Supersymmetry

From the Basics to Exact Results in Gauge Theories

World Scientific Lecture Notes in Physics

ISSN: 1793-1436

---

*Published titles**

Vol. 70: Noise Sustained Patterns: Fluctuations and Nonlinearities
Markus Loecher

Vol. 71: The QCD Vacuum, Hadrons and Superdense Matter (2nd ed.)
Edward V Shuryak

Vol. 72: Massive Neutrinos in Physics and Astrophysics (3rd ed.)
R Mohapatra and P B Pal

Vol. 73: The Elementary Process of Bremsstrahlung
W Nakel and E Haug

Vol. 74: Lattice Gauge Theories: An Introduction (3rd ed.)
H J Rothe

Vol. 75: Field Theory: A Path Integral Approach (2nd ed.)
A Das

Vol. 76: Effective Field Approach to Phase Transitions and Some Applications to Ferroelectrics (2nd ed.)
J A Gonzalo

Vol. 77: Principles of Phase Structures in Particle Physics
H Meyer-Ortmanns and T Reisz

Vol. 78: Foundations of Quantum Chromodynamics: An Introduction to Perturbation Methods in Gauge Theories (3rd ed.)
T Muta

Vol. 79: Geometry and Phase Transitions in Colloids and Polymers
W Kung

Vol. 80: Introduction to Supersymmetry (2nd ed.)
H J W Müller-Kirsten and A Wiedemann

Vol. 81: Classical and Quantum Dynamics of Constrained Hamiltonian Systems
H J Rothe and K D Rothe

Vol. 82: Lattice Gauge Theories: An Introduction (4th ed.)
H J Rothe

Vol. 83: Field Theory: A Path Integral Approach (3rd ed.)
Ashok Das

Vol. 84: Foundations of Quantum Field Theory
Klaus Dieter Rothe

Vol. 85: Quark-Gluon Plasma, Heavy Ion Collisions and Hadrons
Edward Shuryak

Vol. 86: Supersymmetry: From the Basics to Exact Results in Gauge Theories
Matteo Bertolini

*For the complete list of published titles, please visit
http://www.worldscientific.com/series/wslnp

World Scientific Lecture Notes in Physics – Vol. 86

Supersymmetry

From the Basics to Exact Results in Gauge Theories

Matteo Bertolini
SISSA, Trieste, Italy

World Scientific

NEW JERSEY • LONDON • SINGAPORE • BEIJING • SHANGHAI • HONG KONG • TAIPEI • CHENNAI

Recommended Titles in Related Topics

Introduction to Supersymmetry and Supergravity:
Revised and Extended Second Edition
by Peter West
ISBN: 978-981-02-0098-5
ISBN: 978-981-02-0099-2 (pbk)

The Supersymmetric World: The Beginnings of the Theory
edited by G Kane and M Shifman
ISBN: 978-981-02-4522-1
ISBN: 978-981-02-4539-9 (pbk)

Supersymmetry in Quantum Mechanics
by Fred Cooper, Avinash Khare and Uday Sukhatme
ISBN: 978-981-02-4605-1
ISBN: 978-981-02-4612-9 (pbk)

Introduction to Supersymmetry
Second Edition
by Harald J W Müller-Kirsten and Armin Wiedemann
ISBN: 978-981-4293-41-9
ISBN: 978-981-4293-42-6 (pbk)

A Supergravity Primer:
From Geometrical Principles to the Final Lagrangian
by Michel Rausch de Traubenberg and Mauricio Valenzuela
ISBN: 978-981-12-1051-8

Published by

World Scientific Publishing Co. Pte. Ltd.
5 Toh Tuck Link, Singapore 596224
USA office: 27 Warren Street, Suite 401-402, Hackensack, NJ 07601
UK office: 57 Shelton Street, Covent Garden, London WC2H 9HE

Library of Congress Control Number: 2024043679

British Library Cataloguing-in-Publication Data
A catalogue record for this book is available from the British Library.

World Scientific Lecture Notes in Physics — Vol. 86
SUPERSYMMETRY
From the Basics to Exact Results in Gauge Theories

ISBN 978-981-98-0024-7 (hardcover)
ISBN 978-981-98-0025-4 (ebook for institutions)
ISBN 978-981-98-0026-1 (ebook for individuals)

For any available supplementary material, please visit
https://www.worldscientific.com/worldscibooks/10.1142/14026#t=suppl

Typeset by Stallion Press
Email: enquiries@stallionpress.com

To the memory of my father

Preface

This book emerged from a graduate course on Supersymmetry that I have been teaching for several years at SISSA, the International School for Advanced Studies of Trieste. A draft version of my lecture notes has been used also by many students outside SISSA. Quite often, I have met young researchers at a conference or workshop telling me, "I learned supersymmetry from your lecture notes". Eventually, encouraged by some colleagues — and by the editors at World Scientific — I decided to turn these notes into a book. What you have in your hands (or on your PC in its e-book version) is the result of my effort. I hope it could be of some help to you too.

While primarily aimed at graduate students, advanced undergraduate students in high-energy or mathematical physics may also find the book useful, as well as researchers working in this or related areas. Its goal is to provide a pedagogical presentation not only of the basics but also of several advanced topics. Basic notions of quantum field theory, differential geometry and group theory are assumed as prerequisites.

To make the book meaningful as a textbook for a one-semester course, some choices have been made. In particular, only rigid supersymmetry is discussed and the focus is on four space-time dimensions. Moreover, the book is mostly theoretically oriented, meaning that phenomenological aspects are only briefly sketched to the extent that they serve the logical flow. This choice is influenced by my own taste and expertise but also by the status of present day research where supersymmetry phenomenology is somewhat sleepy, waiting for experimental signals at colliders (or elsewhere), and by the fact that there exist excellent reviews on supersymmetry phenomenology on the market to which I could have added very few. On the other hand, remarkable progress on more formal aspects has been made recently, resurrecting topics which were popular in the 1990s/early 2000s and which have become even more so today. My ambition is that by the end of the course readers will have acquired the sufficient background to be able to tackle, on their own, topics at the forefront of current research in the field.

The book can be roughly divided into three parts.

The first part, Chapters 1-4, provides the bases of supersymmetry, introduces the supersymmetry algebra and discusses its representations both on states (on-shell) and on fields (off-shell). As for the latter, for both minimal and extended supersymmetry, we will make use of the $\mathcal{N} = 1$ superspace formalism for which a thorough presentation is made. Besides the basic superfields needed to describe (non-gravitational) supersymmetric Lagrangians, current superfields, which accommodate conserved current operators and the supersymmetry current, are also constructed. Often, these are not included in standard supersymmetry textbooks but are actually needed whenever one wants to discuss non-perturbative aspects of field theories and their dynamics, existence of interacting fixed points, etc... and are relevant in many contexts as, for instance, the celebrated AdS/CFT correspondence.

In the second part, Chapters 5–8, supersymmetric invariant Lagrangians in four dimensions with $\mathcal{N} = 1, 2, 4$ supersymmetry are constructed. Both renormalizable and effective theories (*i.e.* supersymmetric sigma models) are presented along with several examples. Then, an overview is given on perturbative non-renormalization theorems and their consequences — as they were understood before the progress in early nineties in which holomorphy has played a central role. An entire chapter is devoted to discussing several models of spontaneous tree-level supersymmetry breaking, both in stable and metastable vacua, as well as some indirect criteria for supersymmetry breaking, which is helpful when direct tools to understand the vacuum structure of the theory one wants to study are lacking. Finally, we review a few aspects of supersymmetry phenomenology, giving a smattering of the structure of supersymmetric models of beyond the Standard Model physics, including the basics of gauge mediation and gravity mediation scenarios.

The third part, Chapters 9-12, focuses on the quantum behavior of supersymmetric (gauge) theories, for the understanding of which holomorphy, symmetries and dualities play a central role. Non-renormalization theorems are re-derived (and extended) using holomorphy, and the quantum behavior of gauge theories with minimal and extended supersymmetry are analyzed. We will use $\mathcal{N} = 1, 2$ SCQD as a prototype to introduce concepts like gaugino condensation, non-perturbatively generated superpotentials, chiral symmetry breaking, Seiberg duality, interacting superconformal field theories (e.g. the $\mathcal{N} = 1$ conformal window as well as $\mathcal{N} = 2$ interacting abelian Coulomb phases), Seiberg-Witten theory, confinement via monopole condensation and more. Furthermore, Montonen-Olive duality of $\mathcal{N} = 4$ Super-Yang-Mills is discussed, including also those properties in which the global structure of the gauge group makes a difference. Finally, one chapter is devoted to present models where supersymmetry is broken dynamically, by strong coupling effects. This includes some sample dynamical supersymmetry breaking (DSB) models into stable vacua as well as the prototype of DSB into metastable vacua, namely massive $\mathcal{N} = 1$ SQCD. The latter has been the first model of this kind which has appeared in the literature and has generated an intense activity in the last two decades within high energy physics as well as overlapping fields, like cosmology. In the economy of the book, this model also turns out to be an excellent playground for the reader to see how many of the concepts and technical tools learned during the entire course have been absorbed.

Each chapter contains a number of exercises to illustrate special cases, discuss some extensions or prove statements from the main text. There is no bibliography at the end of the book, but each chapter contains its own (biased and incomplete) collection of references, where I refer to some original papers, but mostly to review articles or textbooks on which I have primarily relied upon in preparing the material. This way of referencing will also help readers to be guided if they want to delve into a specific topic.

Acknowledgements

I was fortunate to have great guides early in my career who contributed, each in their own style, in shaping the way I think about physics: Riccardo D'Auria and Sergio Ferrara during my Master's; Pietro Fré and Roberto Iengo during my PhD; Paolo Di Vecchia, Gabriele Ferretti and Alberto Lerda during my early years as a postdoc. Almost my entire scientific career has been concerned with supersymmetry and each of them has taught me something about this fascinating subject. I am particularly grateful to Gabriele for introducing me to the realm of holomorphy and related topics, which gave me a twist in the way I think about (and understand) quantum field theories. Gabriele is also the one, with Andrea Romanino and Alberto Mariotti, from whom I learned anything I know (which is very little compared to them!) about supersymmetry phenomenology.

Generations of students have attended my lectures on supersymmetry at SISSA over the last two decades. Their questions and comments forced me to think deeper and deeper about supersymmetry and its implications, and to improve my notes. I would like to thank them all. A special mention is due to Stefano Cremonesi for sharing his notes on Seiberg-Witten theory, to Lorenzo Di Pietro for countless feedbacks on many of the topics discussed in the book (and for his patience!), to Marco Gorghetto for his incredible persistence in finding typos in my original handouts and, finally, to Andrea Antinucci, Giovanni Galati and Giovanni Rizi for our last-minute discussions about generalized symmetries and related topics.

I am grateful to all colleagues whom I have had the pleasure to collaborate or just discuss with in the last thirty years or so, and which are too many to be mentioned here. Let me just express special gratitude to those who — at different stages of my career — have significantly contributed to improving my understanding of supersymmetry, namely Laura Andrianopoli, Francesco Bigazzi, Aldo Cotrone, Sebastian Franco, Shamit Kachru, Rodolfo Russo and Mario Trigiante. I am also particularly grateful to my SISSA colleagues: to Francesco Benini, Giulio Bonelli and Marco Serone for their patience every time I knocked at the door of their offices and ask about this and that, even though I had to be the one who should have known supersymmetry better; and to Sergio Cecotti, for teaching me that whenever you think you understand something, there is always a more elegant way to go.

Above all, however, I want to thank my friend and longtime collaborator Riccardo Argurio. It is impossible to overestimate how much Riccardo contributed to anything I know about the content of this book. None of its chapters could have been written if not for the endless discussions and exchange of ideas I had with Riccardo over all these years, on supersymmetry and more. This book is also his.

Finally, I would like to thank the love of my life, Giulietta, for her encouragement while writing this book and, more importantly, for the long and exciting journey which she has given me the opportunity to share with her in this world.

Trieste, Italy *Matteo Bertolini*
July 2024

Foreword

Why should a theoretical particle physicist ever know about supersymmetry (and embark on reading a book as long as this)? There are several reasons why it is worth it. Let me mention the ones I think are the most relevant.

The first of such reasons, at least from a chronological point of view, is also the more phenomenological in nature. Back in 2012 the missing building block of the Standard Model, the Higgs particle, has been discovered at CERN's Large Hadron Collider (LHC). This has been an impressive achievement, one of the greatest successes of how we think Nature works at short distances and of the tool we use to describe it, *i.e.* Quantum Field Theory (QFT). On the other hand, there exist many reasons — some of which we will review in the following — which suggest that this cannot be the end of the story: new physics should show up at energy scales higher than those we have so far been able to have access to (lower than the Planck scale, in general). It turns out that of all possible options, the most compelling and motivated scenario for such beyond the Standard Model physics is supersymmetry. So, when it comes to understanding how particles behave at high energies, equivalently at shorter and shorter distances, supersymmetry is a piece of basic knowledge any particle physicist should have. It should be said that at the time of writing, no supersymmetric particles have been discovered yet, nor do we have any indirect evidence for their existence. While I think it is hard to doubt that supersymmetry is actually realized in Nature, this lack of experimental signature is challenging the very idea of low energy supersymmetry, suggesting at least some twist in the way we think about it. Things might be slightly more involved than we imagined; supersymmetric particles might not be around the corner but actually a few more steps ahead, implying that the way supersymmetry tackles the different phenomenological problems it is expected to solve might be more tricky than we thought. However, I do not think we are yet at a stage to declare supersymmetry phenomenology dead, and I keep on thinking that there is still room for such a phenomenological motivation for supersymmetry.

Supersymmetry is believed to play an even more profound role in the dynamics and ultimate structure of space-time, in the way gravity behaves at very high energies, as high as the Planck scale, via string theory. The latter is the most successful framework for describing all interactions, including gravity, in a way consistent with quantum mechanics. However, different from an ordinary quantum field theory, string theory is inherently supersymmetric. From this point of view, no matter the scale at which it might show up, supersymmetry appears as a crucial ingredient in our understanding of the ultimate laws of Nature.

Supersymmetry is also at the core of what is probably the more amazing and far-reaching discovery in theoretical physics in the last decades — the celebrated AdS/CFT correspondence. In short, this correspondence predicts that a (non-gravitational) QFT in d space-time dimensions can actually be dual to a theory of quantum gravity in one dimension higher. This means that the two theories are equivalent at the full quantum level and, upon using a proper dictionary, all observables agree. The best studied (and solid) examples of such remarkable duality involve supersymmetric QFTs in d-dimensional Minkowski space and (super)string theory in $d+1$-dimensional anti-de Sitter space. This is why supersymmetric quantum field theories have now also become a tool to study quantum gravity.

One other thing we, theoretical physicists, want to understand is the behavior of quantum field theories at strong coupling. This is a regime where usual perturbative techniques fail and we lack analytical tools. However, many phenomena we observe in Nature are described by the behavior of quantum field theories in such a regime, the most notable example being the way phenomena like confinement, dynamical mass generation and chiral symmetry breaking are realized in Nature. One spectacular property of supersymmetry is that it makes these phenomena more accessible: supersymmetric quantum field theories turn out to have much more constrained dynamics with respect to non-supersymmetric ones — so constrained that it is often possible to understand their strong coupling regime analytically. In this regard, supersymmetry is seen (and is being used) as a theoretical laboratory to study quantum field theories at strong coupling and get some intuition on how phenomena like those mentioned above are realized in non-supersymmetric field theories (as, for example, QCD). Remarkably, several ideas that had been proposed to account for such phenomena and which could only be conjectural as far as ordinary quantum field theories are concerned, have been analytically proven in the supersymmetric context. Notable examples include confinement as a result of monopole condensation, or that at strong coupling fermion bilinears condense. From this point of view, even setting aside its phenomenological or formal applications, supersymmetry is useful to deepen our understanding of QFT in general, seeing all of its features at work in a well-controlled setting.

I will not be able to discuss all above aspects in detail. The aim of this book is just to provide the minimum foundation you need to get into this fascinating subject and to give you some taste of some advanced topics. What to do with it then... will be your choice.

I hope you will enjoy the journey!

Contents

Chapter 1

Supersymmetry: A Bird's Eye View

In this first chapter, which is introductory/motivational, we want to give an idea of *what* supersymmetry is and *why* it is interesting to study it. This should be seen as an invitation for what comes next. The rest of the book will try to provide much more detailed answers to these two basic questions.

1.1 What is supersymmetry?

Supersymmetry (SUSY) is a *space-time symmetry* mapping particles and fields of integer spin (bosons) into particles and fields of half integer spin (fermions), and vice versa. The generators Q act as

$$Q\,|Fermion\rangle = |Boson\rangle \quad \text{and vice versa} \tag{1.1}$$

From its very definition, this operator has two obvious but far-reaching properties that can be summarized as follows:

- It changes the spin of a particle and hence its space-time properties. This is why supersymmetry is not an internal symmetry but a space-time symmetry.
- In a theory where supersymmetry is realized, each one-particle state has at least a superpartner. Therefore, in a supersymmetric world, instead of single particle states, one has to deal with (super)multiplets of particle states with different spins.

Supersymmetry generators have specific commutation properties with other generators. In particular:

- Q commutes with translations and internal quantum numbers (e.g. gauge and global symmetries), but it does not commute with Lorentz generators

$$[Q, P_\mu] = 0, \quad [Q, G] = 0, \quad [Q, M_{\mu\nu}] \neq 0. \tag{1.2}$$

This implies that particles belonging to the same supermultiplet have different spin but same mass and same quantum numbers.

A supersymmetric field theory is a set of fields and a Lagrangian which exhibits such a symmetry. As ordinary field theories, supersymmetric theories describe particles and interactions between them: supersymmetry manifests itself in the specific particle spectrum a theory enjoys and in the way particles interact between themselves.

A supersymmetric model which is covariant under *general coordinate transformations* is called supergravity (SUGRA) model. In this respect, a non-trivial fact, which again comes from the algebra, in particular from the (anti)commutation relation

$$\{Q, \overline{Q}\} \sim P_\mu, \tag{1.3}$$

is that having general coordinate transformations is equivalent to having *local* supersymmetry, the gauge mediator being a spin 3/2 particle, the gravitino. Hence, local supersymmetry and General Relativity are intimately tied together.

One can have theories with different number of supersymmetry generators Q: Q^I; $I = 1, \ldots, N$. The number of supersymmetry generators, however, cannot be arbitrarily large. The reason is that any supermultiplet contains particles with spin at least as large as $\frac{1}{4}N$. Therefore, N can be at most as large as 4 for theories with maximal spin 1 (gauge theories) and as large as 8 for theories with maximal spin 2 (gravity). Thus stated, this statement is true in four space-time dimensions. Equivalent statements can be made in higher/lower dimensions, where the dimension of the spinor representation of the Lorentz group is larger/smaller (for instance, in ten dimensions, which is the natural dimension where superstring theory lives, the maximum allowed N is 2, which corresponds to $N = 8$ in four dimensions). What really matters is the number of single state supersymmetry generators, which is a dimension-independent statement.

Finally, notice that since supersymmetric theories automatically accommodate both bosons and fermions, supersymmetry looks like the most natural framework to formulate a theory that is able to describe matter and interactions in a unified way.

1.2 Why supersymmetry?

Let us outline a number of reasons why it might be reasonable (and useful) to have such a bizarre and unconventional symmetry actually realized in Nature. They can be roughly divided into three classes.

1.2.1 *Theoretical reasons*

Coleman-Mandula theorem. A central role in quantum field theory is played by the S-matrix, which encodes the (exact) information about physical processes between asymptotic states. A natural question one might ask is what are the more general allowed continuous symmetries of the S-matrix for a theory defined, say, in Minkowski space. More precisely, what are the possible symmetry generators that commute with the S-matrix, that take single-particle states into single-particle states, and whose action on multiparticle states is the direct sum of their action on single-particle states. In 1967 Coleman and Mandula proved a theorem which says that in a generic quantum field theory, under a number of (very reasonable and physical) assumptions, like *locality*, *causality*, *positivity of energy* and *finiteness of number of particles*, the only possible such continuous symmetries are those generated by Poincaré group generators, P_μ and $M_{\mu\nu}$, plus some internal symmetry group generators $\mathbb{G}$ commuting with them

$$[\mathbb{G}, P_\mu] = [\mathbb{G}, M_{\mu\nu}] = 0, \tag{1.4}$$

where the group G is a semi-simple group times abelian factors.

In other words, the most general symmetry group enjoyed by the S-matrix is

$$\text{Poincaré} \times \text{Internal Symmetries}$$

The Coleman-Mandula theorem can be evaded by weakening one or more of its assumptions. One such assumption is that the symmetry algebra only involves *commutators*, all generators being *bosonic* generators. This assumption does not have any particular physical reason not to be relaxed. Allowing for *fermionic* generators, which satisfy anti-commutation relations, it turns out that the set of allowed symmetries can be enlarged. More specifically, in 1975, Haag, Lopuszanski and Sohnius showed that supersymmetry (which, as we will see, is a very specific way to add fermionic generators to a symmetry algebra) is the only possible such option. This makes the Poincaré group become SuperPoincaré. Therefore, the most general symmetry group the S-matrix can enjoy turns out to be

$$\text{SuperPoincaré} \times \text{Internal Symmetries}$$

From a purely theoretical view point, one could then well expect that Nature might have realized all possible kinds of allowed symmetries, given that we already know this is indeed the case for all known symmetries, but supersymmetry, *i.e.* the Standard Model.

A history of unification. The history of our understanding of physical laws can be seen as a long sequence of successful unifications, the effort to describe different phenomena using the same laws and different laws within a common framework (suggesting that physics is reductionist, as somebody might want to say). A famous example is Newton's law of universal gravitation, which says that one and the same equation describes the attraction a planet exert on another planet and on... an apple! Maxwell equations unify electric and magnetic phenomena, and so do following the principles of special relativity. Quantum Electrodynamics shows that electrodynamics is compatible with the principles of quantum mechanics. And so on and so forth, till the formulation of the Standard Model which describes in a unified way all known non-gravitational interactions. Supersymmetry (and its local version, supergravity) is the most natural candidate to complete this long journey. It is a way not just to describe in a unified way all known interactions, including gravity, but in fact to describe matter *and* radiation in a common framework. This sounds compelling, and from this view point it seems natural studying supersymmetry and its consequences.

Quantum gravity and String theory. There is one more reason as to why one could expect that supersymmetry is out there after all. As emphasized, string theory today stands as the most satisfactory theory in describing quantum gravity in a consistent way and, also, in describing all known interactions in a unified framework. So, it might very well be that Nature, at sufficiently high energies, is described by string theory. Unlike a theory of fields, a theory of strings can only be made consistent if it is supersymmetric. So, in this sense, supersymmetry is predicted to be realized in Nature, if string theory is correct. Supersymmetry is in fact one of the two more striking predictions of string theory (the other being the existence of extra-dimensions). In this respect, it is worth noticing that supersymmetry did not first appear in ordinary four-dimensional quantum field theories but as a symmetry of the string world-sheet, at the very beginning of the seventies. Only later, it was shown to be possible to have supersymmetry in space-time.

All arguments reviewed above suggest that supersymmetry may be realized in Nature. However, none of such arguments give any obvious indication on the energy scale at which supersymmetry might show up. In principle, this scale can be very high, as high as the Planck scale. Below, we will present few arguments, more phenomenological in nature, which suggest that low energy supersymmetry (as low as TeV scale or slightly higher) would be the preferred option.

1.2.2 *Phenomenological reasons*

Naturalness and the hierarchy problem. Three out of four of the fundamental interactions among elementary particles (strong, weak and electro-magnetic) are described by the Standard Model (SM). The typical scale of the SM, the electroweak scale, is

$$M_{\rm ew} \sim 250\ {\rm GeV} \iff L_{\rm ew} \sim 10^{-16}\ {\rm mm}, \tag{1.5}$$

The theory is very well tested, experimentally, up to such energies and slightly more, say up to the TeV. This cannot be the end of the story, though: for one thing, at high enough energies, as high as the Planck scale $M_{\rm Pl}$, gravity becomes comparable with other forces and cannot be neglected in elementary particle interactions. At some point, we need a quantum theory of gravity. In fact, that $M_{\rm ew}/M_{\rm Pl} \ll 1$ calls for new physics at a much lower scale. One way to see this is as follows. The Higgs potential reads

$$V(H) \sim \mu^2|H|^2 + \lambda|H|^4 \quad \text{where} \quad \mu^2 < 0. \tag{1.6}$$

Experimentally, the minimum of such potential, $\langle H\rangle = \sqrt{-\mu^2/2\lambda}$, is at around 174 GeV. This implies that the bare mass of the Higgs particle is roughly around 100 GeV or so, $m_H^2 = -\mu^2 \sim (100\,{\rm GeV})^2$. What about radiative corrections? Scalar masses are subject to quadratic divergences in perturbation theory. The SM fermion coupling $-\lambda_f H\overline{f}f$ induces a one-loop correction to the Higgs mass as

$$\Delta m_H^2 \sim -2\,\lambda_f^2\,\Lambda^2 \tag{1.7}$$

due to diagrams such as the one in Figure 1.1. A natural physical UV cut-off Λ should then be at around the TeV scale in order to protect the Higgs

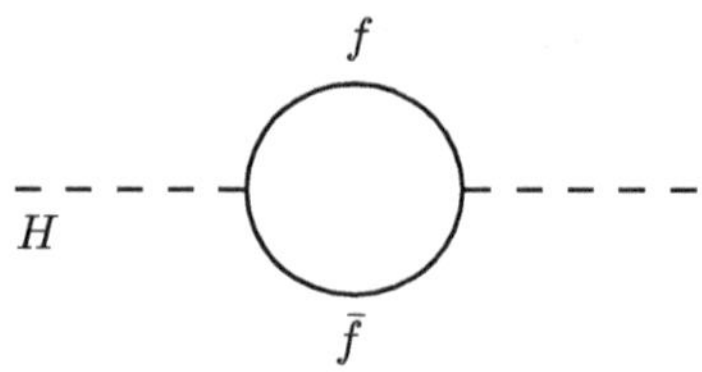

Fig. 1.1 One-loop radiative correction to the Higgs mass due to fermion couplings.

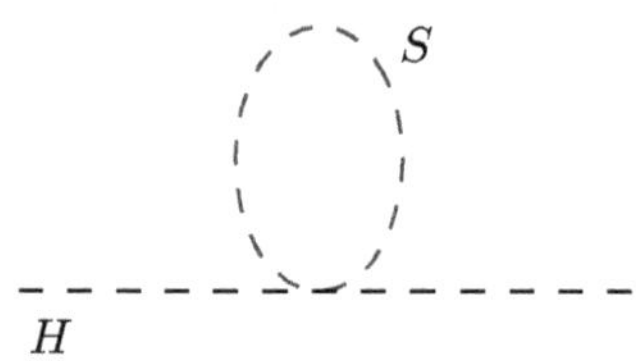

Fig. 1.2 One-loop radiative correction to the Higgs mass due to scalar couplings.

mass, and the SM should then be seen as an effective theory valid up to energies $E \leq M_{\text{eff}} \sim$ TeV, well below the Planck scale.

What can be the new physics beyond such scale and how can such new physics protect the otherwise perturbative divergent Higgs mass? New physics, if any, may include new fermionic and bosonic fields, possibly coupling to the SM Higgs. Each of these fields will give radiative contributions to the Higgs mass of the kind above. Hence, no matter what this new physics is, the natural mass for the the Higgs field would always be of the order of the UV cut-off of the theory, generically around $\sim M_{\text{Pl}}$. We would need a huge fine-tuning to get it stabilized at $\sim$100 GeV (we know that the physical Higgs mass is in fact at 125 GeV)! This is known as the *hierarchy problem*: the experimental value of the Higgs mass is unnaturally smaller than its natural theoretical value.

In principle, there is a very simple way out of this. This resides in the fact that scalar couplings provide one-loop radiative contributions which are opposite in sign with respect to fermions. Suppose there exists some new scalar, S, with Higgs coupling $-\lambda_S|H|^2|S|^2$. Such coupling would induce corrections to the Higgs mass via the one-loop diagram in Figure 1.2.

Such corrections would have opposite sign with respect to those coming from fermion couplings, that is

$$\Delta m_H^2 \sim \lambda_S \, \Lambda^2. \tag{1.8}$$

Therefore, if the new physics would be such that each quark and lepton of the SM were accompanied by two complex scalars having the same Higgs couplings of the quark and lepton, *i.e.* $\lambda_S = |\lambda_f|^2$, then all Λ^2 contributions would automatically cancel, and the Higgs mass could be stabilized at its tree level value. Such conspiracy, however, would be quite ad hoc, and instead of solving the fine-tuning problem mentioned above, it is merely a rephrasing. A natural thing to invoke to have such magic cancellations would be to have a symmetry protecting m_H, exactly in the same way that gauge symmetry protects the masslessness of spin-1 particles or chiral symmetry that of chiral fermions. Some symmetry imposing to the theory the correct matter content (and couplings) for such cancellations to occur. This is exactly what supersymmetry is: in a supersymmetric theory there are fermions and bosons (and couplings) just in the right way to provide *exact* cancellation between diagrams like the ones above! In summary, supersymmetry is a very natural and economical way (though not the only possible one) to solve the hierarchy problem.

Known fermions and bosons cannot be partners of each other. We do not observe any degeneracy in the masses of elementary particles that we know. Moreover, and this is possibly a stronger reason, the quantum numbers do not match: gauge bosons transform in the adjoint representations of the SM gauge group while quarks and leptons in the fundamental or singlet representations. Hence, in a supersymmetric world, each SM particle should have a (yet not observed!) supersymmetric partner, usually dubbed *sparticle*. Roughly, the spectrum of such supersymmetric Standard Model (SSM) should be as follows

SM particles	**SUSY partners**
gauge bosons	gauginos
quarks, leptons	squarks, sleptons
Higgs	higgsino

Notice: the (down) Higgs has the same quantum numbers as the scalar partner of neutrino and leptons, sneutrino and sleptons respectively, $(H_d^0, H_d^-) \leftrightarrow (\tilde{\nu}, \tilde{e}_L)$. Hence, one can imagine that the Higgs is in fact a sparticle. This cannot be. In such scenario, there would be phenomenological problems, *e.g.* lepton number violation and (at least one) neutrino mass in gross violation of experimental bounds.

In summary, the world we already had direct experimental access to, is not supersymmetric. If at all realized, supersymmetry should be a (spontaneously) broken symmetry in the vacuum state chosen by Nature.

However, in order to solve the hierarchy problem without too much fine-tuning this scale should be not much higher than 1 TeV. Including lower bounds from present day experiments, it turns out that the supersymmetry breaking scale should be in the following energy range

$$10^2\,\text{GeV} \leq \text{SUSY breaking scale} \leq 10^3 - 10^4\,\text{GeV}.$$

Let us emphasize that these bounds are just a crude and rough estimate, as they depend very much on the specific supersymmetric Standard Model extension one is actually considering. In particular, the upper bound can be made higher by enriching the structure of the SSM in various ways, while keeping naturalness as a guiding principle. In any event, these bounds are the basic reason why it was believed supersymmetry would show up at CERN's Large Hadron Collider (LHC).

It is worth stressing that, at the time of writing of this book, no signal of supersymmetry has been found at LHC or elsewhere and this has made the above upper bounds more and more in tension with experimental data and, in turn, the very idea of naturalness being reconsidered, at least in this context. There are ongoing discussions on these aspects, including the idea that the resolution of the hierarchy problem should not use naturalness as a guiding principle and that it should be explained by something different, as for instance anthropic arguments or something we do not yet fully understand.

Gauge coupling unification. There is another reason to believe in (low energy) supersymmetry; possibly stronger, from a phenomenological point of view, than that provided by the hierarchy problem. Forget about supersymmetry for a while, and consider the $SU(3) \times SU(2)_L \times U(1)_Y$ Standard Model as it stands. Interesting enough, besides the electro-weak scale, the SM contains in itself a new scale of order 10^{15} GeV. The three SM gauge couplings run according to Renormalization Group (RG) equations like

$$\frac{4\pi}{g_i^2(\mu)} = \frac{b_i}{2\pi} \ln \frac{\mu}{\Lambda_i} \qquad i = 1, 2, 3, \tag{1.9}$$

where Λ_i are the scales at which the one-loop couplings diverge (an IR scale or a UV cut-off for b_i positive or negative, respectively). At the electro-weak scale, $\mu = M_Z$, there is a hierarchy between the three gauge couplings, *i.e.* $g_1(M_Z) < g_2(M_Z) < g_3(M_Z)$. But RG equations make this hierarchy change with the energy scale. In particular, supposing there are no particles other than the SM ones, at a much higher scale, $M_{\text{GUT}} \sim 10^{15}$GeV, the three couplings tend to meet. This naturally calls for a Grand Unified Theory (GUT), where the three interactions are unified into a single one, two possible GUT gauge groups being $SU(5)$ and $SO(10)$. The symmetry breaking pattern one should have in mind would then be as follows

$$SU(5) \underset{\phi}{\rightarrow} SU(3) \times SU(2)_L \times U(1)_Y \underset{H}{\rightarrow} SU(3) \times U(1)_{\text{em}}$$

where ϕ is a *heavy* Higgs field inducing spontaneous symmetry breaking at energies $M_{GUT} \sim 10^{15}\,\text{GeV}$, and H the SM *light* Higgs, inducing electroweak spontaneous symmetry breaking around the TeV scale.

This idea makes a lot of sense but poses several problems. First, there is a new hierarchy problem (generically, the SM Higgs mass is expected to get corrections from the heavy Higgs ϕ). Second, there is a proton decay problem: some of the additional gauge bosons predicted by the GUT group can mediate baryon number violating transitions, allowing processes as $p \rightarrow e^+ + \pi_0$. This makes the proton not fully stable and it turns out that its expected lifetime in such GUT frameworks is violated by present experimental bounds. Finally, on a more theoretical side, if we do not allow for new particles besides the SM ones to be there at some intermediate scale, the three gauge couplings only *approximately* meet and it turns out that this cannot be taken care of just by experimental uncertainties. The latter is an unpleasant feature: small numbers are unnatural from a theoretical view point, unless there are specific reasons (as symmetries) justifying their otherwise unnatural smallness.

Remarkably, making the GUT supersymmetric (SGUT) solves all of these problems at a glance! As already emphasized, with supersymmetry the Higgs mass and that of any scalar is in fact automatically protected. Moreover, just allowing for the minimal supersymmetric extension of the SM spectrum, known as MSSM, the three gauge couplings exactly meet (more precisely, they still miss but now well within experimental uncertainties). Finally, the GUT scale is raised enough, up to around 10^{16} GeV, so that proton decay rate becomes compatible with experimental bounds. So, supersymmetry makes the very natural idea of gauge coupling unification via a Grand Unified Theory free of any apparent drawbacks. A qualitative picture of SM gauge coupling running with and without supersymmetry is reported in Figure 1.3.

An important disclaimer is worth making at this point. The MSSM is not the only possible option for supersymmetry beyond the SM, just the most economic one. In the MSSM one just adds a superpartner to each

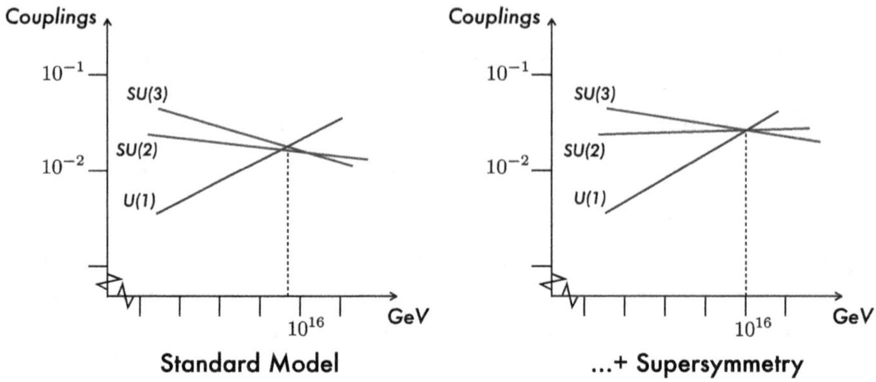

Fig. 1.3 On the left a qualitative picture describing the running of the three SM couplings, which approximately meet at a scale of order 10^{15} GeV. On the right, the same picture in a minimal supersymmetric extension of the SM, where the couplings exactly meet (within experimental uncertainties) at a scale of order 10^{16} GeV.

SM particle, therefore introducing the higgsino, the wino, the zino, several gluinos, together with all squarks and sleptons, and no more. [There is in fact an exception. To have a consistent model one has to double the Higgs sector, and have two Higgs doublets. One reason for that is gauge anomaly cancellation: the higgsinos are fermions in the fundamental representation of $SU(2)_L$ hence two of them are needed, with opposite hypercharge, so to preserve the anomaly-free properties of the SM. A second reason is that in the SM the field H gives mass to down quarks and charged leptons while its charge conjugate, $H^c(\sim \overline{H})$ gives mass to up quarks. As we will see, in a supersymmetric model $\overline{H}$ cannot enter in the potential, which is a function of H only. Therefore, in a supersymmetric scenario, to give mass to up quarks, one needs a second, independent Higgs doublet.] There exist many non-minimal supersymmetric extensions of the Standard Model (which, in fact, are in better shape against experimental constraints with respect to the MSSM). One can in principle construct any SSM one likes. In doing so, however, several constraints are to be taken into account. For example, it is not so easy to make such non-minimal extensions keeping the nice exact gauge coupling unification as enjoyed by the MSSM.

It is worth stressing that gauge coupling unification and the hierarchy problem are independent issues. Indeed, for the former to hold one does not need a full supersymmetric spectrum at low energy. Only light fermionic partners are needed. Scalar partners of SM fermions sit in full GUT families so they do not contribute to gauge coupling unification; they just shift all couplings by one and the same constant. At the price of forgetting about naturalness, this observation opened up the idea that the supersymmetric spectrum can be split — with light fermions and heavy scalars — with supersymmetry being realized only at high energy. This scenario goes under the name of Split Supersymmetry.

Supersymmetry and dark matter. Another context where supersymmetry can play an important role is cosmology. There are various evidences which indicate that around 26% of the energy density in the Universe should be made of *dark matter*, *i.e.* non-luminous and non-baryonic matter. The only SM candidates for dark matter are neutrinos, but they are disfavored by available experimental data (basically, neutrinos are too light to account for such an enormous energy density). Supersymmetry provides instead a very natural dark matter candidate: the neutralino. Neutralinos are mass eigenstates of a linear superposition of the supersymmetric partners of the neutral Higgs and of the SU(2) and U(1) neutral gauge bosons

$$\chi_i = \alpha_{i1}\widetilde{B}^0 + \alpha_{i2}\widetilde{W}^0 + \alpha_{i3}\widetilde{H}_u^0 + \alpha_{i4}\widetilde{H}_d^0. \tag{1.10}$$

In most supersymmetric frameworks the neutralino is the lightest supersymmetric particle (LSP) and it is fully stable, as a dark matter candidate should be.

1.2.3 *Supersymmetry as a theoretical lab for strongly coupled gauge dynamics*

What if supersymmetry turns out not to be the correct theory to describe beyond the Standard Model physics? Or, worse, what if supersymmetry turns out not to be realized at all in Nature (something we could hardly ever be able to prove, actually)? Interestingly, there is a remarkable fact that makes it worth studying supersymmetric theories, independently from the role supersymmetry might or might not play as a theory describing high energy physics.

Let us consider non-abelian gauge theories, which strong interactions are an example of. Every time a non-abelian gauge group remains unbroken at low energy, we have to deal with strong coupling. The typical questions one should try and answer (in QCD or similar theories) are:

- The bare Lagrangian is described in terms of quark and gluons, which are UV degrees of freedom. Which are the IR (light) degrees of freedom

of QCD? What is the effective Lagrangian in terms of such degrees of freedom?
- Strong coupling physics is very rich. Typically, one has to deal with phenomena like confinement, charge screening, the generation of a mass gap, etc... Is there any theoretical understanding of such phenomena?
- It is believed that the QCD vacuum is populated by vacuum condensates of fermion bilinears, which induce chiral symmetry breaking. What is the microscopic mechanism behind this phenomenon?

Most of the IR properties of QCD have eluded so far a clear understanding, since we lack analytical tools to deal with strong coupling dynamics. Most results come from lattice computations, but these do not furnish a first principle understanding of the above phenomena.

Because of their nice renormalization properties, supersymmetric theories are more constrained than ordinary field theories and let one have a better control on strong coupling regimes. Therefore, one might hope to use them as toy models to study properties of more realistic theories, such as QCD, in a more controlled way. As we will see, supersymmetric theories do provide examples where some of the above strong coupling effects can be studied exactly. This is possible due to powerful non-renormalizations theorems that supersymmetric theories enjoy, a very special property known as *holomorphy* and remarkable strong-weak coupling dualities, which in certain circumstances let one compute several non-perturbative contributions to the Lagrangian exactly. The third part of this book, Chapters 9 to 12, will focus on many such issues.

1.3 Some useful textbooks

Supersymmetry has now been around for more than 50 years and the list of references in the literature is endless. In preparing the materials presented in this book I have drawn upon textbooks, reviews available on the high energy physics archive (to be found at https://arxiv.org) and occasionally original papers, which I refer to at the end of each chapter.

Below, I list a few books that may be useful to consult. Some may be better than others depending on the specific topic of interest (and on personal taste). Immersing yourself in their reading is also a very instructive exercise to see that what has been perceived as relevant, more or less important, more or less worth pointing out is a time-dependent function.

S. J. Gates, M. T. Grisaru, M. Rocek and W. Siegel, Superspace Or One Thousand and One Lessons in Supersymmetry, *Front. Phys.* **58** (1983).

M. F. Sohnius, Introducing Supersymmetry, *Phys. Rep.* **128** (1985).

P. C. West, *Introduction to Supersymmetry and Supergravity*, World Scientific Singapore (1990)

Wess and J. Bagger, *Supersymmetry and Supergravity*, Princeton University Press (1992).

S. Weinberg, *The Quantum Thcory of Fields Vol. 3: Supersymmetry*, Cambridge University Press (2000).

J. Terning, *Modern Supersymmetry: Dynamics and Duality*, Oxford University Press (2006).

M. Dine, *Supersymmetry and String Theory: Beyond the Standard Model*, Cambridge University Press (2007).

H. J. Müller-Kirsten and A. Wiedemann, *Introduction to Supersymmetry*, World Scientific Singapore (2010).

S. Cecotti, *Supersymmetric Field Theories*, Cambridge University Press (2015).

Chapter 2

The Supersymmetry Algebra

In this chapter we will introduce the supersymmetry algebra, which is the algebra encoding the set of symmetries a supersymmetric theory should enjoy. The supersymmetry algebra is an extension of the Poincaré algebra so in the first two sections we will start recalling a few facts about Lorentz and Poincaré algebras, the corresponding groups and their representations. In particular, we will emphasize the relation of the Lorentz algebra with the $SU(2)$ and $SL(2,\mathbb{C})$ algebras and define (two component) spinors as the basic representation of the Lorentz group. In the last section we will introduce the concept of graded Lie algebra and, finally, the supersymmetry algebra, which is a specific instance of a graded Lie algebra.

2.1 Lorentz and Poincaré groups

The Lorentz group $SO(1,3)$ is the subgroup of matrices Λ of $GL(4,\mathbb{R})$ with unit determinant, $\det\Lambda = 1$, and which satisfy the following relation

$$\Lambda^T \eta \, \Lambda = \eta, \tag{2.1}$$

where η is the (mostly minus in our conventions) flat Minkowski metric

$$\eta_{\mu\nu} = \mathrm{diag}(+,-,-,-). \tag{2.2}$$

The Lorentz group has six generators (associated to space rotations and boosts) enjoying the following commutation relations

$$[J_i, J_j] = i\epsilon_{ijk} J_k, \quad [J_i, K_j] = i\epsilon_{ijk} K_k, \quad [K_i, K_j] = -i\epsilon_{ijk} J_k. \tag{2.3}$$

Notice that while the J_i are hermitian, the boosts K_i are anti-hermitian, this being related to the fact that the Lorentz group is non-compact (topologically, the Lorentz group is $\mathbb{R}^3 \times S^3/\mathbb{Z}_2$, the non-compact factor corresponding to boosts and the doubly connected $S^3/\mathbb{Z}_2$ corresponding to rotations). In order to construct representations of this algebra it is useful to introduce the following complex linear combinations of the generators J_i and K_i

$$J_i^{\pm} = \frac{1}{2}(J_i \pm iK_i), \tag{2.4}$$

where now the $J_i^{\pm}$ are hermitian. In terms of $J_i^{\pm}$ the algebra (2.3) becomes

$$[J_i^{\pm}, J_j^{\pm}] = i\epsilon_{ijk}J_k^{\pm}, \quad [J_i^{\pm}, J_j^{\mp}] = 0. \tag{2.5}$$

This shows that the Lorentz algebra is equivalent to two $SU(2)$ algebras. As we will see later, this simplifies a lot the study of representations of the Lorentz group, which can be organized into (couples of) $SU(2)$ representations. This isomorphism comes from the theory of Lie Algebra which says that at the level of complex algebras

$$SO(4) \simeq SU(2) \times SU(2). \tag{2.6}$$

The Lorentz algebra is a specific real form of that of $SO(4)$. This difference can be seen from the defining commutation relations (2.3): for $SO(4)$ one would have had a plus sign on the right hand side of the third such commutation relations. This difference has some consequences when it comes to studying representations. In particular, while all representations are real or pseudoreal in Euclidean space, complex conjugation interchanges the two $SU(2)$'s in Minkowski space. This can also be seen at the level of the generators $J_i^{\pm}$. In order for all rotation and boost parameters to be real, one must take all the J_i and K_i to be imaginary and hence from eq. (2.4) one sees that

$$(J_i^{\pm})^* = -J_i^{\mp}. \tag{2.7}$$

In terms of algebras, this discussion can be summarized noticing that for the Lorentz algebra the isomorphism (2.6) changes into

$$SO(1,3) \simeq SU(2) \times SU(2)^*. \tag{2.8}$$

For later purpose, let us introduce a four-vector notation for the Lorentz generators, in terms of an anti-symmetric tensor $M_{\mu\nu}$ defined as

$$M_{\mu\nu} = -M_{\nu\mu} \quad \text{with} \quad M_{0i} = K_i \quad \text{and} \quad M_{ij} = \epsilon_{ijk}J_k, \tag{2.9}$$

where $\mu = 0, 1, 2, 3$. In terms of such matrices, the Lorentz algebra reads

$$[M_{\mu\nu}, M_{\rho\sigma}] = -i\eta_{\mu\rho}M_{\nu\sigma} - i\eta_{\nu\sigma}M_{\mu\rho} + i\eta_{\mu\sigma}M_{\nu\rho} + i\eta_{\nu\rho}M_{\mu\sigma}. \tag{2.10}$$

Another useful relation one should bear in mind is the relation between the Lorentz group and $SL(2,\mathbb{C})$, the group of 2×2 complex matrices with unit determinant. More precisely, there exists a homomorphism between $SL(2,\mathbb{C})$ and $SO(1,3)$, which means that for any matrix $A \in SL(2,\mathbb{C})$ there exists an associated Lorentz matrix Λ, and that

$$\Lambda(A)\,\Lambda(B) = \Lambda(AB), \tag{2.11}$$

where A and B are $SL(2,\mathbb{C})$ matrices. This can be proved as follows. Lorentz transformations act on four-vectors as

$$x'^{\mu} = \Lambda^{\mu}{}_{\nu}\, x^{\nu}, \tag{2.12}$$

where the matrices Λ's are a representation of the generators $M_{\mu\nu}$ defined above. Let us introduce 2×2 matrices σ_μ where σ_0 is the identity matrix and σ_i are the Pauli matrices defined as

$$\sigma_1 = \begin{pmatrix} 0 & 1 \\ 1 & 0 \end{pmatrix}, \quad \sigma_2 = \begin{pmatrix} 0 & -i \\ i & 0 \end{pmatrix}, \quad \sigma_3 = \begin{pmatrix} 1 & 0 \\ 0 & -1 \end{pmatrix}. \tag{2.13}$$

Let us also define the matrices with upper indices, σ^μ, as

$$\sigma^\mu = (\sigma^0, \sigma^i) = (\sigma_0, -\sigma_i). \tag{2.14}$$

The matrices σ_μ are a complete set, in the sense that any 2×2 complex matrix can be written as a linear combination of them. For every four-dimensional vector x^μ, let us construct the 2×2 complex matrix

$$\rho : x^\mu \to x^\mu \sigma_\mu = X. \tag{2.15}$$

The matrix X is hermitian, since the Pauli matrices are hermitian, and has determinant equal to $x^\mu x_\mu$, which is a Lorentz invariant quantity. Therefore, ρ is a map from Minkowski space to H, the space of 2×2 hermitian complex matrices

$$M_4 \underset{\rho}{\longrightarrow} H. \tag{2.16}$$

Let us now act on X with a $SL(2,\mathbb{C})$ transformation A

$$A : X \to AXA^\dagger = X'. \tag{2.17}$$

This transformation preserves the determinant since $\det A = 1$ and also preserves the hermicity of X since

$$X'^\dagger = (AXA^\dagger)^\dagger = AX^\dagger A^\dagger = AXA^\dagger = X'. \tag{2.18}$$

Therefore A is a map between H and itself

$$H \underset{A}{\longrightarrow} H. \tag{2.19}$$

We finally apply the inverse map ρ^{-1} to X' and get a four-vector x'^{μ}. The inverse map is defined as

$$\rho^{-1} = \frac{1}{2}\,\mathrm{Tr}[\bullet\ \bar{\sigma}^{\mu}] \tag{2.20}$$

(where, as we will later see more rigorously, a complex 2×2 matrix $\bar{\sigma}^{\mu}$ is the same as σ_{μ}). Indeed

$$\rho^{-1}X = \frac{1}{2}\,\mathrm{Tr}[X\bar{\sigma}^{\mu}] = \frac{1}{2}\,\mathrm{Tr}[x_{\nu}\sigma^{\nu}\bar{\sigma}^{\mu}] = \frac{1}{2}\,\mathrm{Tr}[\sigma^{\nu}\bar{\sigma}^{\mu}]x_{\nu} = \frac{1}{2}\,2\,\eta^{\mu\nu}x_{\nu} = x^{\mu}. \tag{2.21}$$

Assembling everything together we then get a map from Minkowski space into itself via the following chain

$$\begin{array}{ccccccc} M_4 & \underset{\rho}{\longrightarrow} & H & \underset{A}{\longrightarrow} & H & \underset{\rho^{-1}}{\longrightarrow} & M_4 \\ x_{\nu} & \underset{\rho}{\longrightarrow} & x_{\nu}\sigma^{\nu} & \underset{A}{\longrightarrow} & Ax_{\nu}\sigma^{\nu}A^{\dagger} & \underset{\rho^{-1}}{\longrightarrow} & \frac{1}{2}\,\mathrm{Tr}[Ax_{\nu}\sigma^{\nu}A^{\dagger}\bar{\sigma}^{\mu}] = x'^{\mu}. \end{array} \tag{2.22}$$

This is nothing but a Lorentz transformation obtained by the $SL(2,\mathbb{C})$ transformation A as

$$\Lambda^{\mu}{}_{\nu}(A) = \frac{1}{2}\mathrm{Tr}[\bar{\sigma}^{\mu}A\,\sigma_{\nu}A^{\dagger}]. \tag{2.23}$$

It is now a trivial exercise, given eq. (2.23), to prove the homomorphism (2.11).

Notice that the relation (2.23) can in principle be inverted, in the sense that for a given Λ one can find a corresponding $A \in SL(2,\mathbb{C})$. However, the relation is not an isomorphism, since it is double valued. The isomorphism holds between the Lorentz group and $SL(2,\mathbb{C})/\mathbb{Z}_2$ (in other words $SL(2,\mathbb{C})$ is a double cover of the Lorentz group). This can be seen as follows. Consider the 2×2 matrix

$$M(\theta) = \begin{pmatrix} e^{-i\theta/2} & 0 \\ 0 & e^{i\theta/2} \end{pmatrix} \tag{2.24}$$

which corresponds to a Lorentz transformation producing a rotation by an angle θ about the z-axis. Taking $\theta = 2\pi$ which corresponds to the identity in the Lorentz group, one gets $M = -\mathbf{1}$ which is a non-trivial element of $SL(2,\mathbb{C})$. It then follows that the elements of $SL(2,\mathbb{C})$ are identified two-by-two under a $\mathbb{Z}_2$ transformation in the Lorentz group. Note that this $\mathbb{Z}_2$ identification holds also in Euclidean space: at the level of groups $SU(2) \times SU(2) = Spin(4)$, where $Spin(4)$ is a double cover of $SO(4)$ as a group (it has an extra $\mathbb{Z}_2$).

The Poincaré group, $ISO(1,3)$, is the Lorentz group augmented by the space-time translation generators P_μ. In terms of the generators P_μ and $M_{\mu\nu}$ the Poincaré algebra reads

$$\begin{aligned}
[P_\mu, P_\nu] &= 0 \\
[M_{\mu\nu}, M_{\rho\sigma}] &= -i\eta_{\mu\rho}M_{\nu\sigma} - i\eta_{\nu\sigma}M_{\mu\rho} + i\eta_{\mu\sigma}M_{\nu\rho} + i\eta_{\nu\rho}M_{\mu\sigma} \\
[M_{\mu\nu}, P_\rho] &= -i\eta_{\rho\mu}P_\nu + i\eta_{\rho\nu}P_\mu .
\end{aligned} \tag{2.25}$$

2.2 Spinors and representations of the Lorentz group

We are now ready to discuss representations of the Lorentz group. Thanks to the isomorphism (2.8), they can be easily organized in terms of those of $SU(2)$ which can be labeled by the spins. In this respect, let us introduce two-component spinors as the objects carrying the basic representations of $SL(2,\mathbb{C})$. There exist two such representations. A spinor transforming in the self-representation $\mathcal{M}$ is a two complex component object

$$\psi = \begin{pmatrix} \psi_1 \\ \psi_2 \end{pmatrix} \tag{2.26}$$

where ψ_1 and ψ_2 are complex Grassmann numbers, which transform under a matrix $\mathcal{M} \in SL(2,\mathbb{C})$ as

$$\psi_\alpha \to \psi'_\alpha = \mathcal{M}_\alpha^{\ \beta}\,\psi_\beta \quad \alpha,\beta = 1,2. \tag{2.27}$$

The complex conjugate representation is defined from $\mathcal{M}^*$, where $\mathcal{M}^*$ means complex conjugation, as

$$\overline{\psi}_{\dot\alpha} \to \overline{\psi}'_{\dot\alpha} = \mathcal{M}^{*\ \dot\beta}_{\ \dot\alpha}\,\overline{\psi}_{\dot\beta} \quad \dot\alpha,\dot\beta = 1,2. \tag{2.28}$$

These two representations are *not* equivalent, that is, it does not exist a matrix C such that $\mathcal{M} = C\mathcal{M}^*C^{-1}$.

There are, however, other representations which are equivalent to the former. Let us first introduce the invariant tensor of $SU(2)$, $\epsilon_{\alpha\beta}$, and similarly for the other $SU(2)$, $\epsilon_{\dot\alpha\dot\beta}$, which one uses to raise and lower spinorial indices as well as to construct scalars and higher spin representations by spinor contractions

$$\epsilon_{\alpha\beta} = \epsilon_{\dot\alpha\dot\beta} = \begin{pmatrix} 0 & -1 \\ 1 & 0 \end{pmatrix} \quad \epsilon^{\alpha\beta} = \epsilon^{\dot\alpha\dot\beta} = \begin{pmatrix} 0 & 1 \\ -1 & 0 \end{pmatrix}. \tag{2.29}$$

We can then define

$$\psi^\alpha = \epsilon^{\alpha\beta}\psi_\beta, \quad \psi_\alpha = \epsilon_{\alpha\beta}\psi^\beta, \quad \overline{\psi}_{\dot\alpha} = \epsilon_{\dot\alpha\dot\beta}\overline{\psi}^{\dot\beta}, \quad \overline{\psi}^{\dot\alpha} = \epsilon^{\dot\alpha\dot\beta}\overline{\psi}_{\dot\beta}. \tag{2.30}$$

The convention here is that adjacent indices are always contracted, putting the epsilon tensor on the left.

Using above conventions, one can easily prove that $\psi'^{\alpha} = (\mathcal{M}^{-1T})^{\alpha}{}_{\beta}\psi^{\beta}$. Since $\mathcal{M}^{-1T} \simeq \mathcal{M}$ (the matrix C being in fact the epsilon tensor $\epsilon_{\alpha\beta}$), it follows that the self-representation ψ_α and the representation ψ^α are equivalent. A similar story holds for $\overline{\psi}^{\dot\alpha}$ which transforms in the representation $\mathcal{M}^{*-1T}$, that is $\overline{\psi}'^{\dot\alpha} = (\mathcal{M}^{*-1T})^{\dot\alpha}{}_{\dot\beta}\overline{\psi}^{\dot\beta}$, which is equivalent to the complex conjugate representation $\overline{\psi}_{\dot\alpha}$ (the matrix C connecting $\mathcal{M}^{*-1T}$ and $\mathcal{M}^*$ is now the epsilon tensor $\epsilon_{\dot\alpha\dot\beta}$). From our conventions, one can easily see that the complex conjugate matrix $(\mathcal{M}_\alpha^{\ \beta})^*$ (that is, the matrix obtained from $\mathcal{M}_\alpha^{\ \beta}$ by taking the complex conjugate of each entry), once expressed in terms of dotted indices, is not $\mathcal{M}^*{}_{\dot\alpha}{}^{\dot\beta}$, but rather $(\mathcal{M}^{*-1T})^{\dot\alpha}{}_{\dot\beta}$. Finally, lower undotted indices are row indices, while upper ones are column indices. Dotted indices follow instead the opposite convention. This implies that $(\psi_\alpha)^* = \overline{\psi}^{\dot\alpha}$, while under hermitian conjugation (which also includes transposition), we have, *e.g.* $\overline{\psi}_{\dot\alpha} = (\psi_\alpha)^\dagger$, as operator identity.

Due to the homomorphism between $SL(2,\mathbb{C})$ and $SO(1,3)$, it turns out that the two spinor representations ψ_α and $\overline{\psi}^{\dot\alpha}$ are representations of the Lorentz group, and because of the isomorphism (2.8), they can be labeled in terms of $SU(2)$ representations as

$$\psi_\alpha \equiv \left(\frac{1}{2}, 0\right) \tag{2.31}$$

$$\overline{\psi}^{\dot\alpha} \equiv \left(0, \frac{1}{2}\right). \tag{2.32}$$

To understand the identifications above just note that $\sum_i (J_i^+)^2$ and $\sum_i (J_i^-)^2$ are Casimir of the two $SU(2)$ algebras (2.5) with eigenvalues $n(n+1)$ and $m(m+1)$ with $n, m = 0, \frac{1}{2}, 1, \frac{3}{2}, \ldots$ being the eigenvalues of J_3^+ and J_3^-, respectively. Hence we can indeed label the representations of the Lorentz group by pairs (n, m) and since $J_3 = J_3^+ + J_3^-$ we can identify the spin of the representation as $n + m$, its dimension being $(2n+1)(2m+1)$. The two spinor representations (2.31) and (2.32) are just the basic such representations. Note, in passing, that the representations ψ_α and ψ^α are nothing but the fundamental and anti-fundamental representations of $SU(2)$. That they are equivalent is specific to $SU(2)$ and does not hold for $SU(N)$ with $N > 2$.

Recalling that Grassmann variables anticommute (that is $\psi_1\chi_2 = -\chi_2\psi_1$, $\psi_1\overline{\chi}_{\dot 2} = -\overline{\chi}_{\dot 2}\psi_1$, etc...) we can define a scalar product for spinors as

$$\psi\chi \equiv \psi^\alpha\chi_\alpha = \epsilon^{\alpha\beta}\psi_\beta\chi_\alpha = -\epsilon^{\alpha\beta}\psi_\alpha\chi_\beta = -\psi_\alpha\chi^\alpha = \chi^\alpha\psi_\alpha = \chi\psi \quad (2.33)$$

$$\overline{\psi}\overline{\chi} \equiv \overline{\psi}_{\dot\alpha}\overline{\chi}^{\dot\alpha} = \epsilon_{\dot\alpha\dot\beta}\overline{\psi}^{\dot\beta}\overline{\chi}^{\dot\alpha} = -\epsilon_{\dot\alpha\dot\beta}\overline{\psi}^{\dot\alpha}\overline{\chi}^{\dot\beta} = -\overline{\psi}^{\dot\alpha}\overline{\chi}_{\dot\alpha} = \overline{\chi}_{\dot\alpha}\overline{\psi}^{\dot\alpha} = \overline{\chi}\overline{\psi}. \quad (2.34)$$

Under hermitian conjugation we have

$$(\psi\chi)^\dagger = (\psi^\alpha\chi_\alpha)^\dagger = \chi_\alpha^{\ \dagger}\psi^{\alpha\dagger} = \overline{\chi}_{\dot\alpha}\overline{\psi}^{\dot\alpha} = \overline{\chi}\overline{\psi}. \quad (2.35)$$

In our conventions, undotted indices are contracted from upper left to lower right while dotted indices from lower left to upper right (this rule does not apply when raising or lowering indices with the epsilon tensor). Recalling eq. (2.17), namely that under $SL(2,\mathbb{C})$ the matrix $X = x^\mu\sigma_\mu$ transforms as $AXA^\dagger$ and that the index structure of A and $A^\dagger$ is $A_\alpha^{\ \beta}$ and $A^{*\dot\beta}_{\ \ \dot\alpha}$, respectively, we see that σ_μ naturally has a dotted and an undotted index and can be contracted with an undotted and a dotted spinor as

$$\psi\sigma^\mu\overline{\chi} \equiv \psi^\alpha\sigma^\mu_{\alpha\dot\alpha}\overline{\chi}^{\dot\alpha}. \quad (2.36)$$

Similarly one can define $\bar\sigma^\mu$ as

$$\bar\sigma^{\mu\ \dot\alpha\alpha} = \epsilon^{\dot\alpha\dot\beta}\epsilon^{\alpha\beta}\sigma^\mu_{\beta\dot\beta} = (\sigma_0, \sigma_i), \quad (2.37)$$

and define the product of $\bar\sigma^\mu$ with a dotted and an undotted spinor as

$$\overline{\psi}\bar\sigma^\mu\chi \equiv \overline{\psi}_{\dot\alpha}\bar\sigma^{\mu\ \dot\alpha\beta}\chi_\beta. \quad (2.38)$$

A number of useful identities one can prove are

$$\begin{aligned}
&\psi^\alpha\psi^\beta = -\frac{1}{2}\epsilon^{\alpha\beta}\psi\psi, \quad (\theta\phi)(\theta\psi) = -\frac{1}{2}(\phi\psi)(\theta\theta)\\
&\chi\sigma^\mu\overline{\psi} = -\overline{\psi}\bar\sigma^\mu\chi, \quad \chi\sigma^\mu\bar\sigma^\nu\psi = \psi\sigma^\nu\bar\sigma^\mu\chi\\
&(\chi\sigma^\mu\overline{\psi})^\dagger = \psi\sigma^\mu\overline{\chi}, \quad (\chi\sigma^\mu\bar\sigma^\nu\psi)^\dagger = \overline{\psi}\bar\sigma^\nu\sigma^\mu\overline{\chi}\\
&(\theta\psi)(\theta\sigma^\mu\overline{\phi}) = -\frac{1}{2}(\theta\theta)(\psi\sigma^\mu\overline{\phi}), \quad (\overline{\theta}\overline{\psi})(\overline{\theta}\bar\sigma^\mu\phi) = -\frac{1}{2}(\overline{\theta}\overline{\theta})(\overline{\psi}\bar\sigma^\mu\phi)\\
&(\phi\psi)\cdot\overline{\chi}_{\dot\alpha} = \frac{1}{2}(\phi\sigma^\mu\overline{\chi})(\psi\sigma_\mu)_{\dot\alpha}\,.
\end{aligned} \quad (2.39)$$

As some people might be more familiar with four component spinor notation, let us close this section by briefly mentioning the connection with Dirac spinors. In the Weyl representation Dirac matrices read

$$\gamma^\mu = \begin{pmatrix} 0 & \sigma^\mu \\ \bar\sigma^\mu & 0 \end{pmatrix}, \quad \gamma_5 = i\gamma_0\gamma_1\gamma_2\gamma_3 = \begin{pmatrix} \sigma_0 & 0 \\ 0 & -\sigma_0 \end{pmatrix} \quad (2.40)$$

and a Dirac spinor is

$$\psi = \begin{pmatrix} \psi_\alpha \\ \overline{\chi}^{\dot\alpha} \end{pmatrix} \quad \text{implying} \quad r(\psi) = \left(\frac{1}{2},0\right) \oplus \left(0,\frac{1}{2}\right). \tag{2.41}$$

This shows that a Dirac spinor carries a reducible representation of the Lorentz algebra. Using this four component spinor notation one sees that

$$\begin{pmatrix} \psi_\alpha \\ 0 \end{pmatrix} \quad \text{and} \quad \begin{pmatrix} 0 \\ \overline{\chi}^{\dot\alpha} \end{pmatrix} \tag{2.42}$$

are Weyl (chiral) spinors, with chirality $+1$ and -1, respectively. One can easily show that a Majorana spinor ($\psi^{\mathcal{C}} = \psi$) is a Dirac spinor such that $\chi_\alpha = \psi_\alpha$. To prove this, just recall that in four component notation the conjugate Dirac spinor is defined as $\overline{\psi} = \psi^\dagger \gamma_0$ and the charge conjugate is $\psi^{\mathcal{C}} = \mathcal{C}\overline{\psi}^T$ with the charge conjugate matrix in the Weyl representation being

$$\mathcal{C} = \begin{pmatrix} i(\sigma^2\bar\sigma^0)_\alpha^{\ \beta} & 0 \\ 0 & i(\bar\sigma^2\sigma^0)^{\dot\alpha}_{\ \dot\beta} \end{pmatrix} = \begin{pmatrix} \delta_\alpha^{\ \beta} & 0 \\ 0 & \delta^{\dot\alpha}_{\ \dot\beta} \end{pmatrix}. \tag{2.43}$$

Finally, Lorentz generators are

$$\Sigma^{\mu\nu} = \frac{i}{2}\gamma^{\mu\nu}, \quad \gamma^{\mu\nu} = \frac{1}{2}(\gamma^\mu\gamma^\nu - \gamma^\nu\gamma^\mu) = \frac{1}{2}\begin{pmatrix} \sigma^\mu\bar\sigma^\nu - \sigma^\nu\bar\sigma^\mu & 0 \\ 0 & \bar\sigma^\mu\sigma^\nu - \bar\sigma^\nu\sigma^\mu \end{pmatrix} \tag{2.44}$$

while the 2-index Pauli matrices are defined as

$$(\sigma^{\mu\nu})_\alpha^{\ \beta} = \frac{1}{4}\left(\sigma^\mu_{\alpha\dot\gamma}(\bar\sigma^\nu)^{\dot\gamma\beta} - (\mu \leftrightarrow \nu)\right), \quad (\bar\sigma^{\mu\nu})^{\dot\alpha}_{\ \dot\beta} = \frac{1}{4}\left((\bar\sigma^\mu)^{\dot\alpha\gamma}\sigma^\nu_{\gamma\dot\beta} - (\mu \leftrightarrow \nu)\right) \tag{2.45}$$

From the last equations one then sees that $i\sigma^{\mu\nu}$ acts as a Lorentz generator on ψ_α while $i\bar\sigma^{\mu\nu}$ acts as a Lorentz generator on $\overline{\psi}^{\dot\alpha}$.

2.3 The supersymmetry algebra

As we have already mentioned, a no-go theorem provided by Coleman and Mandula implies that, under certain assumptions (locality, causality, positivity of energy, finiteness of number of particles, etc...), the only possible symmetries of the S-matrix are, besides C, P, T

- Poincaré symmetries, with generators $P_\mu, M_{\mu\nu}$
- An *internal* symmetry group G with generators B_l being Lorentz scalars and with the structure of a compact semi-simple group times $U(1)$ factors.

The full symmetry algebra hence reads

$$[P_\mu, P_\nu] = 0 \tag{2.46}$$

$$[M_{\mu\nu}, M_{\rho\sigma}] = -i\eta_{\mu\rho}M_{\nu\sigma} - i\eta_{\nu\sigma}M_{\mu\rho} + i\eta_{\mu\sigma}M_{\nu\rho} + i\eta_{\nu\rho}M_{\mu\sigma} \tag{2.47}$$

$$[M_{\mu\nu}, P_\rho] = -i\eta_{\rho\mu}P_\nu + i\eta_{\rho\nu}P_\mu \tag{2.48}$$

$$[B_l, B_m] = if_{lm}^{n}\, B_n \tag{2.49}$$

$$[P_\mu, B_l] = 0 \tag{2.50}$$

$$[M_{\mu\nu}, B_l] = 0, \tag{2.51}$$

where f_{lm}^{n} are structure constants and the last two commutation relations simply say that the algebra is the direct sum of the Poincaré algebra and the algebra $\mathbb{G}$ spanned by the generators B_l, that is

$$ISO(1,3) \times G, \tag{2.52}$$

at the level of groups.

The Coleman-Mandula theorem can be evaded by weakening one (or more) of its assumptions. The theorem assumes, in particular, that the symmetry algebra involves only commutators but there are not any specific physical requirements for this to be needed. Haag, Lopuszanski and Sohnius generalized the notion of Lie algebra to include algebraic systems involving, in addition to commutators, also anticommutators. This extended Lie algebra goes under the name of *graded* Lie algebra. Allowing for a graded Lie algebra weakens the Coleman-Mandula theorem enough to allow for supersymmetry, which is nothing but a specific graded Lie algebra.

Let us first define what a graded Lie algebra is. Recall that a Lie algebra is a vector space (over some field, say $\mathbb{R}$ or $\mathbb{C}$) which enjoys an additional composition rule, called product

$$[\quad,\quad] : L \times L \to L, \tag{2.53}$$

with the following properties

$$[v_1, v_2] \in L$$
$$[v_1, (v_2 + v_3)] = [v_1, v_2] + [v_1, v_3]$$
$$[v_1, v_2] = -[v_2, v_1]$$
$$[v_1, [v_2, v_3]] + [v_2, [v_3, v_1]] + [v_3, [v_1, v_2]] = 0,$$

where v_i are elements of the algebra. A *graded* Lie algebra of grade n is a vector space

$$L = \oplus_{i=0}^{i=n} L_i \tag{2.54}$$

where L_i are all vector spaces, and the product

$$[\ ,\ \} : L \times L \to L \tag{2.55}$$

has the following properties

$$[L_i, L_j\} \in L_{i+j} \mod n+1$$

$$[L_i, L_j\} = -(-1)^{ij}[L_j, L_i\}$$

$$[L_i, [L_j, L_k\}\}(-1)^{ik} + [L_j, [L_k, L_i\}\}(-1)^{ij} + [L_k, [L_i, L_j\}\}(-1)^{jk} = 0.$$

From the first such properties it follows that L_0 is a Lie algebra while all other L_i's with $i \neq 0$ are not. The second property is called supersymmetrization while the third one is nothing but the generalization to a graded algebra of the well known Jacobi identity any algebra satisfies.

The supersymmetry algebra is a graded Lie algebra of grade one, namely

$$L = L_0 \oplus L_1, \tag{2.56}$$

where L_0 is the Poincaré algebra and $L_1 = (Q^I_\alpha, \overline{Q}^I_{\dot\alpha})$ with $I = 1, \ldots, N$, where $Q^I_\alpha, \overline{Q}^I_{\dot\alpha}$ is a set of $N + N = 2N$ anticommuting fermionic generators transforming in the representations $\left(\frac{1}{2}, 0\right)$ and $\left(0, \frac{1}{2}\right)$ of the Lorentz group, respectively. Haag, Lopuszanski and Sohnius proved that this is the only possible consistent extension of the Poincaré algebra, given the other (very physical) assumptions one would not like to relax of the Coleman-Mandula theorem. For instance, generators with spin higher than one, like those transforming in representation $\left(\frac{1}{2}, 1\right)$, cannot be there.

The generators of L_1 are spinors and hence they transform non-trivially under the Lorentz group. Therefore, supersymmetry is not an internal symmetry. Rather it is an extension of Poincaré space-time symmetries. In fact, the generators of L_1, *i.e.* the supersymmetry generators, when acting on bosons transform them into fermions (and viceversa). Hence, this symmetry naturally mixes radiation with matter.

The supersymmetry algebra, besides the commutators (2.46)–(2.51), contains the following (anti)commutators

$$\left[P_\mu, Q^I_\alpha\right] = 0 \tag{2.57}$$

$$\left[P_\mu, \overline{Q}^I_{\dot\alpha}\right] = 0 \tag{2.58}$$

$$\left[M_{\mu\nu}, Q^I_\alpha\right] = i(\sigma_{\mu\nu})_\alpha^{\ \beta} Q^I_\beta \tag{2.59}$$

$$\left[M_{\mu\nu}, \overline{Q}^{I\dot{\alpha}}\right] = i(\bar{\sigma}_{\mu\nu})^{\dot{\alpha}}_{\ \dot{\beta}} \overline{Q}^{I\dot{\beta}} \tag{2.60}$$

$$\left\{Q^I_\alpha, \overline{Q}^J_{\dot{\beta}}\right\} = 2\sigma^\mu_{\alpha\dot{\beta}} P_\mu \delta^{IJ} \tag{2.61}$$

$$\left\{Q^I_\alpha, Q^J_\beta\right\} = \epsilon_{\alpha\beta} Z^{IJ}, \quad Z^{IJ} = -Z^{JI} \tag{2.62}$$

$$\left\{\overline{Q}^I_{\dot{\alpha}}, \overline{Q}^J_{\dot{\beta}}\right\} = \epsilon_{\dot{\alpha}\dot{\beta}} (Z^{IJ})^* \tag{2.63}$$

Let us discuss a bit the above structure.

- Eqs. (2.59) and (2.60) follow from the fact that Q^I_α and $\overline{Q}^I_{\dot{\alpha}}$ are spinors of the Lorentz group, recall eq. (2.45). From these same equations, recalling that $M_{12} = J_3$, one also sees that

$$\left[J_3, Q^I_1\right] = \frac{1}{2} Q^I_1, \quad \left[J_3, Q^I_2\right] = -\frac{1}{2} Q^I_2. \tag{2.64}$$

  Taking the hermitian conjugate of the above relations we get

$$\left[J_3, \overline{Q}^I_{\dot{1}}\right] = -\frac{1}{2} \overline{Q}^I_{\dot{1}}, \quad \left[J_3, \overline{Q}^I_{\dot{2}}\right] = \frac{1}{2} \overline{Q}^I_{\dot{2}} \tag{2.65}$$

  and so we see that Q^I_1 and $\overline{Q}^I_{\dot{2}}$ rise the z-component of the spin by half unit while Q^I_2 and $\overline{Q}^I_{\dot{1}}$ lower it by half unit.
- Eq. (2.61) has a very important implication. First notice that given the transformation properties of Q^I_α and $\overline{Q}^J_{\dot{\beta}}$ under Lorentz transformations, their anticommutator should be symmetric under $I \leftrightarrow J$ and should transform as

$$\left(\frac{1}{2}, 0\right) \otimes \left(0, \frac{1}{2}\right) = \left(\frac{1}{2}, \frac{1}{2}\right). \tag{2.66}$$

  The obvious such candidate is P_μ which is the only generator in the algebra with such transformation properties (the δ^{IJ} in eq. (2.61) is achieved by diagonalizing an arbitrary symmetric matrix and rescaling the Q's and the $\overline{Q}$'s). Hence, the commutator of two supersymmetry transformations is a translation. In theories with *local* supersymmetry (*i.e.* where the spinorial infinitesimal parameter of the supersymmetry transformation depends on x^μ), the commutator is an infinitesimal translation whose parameter depends on x^μ. This is nothing but a theory invariant under general coordinate transformation, namely a theory of gravity! The upshot is that theories with local supersymmetry automatically incorporate gravity, the two things are tight together. Such theories are called *supergravity* theories, SUGRA for short.

- Eqs. (2.57) and (2.58) are not at all obvious. Compatibility with Lorentz symmetry would imply the right hand side of eq. (2.57) to transform as

$$\left(\frac{1}{2},\frac{1}{2}\right)\otimes\left(\frac{1}{2},0\right)=\left(0,\frac{1}{2}\right)\oplus\left(1,\frac{1}{2}\right), \tag{2.67}$$

and similarly for eq. (2.58). The second term on the right hand side cannot be there, due to the theorem of Haag, Lopuszanski and Sohnius which says that the only allowed fermionic generators in the algebra are supersymmetry generators, which are spin $\frac{1}{2}$. In other words, there cannot be a consistent extension of the Poincaré algebra including generators transforming in the $(1,\frac{1}{2})$ under the Lorentz group. Still, group theory arguments by themselves do not justify eqs. (2.57) and (2.58) but rather something like

$$[P_\mu, Q^I_\alpha] = C^I{}_J \sigma_{\mu\,\alpha\dot\beta}\,\overline{Q}^{J\dot\beta} \tag{2.68}$$

$$\left[P_\mu, \overline{Q}^I_{\dot\alpha}\right] = (C^I{}_J)^* \bar\sigma_{\mu\,\dot\alpha\beta}\, Q^{J\beta}. \tag{2.69}$$

where $C^I{}_J$ is an undetermined matrix. We want to prove that this matrix vanishes. Let us first consider the generalized Jacobi identity which the supersymmetry algebra should satisfy and let us apply it to the (Q,P,P) system. We get

$$\begin{aligned}
&[[Q^I_\alpha, P_\mu], P_\nu] + [[P_\mu, P_\nu], Q^I_\alpha] + [[P_\nu, Q^I_\alpha], P_\mu] \\
&= -C^I{}_J \sigma_{\mu\,\alpha\dot\beta}\left[\overline{Q}^{J\dot\beta}, P_\nu\right] + C^I{}_J \sigma_{\nu\,\alpha\dot\beta}\left[\overline{Q}^{J\dot\beta}, P_\mu\right] \\
&= C^I{}_J C^{J*}_K \sigma_{\mu\,\alpha\dot\beta}\,\bar\sigma_\nu^{\dot\beta}{}_\gamma Q^{K\gamma} - C^I{}_J C^{J*}_K \sigma_{\nu\,\alpha\dot\beta}\,\bar\sigma_\mu^{\dot\beta}{}_\gamma Q^{K\gamma} \\
&= 4(C\,C^*)^I{}_K\,(\sigma_{\mu\nu})_{\alpha\gamma} Q^{K\gamma} = 0.
\end{aligned}$$

This implies that

$$C\,C^* = 0, \tag{2.70}$$

as a matrix equation. Note that this is not enough to conclude, as we would, that $C = 0$. For that, we also need to show, in addition, that C is symmetric. To this aim we have to consider other equations, as discussed below.

- Let us now consider eqs. (2.62) and (2.63). As for the first, from Lorentz representation theory we would expect

$$\left(\frac{1}{2},0\right)\otimes\left(\frac{1}{2},0\right)=(0,0)\oplus(1,0), \tag{2.71}$$

which explicitly means

$$\{Q^I_\alpha, Q^J_\beta\}=\epsilon_{\alpha\beta}Z^{IJ}+\epsilon_{\beta\gamma}(\sigma^{\mu\nu})_\alpha^{\ \gamma}M_{\mu\nu}Y^{IJ}. \tag{2.72}$$

The Z^{IJ}, being Lorentz scalars, should be some linear combination of the internal symmetry generators B_l and, given the antisymmetric properties of the epsilon tensor under $\alpha\leftrightarrow\beta$, they should be anti-symmetric under $I\leftrightarrow J$. On the contrary, given that $\epsilon_{\beta\gamma}(\sigma^{\mu\nu})_\alpha^{\ \gamma}$ is symmetric in $\alpha\leftrightarrow\beta$, the matrix Y^{IJ} should be symmetric under $I\leftrightarrow J$.

Let us now consider the generalized Jacobi identity between (Q,Q,P), which can be written as

$$[\{Q_{I\alpha},Q_{J\beta}\},P_\mu]=\{Q_{I\alpha},[Q_{J\beta},P_\mu]\}+\{Q_{J\beta},[Q_{I\alpha},P_\mu]\}. \tag{2.73}$$

If one multiplies it by $\epsilon^{\alpha\beta}$, only the anti-symmetric part under $\alpha\leftrightarrow\beta$ of the left hand side survives, which, from eq. (2.72), can be seen to vanish since the matrix Z_{IJ} commutes with P_μ. So we get

$$\begin{aligned}0&=\epsilon^{\alpha\beta}\{Q_{I\alpha},[Q_{J\beta},P_\mu]\}+\epsilon^{\alpha\beta}\{Q_{J\beta},[Q_{I\alpha},P_\mu]\}\\&=\epsilon^{\alpha\beta}C_I^{\ K}\sigma_{\mu\,\beta\dot\beta}\left\{Q_{J\alpha},\overline{Q}^{\dot\beta}_K\right\}-\epsilon^{\alpha\beta}C_J^{\ K}\sigma_{\mu\,\beta\dot\beta}\left\{Q_{I\alpha},\overline{Q}^{\dot\beta}_K\right\}\\&\sim(C_{IJ}-C_{JI})\,\bar\sigma^{\dot\gamma\alpha}_\mu\sigma^\nu_{\alpha\dot\gamma}P_\nu=2\,(C_{IJ}-C_{JI})P_\mu,\end{aligned}$$

which implies that the matrix C is symmetric. So the previously found equation $C\,C^*=0$ can be promoted to $CC^\dagger=0$, which in turn implies $C=0$ and hence eq. (2.57). A similar rationale leads to eq. (2.58).

Let us now come back to eq. (2.62), which we have not yet proven. To do so, we should plug eq. (2.57) into the Jacobi identity (2.73), getting

$$[\{Q_{I\alpha},Q_{J\beta}\},P_\mu]=0, \tag{2.74}$$

which implies, by (2.72), that the matrix Y^{IJ} vanishes because P_μ does not commute with $M_{\mu\nu}$. This finally proves eq. (2.62). Similarly, one can prove eq. (2.63), which is just the hermitian conjugate of (2.62).

What about the commutation relations between supersymmetry generators and internal symmetry generators, if any? In general, the Q's will carry a representation of the internal symmetry group G. So one expects something like

$$\left[Q^I_\alpha, B_l\right] = (b_l)^I{}_J Q^J_\alpha \tag{2.75}$$

$$\left[\overline{Q}_{I\dot\alpha}, B_l\right] = -\overline{Q}_{J\dot\alpha}(b_l)^J{}_I. \tag{2.76}$$

The second commutation relation comes from the first under hermitian conjugation, recalling that the b_l are hermitian, so are the generators B_l. The largest possible internal symmetry group which can act non trivially on the Q's is thus $U(N)$, and this is called the R-symmetry group (the relation between a Lie algebra with generators S and the corresponding Lie group with elements U is $U = e^{iS}$; hence hermitian generators, $S^\dagger = S$, correspond to unitary groups, $U^\dagger = U^{-1}$). In presence of non-vanishing central charges one can prove that the R-symmetry group reduces to $USp(N)$, the compact version of the symplectic group $Sp(N)$, $USp(N) \cong U(N) \cap Sp(N)$.

As already noticed, the operators Z_{IJ} are Lorentz scalars and should then correspond to some linear combination of the internal symmetry group generators B_l of the compact Lie algebra G, say

$$Z^{IJ} = a^{l|IJ} B_l. \tag{2.77}$$

Using the above equation, the supersymmetry algebra (2.57)–(2.63) and eqs. (2.75) and (2.76) one can actually prove that the Z's are *central charges*, that is they commute with the whole supersymmetry algebra, and within themselves. Contrary to what one could naively think, this does not imply they are ineffective. Indeed, central charges are not numbers but quantum operators and their value may vary from state to state. On a state which is annihilated by all supersymmetry generators, they are trivially realized, recall eqs. (2.62) and (2.63). However, they do not need to vanish in general. For instance, as we will see in the subsequent chapter, massive representations are very different if Z^{IJ} vanishes or if it is non-trivially realized on the representation.

We have been using two-component spinor notation in writing the supersymmetry algebra, but one can equivalently use four-component notation. In four component notation the $2N$ supersymmetry generators Q^I_α, $\overline{Q}^I_{\dot\alpha}$ constitute a set of N Majorana spinors

$$Q^I = \begin{pmatrix} Q^I_\alpha \\ \overline{Q}^{I\dot\alpha} \end{pmatrix} \quad \overline{Q}^I = \begin{pmatrix} Q^{I\alpha} & \overline{Q}^I_{\dot\alpha} \end{pmatrix} \tag{2.78}$$

and the supersymmetry algebra reads

$$\{Q^I, \overline{Q}^J\} = 2\delta^{IJ}\gamma^\mu P_\mu - i\,\mathbb{I}\,\mathrm{Im}Z^{IJ} - \gamma_5\,\mathrm{Re}Z^{IJ}$$

$$[Q^I, P_\mu] = 0 \quad [Q^I, M_{\mu\nu}] = \frac{i}{2}\gamma_{\mu\nu}Q^I \quad [Q^I, R] = i\gamma_5 Q^I. \qquad (2.79)$$

Depending on what one needs to do, one notation can be more useful than the other. In the following, we will stick to two component spinor notation, unless otherwise stated.

Let us close this section with a few more comments. First, if $N = 1$ there are only two supersymmetry generators, which correspond to one Majorana spinor in four component notation. In this case we then speak of unextended or minimal supersymmetry (and there are no non-trivial central charges). For $N > 1$ we have extended supersymmetry (and there can now be a central extension of the supersymmetry algebra). From an algebraic point of view there is no limit to N. However, as we will see later, increasing N the theory must contain particles of increasing spin. In particular we have

- $N \leq 4$ for theories without gravity (spin ≤ 1)
- $N \leq 8$ for theories with gravity (spin ≤ 2)

Therefore, to avoid theories with spin higher than two (that is focusing on local, interacting theories) $N = 8$ is an upper bound. As discussed in the previous chapter, this statement is true in four space-time dimensions. Equivalent statements can be made in higher/lower dimensions, where the dimension of the spinor representation of the Lorentz group is bigger/smaller. What matters is the number of single state supersymmetry generators, which is a dimension-independent statement (*e.g.*, $N = 8$ corresponds to 32 supercharges).

For $N = 1$ the R-symmetry group is just $U(1)$ (one can see it from the generalized Jacobi identity between (Q, B, B) which implies that the $f_{lm}^{\ \ n}$ are trivially realized on the supersymmetry generators). In this case the hermitian matrices b_l are just real numbers and by rescaling the generators B_l one gets

$$[R, Q_\alpha] = -Q_\alpha, \quad [R, \overline{Q}_{\dot\alpha}] = +\overline{Q}_{\dot\alpha}. \qquad (2.80)$$

This implies that in minimal supersymmetric models, supersymmetric partners (which are related by the action of the Q's) have different $U(1)$ R-charge. In particular, given eqs. (2.80), if a particle has $R = 0$ then its superpartner has $R = \pm 1$. An important physical consequence of this property is that in a theory where the R-symmetry is preserved, the lightest supersymmetric particle (LSP) is stable.

2.4 Exercises

Exercise 2.1. Prove the following spinor identities

$$\begin{aligned}
\psi^\alpha\psi^\beta &= -\frac{1}{2}\epsilon^{\alpha\beta}\psi\psi & (\theta\phi)(\theta\psi) &= -\frac{1}{2}(\phi\psi)(\theta\theta)\\
\chi\sigma^\mu\overline{\psi} &= -\overline{\psi}\bar{\sigma}^\mu\chi & \chi\sigma^\mu\bar{\sigma}^\nu\psi &= \psi\sigma^\nu\bar{\sigma}^\mu\chi\\
(\chi\sigma^\mu\overline{\psi})^\dagger &= \psi\sigma^\mu\overline{\chi} & (\chi\sigma^\mu\bar{\sigma}^\nu\psi)^\dagger &= \overline{\psi}\bar{\sigma}^\nu\sigma^\mu\overline{\chi}\\
(\theta\psi)(\theta\sigma^\mu\overline{\phi}) &= -\frac{1}{2}(\theta\theta)(\psi\sigma^\mu\overline{\phi}) & (\overline{\theta\psi})(\bar{\theta}\bar{\sigma}^\mu\phi) &= -\frac{1}{2}(\overline{\theta\theta})(\overline{\psi}\bar{\sigma}^\mu\phi)
\end{aligned}$$

$$(\phi\psi)\cdot\overline{\chi}_{\dot{\alpha}} = \frac{1}{2}(\phi\sigma^\mu\overline{\chi})(\psi\sigma_\mu)_{\dot{\alpha}}$$

Exercise 2.2. The operators Z^{IJ} are linear combinations of the internal symmetries generators B_l, eq. (2.77). Hence, they commute with P_μ and $M_{\mu\nu}$. Prove that Z_{IJ} are in fact central charges of the supersymmetry algebra, namely that it also holds that

$$[Z^{IJ}, B_l] = 0, \quad [Z^{IJ}, Z^{KL}] = 0, \quad [Z^{IJ}, Q^K_\alpha] = 0, \quad [Z^{IJ}, \bar{Q}^K_{\dot{\alpha}}] = 0.$$

References

[1] A. Bilal, *Introduction to Supersymmetry*, Section 2, arXiv:hep-th/0101055.

[2] H. J. W. Müller-Kirsten and A. Wiedemann, *Introduction to Supersymmetry*, Chapter 1, World Scientific Singapore (2010).

[3] L. Castellani, R. D'Auria and P. Fre, *Supergravity And Superstrings: A Geometric Perspective. Vol. 1: Mathematical Foundations*, Chapter II.2, World Scientific Singapore (1991).

[4] J. D. Lykken, *Introduction to Supersymmetry*, Appendix, TASI 96, arXiv: hep-th/9612114.

[5] M. F. Sohnius, Introducing Supersymmetry, Section 3, *Phys. Rep.* **128** (1985).

[6] S. Weinberg, *The Quantum Theory of Fields. Vol. 3: Supersymmetry*, Chapter 24.B, Cambridge University Press (2000).

[7] S. R. Coleman and J. Mandula, All Possible Symmetries of the S Matrix, *Phys. Rev.* **159** (1967), 1251.

[8] R. Haag, J. T. Lopuszanski and M. Sohnius, All Possible Generators of Supersymmetries of the s Matrix, *Nucl. Phys. B* **88** (1975), 257.

Chapter 3

Representations of the Supersymmetry Algebra

The goal of this chapter is to construct representations of the supersymmetry algebra. Let us first recall how things go for the Poincaré algebra. The Poincaré algebra (2.25) has two Casimir operators (*i.e.* two operators which commute with all generators)

$$P^2 = P_\mu P^\mu \quad \text{and} \quad W^2 = W_\mu W^\mu , \tag{3.1}$$

where $W^\mu = \frac{1}{2}\epsilon^{\mu\nu\rho\sigma} P_\nu M_{\rho\sigma}$ is the so-called Pauli-Lubanski vector. Casimir operators are useful to classify irreducible representations of a group. In the case of the Poincaré group such representations are nothing but what we usually call *particles*. Let us see how this is realized for massive and massless particles, respectively.

Let us first consider a massive particle with mass m and go to the rest frame, $P_\mu = (m, 0, 0, 0)$. In this frame it is easy to see that the two Casimir reduce to $P^2 = m^2$ and $W^2 = -m^2 j(j+1)$ where j is the spin. The second equality can be proven by noticing that $W_\mu P^\mu = 0$ which implies that in the rest frame $W_0 = 0$. Therefore, in the rest frame $W_\mu = (0, \frac{1}{2}\,\epsilon_{i0jk} m\, M^{jk})$ from which one immediately gets $W^2 = -m^2 \vec{J}^2$. So we see that massive particles are distinguished by their mass and their spin.

Let us now consider massless particles. In the rest frame $P_\mu = (E, 0, 0, E)$. In this case we have that $P^2 = 0$ and $W^2 = 0$, and $W^\mu = M_{12} P^\mu$. In other words, the two operators are proportional for a massless particle, the constant of proportionality being $M_{12} = \pm j$, the helicity. For these representations the spin is then fixed and the different states are distinguished by their energy and by the sign of the helicity (*e.g.* the photon is a massless particle with two helicity states, ± 1).

Now, as a particle is an irreducible representation of the Poincaré algebra, we call superparticle an irreducible representation of the supersymmetry algebra. Since the Poincaré algebra is a subalgebra of the supersymmetry algebra, it follows that any irreducible representation of the supersymmetry algebra is a representation of the Poincaré algebra, which in general will be reducible. This means that a superparticle corresponds to a collection of particles, the latter being related by the action of the supersymmetry generators Q^I_α and $\overline{Q}^I_{\dot\alpha}$ and having spins differing by units of half. Being a multiplet of different particles, a superparticle is often called *supermultiplet.*

Before discussing in detail specific representations of the supersymmetry algebra, let us list three generic properties which any of such representations enjoy, all of them having very important physical implications.

(1) As compared to the Poincaré algebra, in the supersymmetry algebra P^2 is still a Casimir, but W^2 is not anymore (this follows from the fact that $M_{\mu\nu}$ does not commute with the supersymmetry generators). Therefore, particles belonging to the same supermultiplet have the same mass and different spin, since the latter is not a conserved quantum number of the representation. The mass degeneracy between bosons and fermions is something we do not observe in known particle spectra; this implies that supersymmetry, if at all realized, must be broken in Nature.
Note: what's above is true in Minkowski space. If one wants to discuss supersymmetry in, *e.g.* anti-de Sitter space, things are different. In anti-de Sitter space the generators P_μ do not commute with Lorentz generators, nor with the supercharges. The consequence is that P^2 is not anymore a Casimir but rather $C = M^2 + \alpha P^2$ is, with α a dimension-full quantity proportional to the anti-de Sitter radius squared. So, in anti-de Sitter space states belonging to the same multiplet do have the same C-eigenvalue but different P^2 eigenvalue, so different masses.

(2) In a supersymmetric theory the energy of *any* state is always ≥ 0. Consider an arbitrary state $|\phi\rangle$. Using the supersymmetry algebra, we easily get

$$\begin{aligned}\langle\phi|\left\{Q^I_\alpha,\overline{Q}^I_{\dot\alpha}\right\}|\phi\rangle &= 2\sigma^\mu_{\alpha\dot\alpha}\langle\phi|P_\mu|\phi\rangle\,\delta^{II}\\ &= \langle\phi|\left(Q^I_\alpha(Q^I_\alpha)^\dagger + (Q^I_\alpha)^\dagger Q^I_\alpha\right)|\phi\rangle\\ &= ||(Q^I_\alpha)^\dagger|\phi\rangle||^2 + ||Q^I_\alpha|\phi\rangle||^2 \geq 0.\end{aligned}$$

The last inequality follows from positivity of the Hilbert space. Summing over $\alpha = \dot{\alpha} = 1, 2$ and recalling that Tr $\sigma^\mu = 2\delta^{\mu 0}$ we get

$$4 \langle \phi | P_0 | \phi \rangle \geq 0, \tag{3.2}$$

as anticipated.

(3) A supermultiplet contains an equal number of bosonic and fermionic d.o.f., $n_B = n_F$. Define a fermion number operator

$$(-1)^{N_F} = \begin{cases} -1 & \text{fermionic state} \\ +1 & \text{bosonic state} \end{cases} \tag{3.3}$$

N_F can be taken to be twice the spin, $N_F = 2s$. Such an operator, when acting on a bosonic, respectively a fermionic state, gives indeed

$$(-1)^{N_F} |B\rangle = |B\rangle, \quad (-1)^{N_F} |F\rangle = -|F\rangle. \tag{3.4}$$

We want to show that Tr $(-1)^{N_F} = 0$ if the trace is taken over a finite dimensional representation of the supersymmetry algebra. First notice that

$$\{Q^I_\alpha, (-1)^{N_F}\} = 0 \to Q^I_\alpha (-1)^{N_F} = -(-1)^{N_F} Q^I_\alpha. \tag{3.5}$$

Using this property and the cyclicity of the trace one easily sees that

$$\begin{aligned} 0 &= \text{Tr}\left(-Q^I_\alpha (-1)^{N_F} \overline{Q}^J_{\dot{\beta}} + (-1)^{N_F} \overline{Q}^J_{\dot{\beta}} Q^I_\alpha\right) \\ &= \text{Tr}\left((-1)^{N_F} \left\{Q^I_\alpha, \overline{Q}^J_{\dot{\beta}}\right\}\right) = 2\sigma^\mu_{\alpha\dot{\beta}} \text{Tr}\left[(-1)^{N_F}\right] P_\mu \delta^{IJ}. \end{aligned}$$

Summing on I, J and choosing any $P_\mu \neq 0$ it follows that Tr $(-1)^{N_F} = 0$, which implies that $n_B = n_F$.

In the following, we discuss (some) representations in detail. Since the mass is a conserved quantity in a supermultiplet, it is meaningful distinguishing between massless and massive representations. Let us start from the former.

3.1 Massless supermultiplets

Let us first assume that all central charges vanish, i.e. $Z^{IJ} = 0$ (we will see later that this is the only relevant case, for massless representations). Notice that in this case it follows from eqs. (2.62) and (2.63) that all Q's and all $\overline{Q}$'s anticommute among themselves. The steps to construct the irreps are as follows:

(1) Go to the rest frame where $P_\mu = (E, 0, 0, E)$. In such frame we get

$$\sigma^\mu P_\mu = \begin{pmatrix} 0 & 0 \\ 0 & 2E \end{pmatrix} \tag{3.6}$$

Plugging this into eq. (2.61) we get

$$\left\{ Q_\alpha^I, \overline{Q}_{\dot\beta}^J \right\} = \begin{pmatrix} 0 & 0 \\ 0 & 4E \end{pmatrix}_{\alpha\dot\beta} \delta^{IJ} \quad \longrightarrow \quad \left\{ Q_1^I, \overline{Q}_{\dot 1}^J \right\} = 0. \tag{3.7}$$

Due to the positiveness of the Hilbert space, this implies that both Q_1^I and $\overline{Q}_{\dot 1}^I$ are trivially realized. Indeed, from the equation above we get

$$0 = \langle \phi | \left\{ Q_1^I, \overline{Q}_{\dot 1}^I \right\} | \phi \rangle = || Q_1^I | \phi \rangle ||^2 + || \overline{Q}_{\dot 1}^I | \phi \rangle ||^2, \tag{3.8}$$

whose only solution is $Q_1^I = \overline{Q}_{\dot 1}^I = 0$. We are then left with just Q_2^I and $\overline{Q}_{\dot 2}^I$, hence only half of the generators.

(2) From the non-trivial generators we can define

$$a_I \equiv \frac{1}{\sqrt{4E}} Q_2^I, \quad a_I^\dagger \equiv \frac{1}{\sqrt{4E}} \overline{Q}_{\dot 2}^I. \tag{3.9}$$

These operators satisfy the anticommutation relations of a set of N creation and N annihilation operators

$$\left\{ a_I, a_J^\dagger \right\} = \delta^{IJ}, \quad \{ a_I, a_J \} = 0, \quad \left\{ a_I^\dagger, a_J^\dagger \right\} = 0. \tag{3.10}$$

These are the basic tools we need in order to construct representations of the supersymmetry algebra. Recall that in our conventions the operators Q_2^I and $\overline{Q}_{\dot 2}^I$ (and hence a_I and $a_I^\dagger$) lower respectively raise the helicity of half unit on the state they act on.

(3) To construct a representation, one can start by choosing a state annihilated by all a_I's (known as the Clifford vacuum): such state will carry some irrep of the Poincaré algebra. Besides having $m = 0$, it will carry some helicity λ_0, and we call it $|E, \lambda_0\rangle$ ($|\lambda_0\rangle$ for short). For this state

$$a_I |\lambda_0\rangle = 0. \tag{3.11}$$

Note that this state can be either bosonic or fermionic, and should not be confused with the actual vacuum of the theory, which is the state of minimal energy: the Clifford vacuum is a state with quantum numbers (E, λ_0) and which satisfies eq. (3.11).

(4) The full representation (aka supermultiplet) is obtained acting on $|\lambda_0\rangle$ with the creation operators $a_I^\dagger$ as follows

$$|\lambda_0\rangle,\ a_I^\dagger|\lambda_0\rangle \equiv |\lambda_0 + \frac{1}{2}\rangle_I,\ a_I^\dagger a_J^\dagger|\lambda_0\rangle \equiv |\lambda_0 + 1\rangle_{IJ},$$
$$\ldots,\ a_1^\dagger a_2^\dagger \ldots a_N^\dagger|\lambda_0\rangle \equiv |\lambda_0 + \frac{N}{2}\rangle.$$

Hence, starting from a Clifford vacuum with helicity λ_0, the state with highest helicity in the representation has helicity $\lambda = \lambda_0 + \frac{N}{2}$. Due to the antisymmetry in $I \leftrightarrow J$, at helicity level $\lambda = \lambda_0 + \frac{k}{2}$ we have

$$\#\text{ of states with helicity } \lambda_0 + \frac{k}{2} = \binom{N}{k}, \tag{3.12}$$

where $k = 0, 1, \ldots, N$. The total number of states in the irrep will then be

$$\sum_{k=0}^{N} \binom{N}{k} = 2^N = \left(2^{N-1}\right)_B + \left(2^{N-1}\right)_F, \tag{3.13}$$

half of them having integer helicity (bosons), half of them half-integer helicity (fermions).

(5) CPT flips the sign of the helicity. Therefore, unless the helicity is distributed symmetrically around 0, which is not the case in general, a supermultiplet is not CPT-invariant. This means that in order to have a CPT-invariant theory one should in general double the supermultiplet we have just constructed adding its CPT conjugate. This is not needed if the supermultiplet is self-CPT conjugate, which can happen only if $\lambda_0 = -\frac{N}{4}$ (in this case the helicity is indeed distributed symmetrically around 0).

Let us now apply the above procedure and construct several (physically interesting) irreps of the supersymmetry algebra.

N = 1 supersymmetry

- Matter multiplet (aka chiral multiplet):

$$\lambda_0 = 0 \ \rightarrow\ \left(0, +\frac{1}{2}\right) \underset{CPT}{\oplus} \left(-\frac{1}{2}, 0\right). \tag{3.14}$$

The degrees of freedom of this representation are those of one Weyl fermion and one complex scalar (on shell; recall we are constructing supersymmetry representations on states!). Note that since the two representations above are exchanged by CPT, the two spin 0 states have opposite parity, so if one corresponds to a scalar the other is a pseudoscalar. In a $N = 1$ supersymmetric theory this is the representation where matter sits; this is why such multiplets are called matter multiplets.

- Gauge (or vector) multiplet:

$$\lambda_0 = \frac{1}{2} \rightarrow \left(+\frac{1}{2}, +1\right) \underset{CPT}{\oplus} \left(-1, -\frac{1}{2}\right). \tag{3.15}$$

The degrees of freedom are those of one vector and one Weyl fermion. This is the representation one needs to describe gauge fields in a supersymmetric theory. Notice that since internal symmetries (but the R-symmetry) commute with the supersymmetry algebra, the representation the Weyl fermion should transform under gauge transformations should be the same as the vector field, *i.e.* the adjoint. Hence, usual Standard Model matter (quarks and leptons) cannot be accommodated in these multiplets.

Although in this course we will focus on rigid supersymmetry and hence not consider supersymmetric theories with gravity, let us list for completeness (and future reference) also representations containing states with higher helicity.

- Spin 3/2 multiplet:

$$\lambda_0 = 1 \rightarrow \left(1, +\frac{3}{2}\right) \underset{CPT}{\oplus} \left(-\frac{3}{2}, -1\right). \tag{3.16}$$

The degrees of freedom are those of a spin 3/2 particle and one vector.

- Graviton multiplet:

$$\lambda_0 = \frac{3}{2} \rightarrow \left(+\frac{3}{2}, +2\right) \underset{CPT}{\oplus} \left(-2, -\frac{3}{2}\right). \tag{3.17}$$

The degrees of freedom are those of a graviton, which has helicity 2, and a particle of helicity 3/2, known as the gravitino (the supersymmetric partner of the graviton).

Representations constructed from a Clifford vacuum with higher helicity will inevitably include states with helicity higher than 2. Hence, if one is interested in *interacting local field theories*, the story stops here. Recall that in an interacting local field theory massless particles with helicity higher than $\frac{1}{2}$ should couple to conserved quantities at low momentum. The latter are: conserved internal symmetry generators for (soft) massless vectors, supersymmetry generators for (soft) gravitinos and four-vector P_μ for (soft) gravitons. The supersymmetry algebra does not allow for generators other than these ones. Hence, supermultiplets with helicity $\lambda \geq \frac{5}{2}$ are ruled out: they may exist, but they cannot have couplings that survive in the low energy limit.

The above discussion also implies that in a local interacting field theory a spin 3/2 particle is inevitably associated to local supersymmetry and hence, in turn, with gravity. Therefore, there is no much meaning for a theory without the graviton multiplet and a spin 3/2 multiplet, which would be a non-interacting one in fact. In other words, the physical gravitino is the one sitting in the graviton multiplet.

N = 2 supersymmetry

- Matter multiplet (aka hypermultiplet):

$$\lambda_0 = -\frac{1}{2} \to \left(-\frac{1}{2}, 0, 0, +\frac{1}{2}\right) \underset{CPT}{\oplus} \left(-\frac{1}{2}, 0, 0, +\frac{1}{2}\right). \tag{3.18}$$

  The degrees of freedom are those of two Weyl fermions and two complex scalars. This is where matter sits in a $N = 2$ supersymmetric theory. In $N = 1$ language this representation corresponds to two chiral multiplets with opposite chirality (CPT flips the chirality).
  Note: in principle this representation enjoys the necessary condition to be CPT self-conjugate, $\lambda_0 = -\frac{N}{4}$. However, a closer look shows that an hypermultiplet cannot be self-conjugate (that's why we added the CPT conjugate representation). The way the various states are constructed out of the Clifford vacuum shows that under the compact part of the R-symmetry group, $SU(2)$, the helicity 0 states behave as a doublet while the fermionic states are singlets. If the representation were CPT self-conjugate the two scalar degrees of freedom would have been both real. Such states cannot form a $SU(2)$ doublet since a two-dimensional representation of $SU(2)$ is pseudoreal, and hence the doublet should be complex.

- Gauge (or vector) multiplet:

$$\lambda_0 = 0 \to \left(0, +\frac{1}{2}, +\frac{1}{2}, +1\right) \underset{CPT}{\oplus} \left(-1, -\frac{1}{2}, -\frac{1}{2}, 0\right). \tag{3.19}$$

The degrees of freedom are those of one vector, two Weyl fermions and one complex scalar. In $N = 1$ language this is just a vector and a matter multiplet (both transforming in the adjoint representation of the gauge group).

- Spin 3/2 multiplet:

$$\lambda_0 = -\frac{3}{2} \to \left(-\frac{3}{2}, -1, -1, -\frac{1}{2}\right) \underset{CPT}{\oplus} \left(+\frac{1}{2}, +1, +1, +\frac{3}{2}\right). \quad (3.20)$$

The degrees of freedom are those of a spin 3/2 particle, two vectors and one Weyl fermion.

- Graviton multiplet:

$$\lambda_0 = -2 \to \left(-2, -\frac{3}{2}, -\frac{3}{2}, -1\right) \underset{CPT}{\oplus} \left(+1, +\frac{3}{2}, +\frac{3}{2}, +2\right). \quad (3.21)$$

The degrees of freedom are those of a graviton, two gravitini and a vector, which is usually called graviphoton in the supergravity literature.

N = 4 supersymmetry

- Gauge (or vector) multiplet:

$$\lambda_0 = -1 \to \left(-1, \mathbf{4} \times -\frac{1}{2}, \mathbf{6} \times 0, \mathbf{4} \times +\frac{1}{2}, +1\right). \quad (3.22)$$

The degrees of freedom are those of a vector, four Weyl fermions and three complex scalars. In $N = 1$ language this corresponds to one vector multiplet and three matter multiplets (all transforming in the adjoint). Notice that this multiplet is CPT self-conjugate. This time there are no issues with R-symmetry transformations. The vector is a singlet under $SU(4)$, fermions transform under the fundamental representation, and scalars under the two times anti-symmetric representation, which is the fundamental of $SO(6)$, and is real. The fact that the representation under which scalars transform is real also explains why for $\mathcal{N} = 4$ supersymmetry, the R-symmetry group is not $U(4)$ but just $SU(4)$.

For $N = 4$ it is not possible to have matter in the usual sense, since the number of supersymmetry generators is too high to avoid helicity one states. Therefore, $N = 4$ supersymmetry cannot accommodate fermions transforming in fundamental representations. Besides the vector multiplet there are of course also representations with higher helicity, but we refrain to report them here.

One might wonder why we did not discuss $N = 3$ representations. This is just because as far as non-gravitational theories are concerned, $N = 3$ and $N = 4$ are physically equivalent: when constructing $N = 3$ representations with maximal spin one, once the CPT conjugate representation is added (in this case we cannot satisfy the condition $\lambda_0 = -\frac{N}{4}$) one ends up with a multiplet which is the same as the $N = 4$ vector multiplet. $N = 4$ and $N = 3$ differ only for representations including states with spin higher than one.

N > 4 supersymmetry

In this case one can easily get convinced that it is not possible to avoid gravity since there do not exist representations with helicity smaller than $\frac{3}{2}$ when $N > 4$. Hence, theories with $N > 4$ are all supergravity theories. It is interesting to note that $N = 8$ supergravity allows only one possible representation with highest helicity smaller than $\frac{5}{2}$ and that for higher N one cannot avoid states with helicity $\frac{5}{2}$ or higher. Therefore, $N = 8$ is an upper bound on the number of supersymmetry generators, as far as interacting local field theories are concerned. Beware: as stated, all these statements are valid in four space-time dimensions. The way to count supersymmetries depends on the dimension of space-time, since spinorial representations get bigger, the more the dimensions. Obviously, completely analogous statements can be made in higher dimensions. For instance, in ten space-time dimensions the maximum allowed supersymmetry to avoid states with helicity $\frac{5}{2}$ or higher is $N = 2$. A dimension-independent statement can be made counting the number of single component supersymmetry generators. Using this language, the maximum allowed number of supersymmetry generators for non-gravitational theories is 16 (which is indeed $N = 4$ in four dimensions) and 32 for theories with gravity (which is $N = 8$ in four dimensions).

Table 3.1 summarizes all results we have discussed.

A final, very important comment regards chirality. The Standard Model is a chiral theory, in the sense that there exist particles in the spectrum whose chiral and anti-chiral components transform differently under the gauge group (weak interactions are chiral). When it comes to supersymmetric extensions, it is easy to see that only $N = 1$ theories allow for chiral matter. That $N = 1$ irreps can be chiral is obvious:

Table 3.1 The numbers in parenthesis represent the helicity, while others represent the multiplicity of states with given helicity. Notice that, as anticipated, any supermultiplet contains particles with spin at least as large as $\frac{1}{4}N$. The $N = 7$ theory has not been reported since that allows only the graviton multiplet which, once CPT invariance is required, is identical to the $N = 8$ graviton multiplet. In other words, at the interacting level, $N = 7$ supergravity is the same as $N = 8$ supergravity (this is the same argument we used for not discussing $N = 3$ representations as far as maximal spin one is concerned).

N	$\lambda_{max} = 1$	$\lambda_{max} = \frac{1}{2}$	$\lambda_{max} = 2$	$\lambda_{max} = \frac{3}{2}$
8	none	none	$\left[(2), 8\left(\frac{3}{2}\right), 28(1), 56\left(\frac{1}{2}\right), 70(0)\right]$	none
6	none	none	$\left[(2), 6\left(\frac{3}{2}\right), 16(1), 26\left(\frac{1}{2}\right), 30(0)\right]$	$\left[\left(\frac{3}{2}\right), 6(1), 15\left(\frac{1}{2}\right), 20(0)\right]$
5	none	none	$\left[(2), 5\left(\frac{3}{2}\right), 10(1), 11\left(\frac{1}{2}\right), 10(0)\right]$	$\left[\left(\frac{3}{2}\right), 6(1), 15\left(\frac{1}{2}\right), 20(0)\right]$
4	$\left[(1), 4\left(\frac{1}{2}\right), 6(0)\right]$	none	$\left[(2), 4\left(\frac{3}{2}\right), 6(1), 4\left(\frac{1}{2}\right), 2(0)\right]$	$\left[\left(\frac{3}{2}\right), 4(1), 7\left(\frac{1}{2}\right), 8(0)\right]$
3	$\left[(1), 4\left(\frac{1}{2}\right), 6(0)\right]$	none	$\left[(2), 3\left(\frac{3}{2}\right), 3(1), \left(\frac{1}{2}\right)\right]$	$\left[\left(\frac{3}{2}\right), 3(1), 3\left(\frac{1}{2}\right), 2(0)\right]$
2	$\left[(1), 2\left(\frac{1}{2}\right), 2(0)\right]$	$\left[2\left(\frac{1}{2}\right) 4(0)\right]$	$\left[(2), 2\left(\frac{3}{2}\right), (1)\right]$	$\left[\left(\frac{3}{2}\right), 2(1), \left(\frac{1}{2}\right)\right]$
1	$\left[(1), \left(\frac{1}{2}\right)\right]$	$\left[\left(\frac{1}{2}\right) 2(0)\right]$	$\left[(2), \left(\frac{3}{2}\right)\right]$	$\left[\left(\frac{3}{2}\right), (1)\right]$

chiral multiplets contain one single Weyl fermion. Therefore, in $N = 1$ supersymmetric extensions of the Standard Model one can accommodate left and right components of leptons and quarks in different multiplets, which can then transform differently under the $SU(2)$ gauge group. What about $N > 1$ supersymmetry? First notice that all helicity $\frac{1}{2}$ states belonging to multiplets containing vector fields should transform in the adjoint representation of the gauge group, which is real. Therefore, the only other representation which might allow for helicity $\frac{1}{2}$ states transforming in fundamental representations is the $N = 2$ hypermultiplet. However, as already noticed, a hypermultiplet contains two $N = 1$ chiral multiplets with opposite chirality. Since for any internal symmetry group G, we have that the corresponding algebra commutes with the SuperPoincaré algebra, $[\mathbb{G}, \text{SuperPoincaré}] = 0$, these two chiral multiplets transform in the same representation under G. Therefore, $N = 2$ is non-chiral: left and right components of leptons and quarks would belong to the same matter multiplet and could not transform differently under the $SU(2)$ Standard Model gauge group. Summarizing, if extended supersymmetry is realized in Nature, it should be broken at some high enough energy scale to an effective $N = 1$ model. This is why at low energy people typically focus just on $N = 1$ extensions of the Standard Model.

3.2 Massive supermultiplets

The logical steps one should follow for massive representations are similar to previous ones. There is however one important difference. Let us consider a state with mass m in its rest frame, $P_\mu = (m, 0, 0, 0)$. One can easily see that, differently from the massless case, the number of non-trivial generators gets not diminished: there remain the full set of $2N$ creation and $2N$ annihilation operators. Indeed, eq. (2.61) is now

$$\left\{Q^I_\alpha, \overline{Q}^J_{\dot\beta}\right\} = 2m\, \delta_{\alpha\dot\beta}\, \delta^{IJ} \tag{3.23}$$

and no supersymmetric generators are trivially realized. This means that, generically, massive representations are *longer* than massless ones. Another important difference is that we better speak of spin rather than helicity, now. A given Clifford vacuum will be defined by mass m and spin j, with $j(j+1)$ being the eigenvalue of $\vec{J}^2$. Hence, the Clifford vacuum will have itself degeneracy $2j+1$ since j_3 takes values from $-j$ to $+j$.

N = 1 supersymmetry

The annihilation and creation operators, satisfying the usual oscillator algebra, now read

$$a_{1,2} \equiv \frac{1}{\sqrt{2m}} Q_{1,2}, \quad a^{\dagger}_{1,2} \equiv \frac{1}{\sqrt{2m}} \overline{Q}_{\dot{1},\dot{2}}. \tag{3.24}$$

As anticipated these are twice those for the massless case. Notice that $a_1^{\dagger}$ lowers the spin by half unit while $a_2^{\dagger}$ raises it. We can now define a Clifford vacuum as a state with mass m and spin j_0 which is annihilated by both a_1 and a_2 and act with the creation operators to construct the corresponding massive representations.

- Matter multiplet:

$$j = 0 \;\rightarrow\; \left(-\frac{1}{2}, 0, 0', +\frac{1}{2}\right). \tag{3.25}$$

The number of degrees of freedom is the same as the massless case (but with no need to add any CPT conjugate, now). It is worth noticing that also in the present case the two spin 0 states, $0'$ and 0, as one can easily proof playing a bit with the operator algebra. Summarizing, the multiplet is made of a massive complex scalar and a massive Majorana fermion.

- Gauge (or vector) multiplet:

$$j = \frac{1}{2} \;\rightarrow\; \left(-1, \mathbf{2} \times -\frac{1}{2}, \mathbf{2} \times 0, \mathbf{2} \times +\frac{1}{2}, 1\right). \tag{3.26}$$

The degrees of freedom one ends-up with are those of one massive vector, one massive Dirac fermion and one massive real scalar (recall the comment after eq. (3.23), which in this case implies that the Clifford vacuum, $j = \frac{1}{2}$ contains two single particle states, with j_3 component $|1/2\rangle$ and $|-1/2\rangle$, respectively). The representation is longer than that of a massless vector supermultiplet, as expected. Notice, though, that the number of these degrees of freedom is the same as those of a massless vector multiplet plus one massless matter multiplet. This is reassuring, since we do not like massive vectors to start with, and only allow Higgs-like mechanisms to generate masses for vector fields in a unitary and renormalizable theory. One can generate massive vector multiplets by a generalization of the Higgs mechanism, in which a massless vector multiplet eats-up a chiral multiplet while preserving supersymmetry.

Since we cannot really make sense of massive particles with spin higher than one (and we are not much interested in supergravity theories in this course, anyway), we stop here and move to extended supersymmetry representations.

Extended supersymmetry

Let us then consider $N > 1$ and allow also for non-trivial central charges. A change of basis in the space of supersymmetry generators turns out to be useful for the following analysis. Since the central charge matrix Z^{IJ} is antisymmetric, with a $U(N)$ rotation one can put it in the standard block-diagonal form

$$Z^{IJ} = \begin{pmatrix} 0 & Z_1 & & & & & & \\ -Z_1 & 0 & & & & & & \\ & & 0 & Z_2 & & & & \\ & & -Z_2 & 0 & & & & \\ & & & & \cdots & \cdots & & \\ & & & & \cdots & \cdots & & \\ & & & & & & 0 & Z_{N/2} \\ & & & & & & -Z_{N/2} & 0 \end{pmatrix} \tag{3.27}$$

(we have assumed for simplicity that N is even). One can now define

$$\begin{aligned}
a^1_\alpha &= \frac{1}{\sqrt{2}} \left(Q^1_\alpha + \epsilon_{\alpha\beta}(Q^2_\beta)^\dagger\right) \\
b^1_\alpha &= \frac{1}{\sqrt{2}} \left(Q^1_\alpha - \epsilon_{\alpha\beta}(Q^2_\beta)^\dagger\right) \\
a^2_\alpha &= \frac{1}{\sqrt{2}} \left(Q^3_\alpha + \epsilon_{\alpha\beta}(Q^4_\beta)^\dagger\right) \\
b^2_\alpha &= \frac{1}{\sqrt{2}} \left(Q^3_\alpha - \epsilon_{\alpha\beta}(Q^4_\beta)^\dagger\right) \\
\ldots &= \ldots \\
\ldots &= \ldots \\
a^{N/2}_\alpha &= \frac{1}{\sqrt{2}} \left(Q^{N-1}_\alpha + \epsilon_{\alpha\beta}(Q^N_\beta)^\dagger\right) \\
b^{N/2}_\alpha &= \frac{1}{\sqrt{2}} \left(Q^{N-1}_\alpha - \epsilon_{\alpha\beta}(Q^N_\beta)^\dagger\right)
\end{aligned}$$

which satisfy the oscillator algebra

$$\begin{aligned}\{a^r_\alpha, (a^s_\beta)^\dagger\} &= (2m + Z_r)\,\delta_{rs}\,\delta_{\alpha\beta}\\ \{b^r_\alpha, (b^s_\beta)^\dagger\} &= (2m - Z_r)\,\delta_{rs}\,\delta_{\alpha\beta}\\ \{a^r_\alpha, (b^s_\beta)^\dagger\} &= \{a^r_\alpha, a^s_\beta\} = \cdots = 0.\end{aligned}$$

where $r, s = 1, \ldots, N/2$. As anticipated, we have now $2N$ creation operators

$$(a^r_\alpha)^\dagger, \quad (b^r_\alpha)^\dagger \quad r = 1, \ldots, N/2, \quad \alpha = 1, 2 \tag{3.28}$$

which we can use to construct massive representations starting from some given Clifford vacuum. Notice that, from their very definition, it follows that creation operators with spinorial index $\alpha = 1$ lower the spin by half unit, while those with spinorial index $\alpha = 2$ raise it.

Several important comments are in order. Due to the positiveness of the Hilbert space, from the oscillator algebra above one can show that

$$2m \geq |Z_r|, \quad r = 1, \ldots, \frac{N}{2}. \tag{3.29}$$

This means that the mass of a given representation is always larger (or equal) than (half) the modulus of any central charge eigenvalue. The first important consequence of the bound (3.29) is that for massless representations (for which the left hand side is identically 0) the central charges are always trivially realized, i.e. $Z^{IJ} = 0$. That's why we did not discuss massless multiplets with non vanishing central charges in the previous section.

There is another important consequence of the bound (3.29). Suppose none of the central charge eigenvalues saturate it, namely $2m > |Z_r|$, $\forall r$. Proceeding as before, starting from a Clifford vacuum λ_0 annihilated by all operators a^r_α, b^r_α and acting on it with the creation operators (3.28) one creates 2^{2N} states, 2^{2N-1} bosonic and 2^{2N-1} fermionic, with spin going from $\lambda_0 - N/2$ to $\lambda_0 + N/2$. Therefore, the representation has states with spins spanning $2N + 1$ half-integer values.

Suppose instead that some Z_r saturate the bound (3.29), say $k \leq N/2$ of them do so. Looking at the oscillator algebra one immediately sees that k b-type operators become trivial (we are supposing, without loss of generality, that all Z_r are positive), and the dimension of the representation diminishes accordingly. The multiplet contains only $2^{2(N-k)}$ states now. These are called *short multiplets*. The extreme case is when all Z_r saturate the bound ($k = N/2$). In this case half the creation operators trivialize and we get a multiplet, known as *ultrashort*, whose dimension is identical to that of a massless one: the number of states is indeed 2^N, 2^{N-1} bosonic and 2^{N-1} fermionic.

The upshot of the discussion above is that in theories with extended supersymmetry one can have massive multiplets with different lengths:

$$\begin{array}{ll} \text{long multiplets} & 2^{2N} = \left(2^{2N-1}\right)_B + \left(2^{2N-1}\right)_F \\ \text{short multiplets} & 2^{2N-2k} = \left(2^{2(N-k)-1}\right)_B + \left(2^{2(N-k)-1}\right)_F \\ \text{ultra-short multiplets} & 2^{2N-2\frac{N}{2}} = 2^N = \left(2^{N-1}\right)_B + \left(2^{N-1}\right)_F. \end{array}$$

As it happens for massless representations, states belonging to some representation of supersymmetry also transform into given representations of the R-symmetry group, since the supercharges do so. One should just remember that the R-symmetry group is $U(N)$ in absence of central charges but reduces to $USp(N)$ if central charges are present.

N = 2 supersymmetry

In this case we have one only central charge eigenvalue, Z, and we have four oscillators a_α and b_α (we have dropped the now inessential upper index r).

Let us first consider the case of long multiplets, namely a situation in which the bound (3.29) is not saturated. In this case we cannot have massive matter since we have too many creation operators to avoid spins higher than $\frac{1}{2}$. So the only possibility are (massive) vector multiplets.

- Gauge (or vector) multiplet:

$$j = 0 \;\to\; \left(-1, \mathbf{4} \times -\frac{1}{2}, \mathbf{6} \times 0, \mathbf{4} \times +\frac{1}{2}, 1\right). \tag{3.30}$$

  The degrees of freedom correspond to a massive vector, two Dirac fermions, and five real scalars. Their number equals that of a massless $N = 2$ vector multiplet and a massless $N = 2$ hypermultiplet. As before, such massive vector multiplet should be thought of as obtained via some supersymmetric Higgs-like mechanism.

Let us now consider shorter representations. Since in this case there is only one central charge eigenvalue, Z, the only possible short representation is in fact the ultrashort, whose length should equal that of the corresponding massless representation. The only non-trivial oscillators are now a_α, since the b_α are trivially realized (we are assuming $Z > 0$).

- Matter multiplet (short rep.):

$$j = 0 \;\to\; \left(\mathbf{2} \times -\frac{1}{2}, \mathbf{4} \times 0, \mathbf{2} \times +\frac{1}{2}\right), \tag{3.31}$$

  (where the doubling of states arises for similar reasons as for the massless hypermultiplet). The degrees of freedom are those of one massive Dirac fermion and two massive complex scalars. As expected the number of degrees of freedom equals those of a massless hypermultiplet.
- Vector multiplet (short rep.):

$$j = \frac{1}{2} \;\to\; \left(-1, \mathbf{2} \times -\frac{1}{2}, \mathbf{2} \times 0, \mathbf{2} \times +\frac{1}{2}, +1\right). \tag{3.32}$$

  The degrees of freedom are those of one massive vector, one massive Dirac fermion and one massive real scalar. While rearranged differently in terms of fields, the number of bosonic and fermionic degrees of freedom equals that of a massless vector multiplet. What's interesting here is that a massive ultrashort vector multiplet can arise dynamically, via some Higgs-like mechanism involving *only* a massless vector multiplet, something peculiar to $N = 2$ supersymmetry and related to the fact that massless vector multiplets contain scalars, and can then self-Higgs.

N = 4 supersymmetry

For $N = 4$ supersymmetry long multiplets are not allowed since the number of states (actually 256!) would include at least spin 2 states; such a theory would then include a massive spin 2 particle which is not believed to be possible in a local quantum field theory. What are are possible are short multiplets, actually only the ultrashort, whose field content amounts to rearrange the fields characterizing a massless vector multiplet into massive states: one would get a massive vector, two Dirac fermions and five real scalars. The construction is left to the reader.

Let us finally notice that all short multiplets are *supersymmetry preserving*, meaning they are annihilated by the supersymmetry generators whose corresponding central charge eigenvalue saturates the bound. In general one can then have $\frac{1}{N}, \frac{2}{N}, \ldots, \frac{N/2}{N}$ supersymmetry preserving multiplets (the numerator is nothing but the integer k previously defined). For instance, ultrashort multiplets, for which $k = N/2$, are $\frac{1}{2}$ supersymmetry preserving states. These multiplets accommodate states which have very important properties at the quantum level; most notably, it turns out that they are more protected against quantum corrections with respect to states belonging to long multiplets. Short multiplets are also called BPS, since the bound (3.29) is very much reminiscent of the famous Bogomonlyi-Prasad-Sommerfeld bound which is saturated by solitons, tipically. This is not just a mere analogy, since the bound (3.29) is in fact not just an algebraic relation but it has a very concrete physical meaning: it is nothing but a specific BPS-like bound. Indeed, short multiplets often arise as solitons in supersymmetric field theories, and central charges correspond to physical (topological) charges. We will see concrete examples of BPS states later in this course.

3.3 Representation on fields: a first try

So far we have discussed supersymmetry representations on states. However, we would like to discuss supersymmetric *field* theories, eventually. Therefore, we need to construct supersymmetric representations in terms of multiplets of fields rather than multiplets of states. In principle, following our previous strategy this can be done quite easily.

Let us start focusing on $N = 1$ supersymmetry. To build a representation of the supersymmetry algebra on fields, we start from some field $\phi(x)$ for which

$$[\overline{Q}_{\dot{\alpha}}, \phi(x)] = 0. \tag{3.33}$$

The field ϕ is the analogous of the Clifford vacuum $|\lambda_0\rangle$ we used previously, the ground state of the representation. Similarly as before, acting on this ground state $\phi(x)$ with the supersymmetry generator Q_α, we can generate new fields out of it, all belonging to the same supermultiplet.

For definiteness, we choose $\phi(x)$ to be a scalar field, but one can also have ground states which are fields with some non-trivial tensor structure, as we will later see. Not much of what we want to say here depends on this choice.

The first thing to notice is that the scalar field $\phi(x)$ is actually complex. Suppose it were real. Then, taking the hermitian conjugate of eq. (3.33) one would have obtained

$$[Q_\alpha, \phi(x)] = 0. \tag{3.34}$$

One can now use the generalized Jacobi identity for $(\phi, Q, \overline{Q})$ and get

$$\begin{aligned}&\left[\phi(x), \{Q_\alpha, \overline{Q}_{\dot{\alpha}}\}\right] + \left\{Q_\alpha, \left[\overline{Q}_{\dot{\alpha}}, \phi(x)\right]\right\} - \left\{\overline{Q}_{\dot{\alpha}}, \left[\phi(x), Q_\alpha\right]\right\} = 0 \\ &\longrightarrow \quad 2\sigma^\mu \left[\phi(x), P_\mu\right] = 0 \quad \longrightarrow \quad \left[P_\mu, \phi(x)\right] \sim \partial_\mu \phi(x) = 0, \end{aligned} \tag{3.35}$$

which should then imply that the field is actually a constant (not a field, really!). So better $\phi(x)$ to be complex. In this case eq. (3.34) does not hold, but rather

$$[Q_\alpha, \phi(x)] \equiv \psi_\alpha(x). \tag{3.36}$$

This defines a new field ψ_α belonging to the same representation (since ϕ is a scalar, ψ is a Weyl spinor). The next step is to see whether acting with supersymmetry generators on ψ_α gives new fields or just derivatives (or combinations) of fields already present in the representation. In principle we have

$$\{Q_\alpha, \psi_\beta(x)\} = F_{\alpha\beta}(x) \tag{3.37}$$

$$\left\{\overline{Q}_{\dot{\alpha}}, \psi_\beta(x)\right\} = X_{\dot{\alpha}\beta}(x). \tag{3.38}$$

Enforcing the generalized Jacobi identity on $(\phi, Q, \overline{Q})$, and using eq. (3.36), after some trivial algebra one gets

$$X_{\dot{\alpha}\beta} = \left\{\psi_\beta(x), \overline{Q}_{\dot{\alpha}}\right\} = 2\sigma^\mu_{\beta\dot{\alpha}} \left[P_\mu, \phi\right] \sim \partial_\mu \phi, \tag{3.39}$$

which implies that $X_{\dot{\alpha}\beta}$ is not a new field but just the space-time derivative of the scalar field ϕ. Let us now enforce the generalized Jacobi identity on (ϕ, Q, Q). Since the Q's anticommute (recall we are considering $N = 1$ supersymmetry and hence there are no central charges) one simply gets

$$\{Q_\alpha, [Q_\beta, \phi]\} - \{Q_\beta, [\phi, Q_\alpha]\} = 0 \quad \longrightarrow \quad F_{\alpha\beta} + F_{\beta\alpha} = 0. \tag{3.40}$$

This says that the field $F_{\alpha\beta}$ is antisymmetric under $\alpha \leftrightarrow \beta$, which implies that

$$F_{\alpha\beta}(x) = \epsilon_{\alpha\beta} F(x). \tag{3.41}$$

No other constraints are imposed on F by other consistency conditions. So we find here a new complex scalar field F. Again, we should now ask whether acting on it with supersymmetry generators produces new fields. We get

$$[Q_\alpha, F] = \lambda_\alpha \tag{3.42}$$

$$\left[\overline{Q}_{\dot{\alpha}}, F\right] = \overline{\chi}_{\dot{\alpha}}. \tag{3.43}$$

Using the generalized Jacobi identities for (ψ, Q, Q) and $(\psi, Q, \overline{Q})$, one can easily prove that λ_α is actually vanishing and that $\overline{\chi}_{\dot{\alpha}}$ is proportional to the space-time derivative of the field ψ. So no new fields in this case: after a certain number of steps the representation closes. The multiplet of fields we have found is then

$$(\phi, \psi, F). \tag{3.44}$$

If ϕ is a scalar field, as we have supposed here, this multiplet is a matter multiplet since it contains particles with spin 0 and 1/2 only. It is called chiral multiplet and it is indeed the field theory counterpart of the chiral multiplet of states we have constructed before. Notice that the equality of the number of fermionic and bosonic states for a given representation still holds: we are now off-shell, and the spinor ψ_α has four degrees of freedom; this is the same number of bosonic degrees of freedom, two coming from the scalar field ϕ and two from the scalar field F

$$(\text{Re}\phi, \text{Im}\phi, \text{Re}F, \text{Im}F)_B\,, \quad (\text{Re}\psi_1, \text{Im}\psi_1, \text{Re}\psi_2, \text{Im}\psi_2)_F\,. \tag{3.45}$$

While we see the expected degeneracy between bosonic and fermonic degrees of freedom, they do not match those of the chiral multiplet of states we have constructed before, which are just $2_B + 2_F$. This is because we are off-shell now. Going on-shell, the 4 fermionic degrees of freedom reduce to just 2 propagating degrees of freedom, due to Dirac equation. The same sort of reduction should occur for the bosonic degrees of freedom, in order to match the $2_B + 2_F$ on-shell condition. But Klein-Gordon equation does not diminish the number of independent degrees of freedom! What happens is that F turns out to be a non-dynamical *auxiliary* field: as we will see when constructing Lagrangians, the equation of motion for F simply tells that this scalar field is not an independent field but rather a (specific) function of other fields, $F = F(\phi, \psi)$. This is not specific to the chiral multiplet we have constructed, but it is in fact a completely general phenomenon. We will come back to this point in next chapters.

The procedure we have followed to construct the multiplet (3.44) can be easily generalized. Modifying the condition (3.33) one can construct other kind of multiplets, like linear multiplets, vector multiplets, etc... And/or construct chiral multiplets with different field content, simply by defining a ground state carrying some space-time index, letting ϕ being a spinor, a vector, etc...

Out of a set of multiplets with the desired field content, one can construct suitable Lagrangian made out of these fields. In order for the theory to be supersymmetric, this Lagrangian should (at most) transform as a total space-time derivative under supersymmetry transformations. Indeed, in this case, the action constructed out of it

$$S = \int d^4x \, \mathcal{L}, \tag{3.46}$$

will be supersymmetric invariant.

In principle, this is a well defined program. In practice, however, to see whether a given action is invariant under supersymmetry is rather cumbersome: one should take any single term in the Lagrangian, act on it with supersymmetry transformations and prove that the variations of all (possibly very many) terms sum up to a total space-time derivative. This turns out to be very involved, in general, but theoretical physicists came up with a brilliant idea to circumvent this problem.

This difficulty is due to the fact that the formulation above is a formulation in which supersymmetry is not manifest. Ordinary field theories are naturally defined in Minkowski space and in such formulation it is easy to construct Lagrangians respecting Poincaré symmetry. It turns out that supersymmetric field theories are naturally defined on an extension of Minkoswki space, known as *superspace*, which, essentially, takes into account the extra space-time symmetries associated to the supersymmetry generators. In such extended space it is much easier to construct supersymmetric Lagrangians, and indeed the superspace formalism is what is most commonly used to discuss supersymmetric field theories. This is the formalism we will use along this course, and next chapter will be devoted to a thorough introduction of superspace.

3.4 Exercises

Exercise 3.1. Prove that P^2 and W^2 are Casimir of the Poincaré algebra.

Exercise 3.2. Prove that CPT flips the sign of the helicity.

Exercise 3.3. Construct explicitly, in terms of creation operators acting on the Clifford vacuum, the massive $N = 1$ vector multiplet (3.26) and the massive $N = 2$ BPS vector multiplet (3.32). For the latter, determine the $SU(2)_R$ representations under which bosonic and fermionic states transform.

Exercise 3.4. Construct explicitly the 1/2 BPS $N = 4$ vector multiplet (Hint: the Clifford vacuum has $j = 0$). Discuss its (massive) content and its relation with the massless vector multiplet. Can one construct a 1/4 BPS $N = 4$ multiplet?

Exercise 3.5. Enforcing the generalized Jacobi identity on (ψ, Q, Q) and $(\psi, Q, \overline{Q})$, using eqs. (3.37), (3.38), (3.42) and (3.43), prove that $\lambda_\alpha = 0$ and $\overline{\chi}_{\dot{\beta}} \sim \partial_\mu(\sigma^\mu \psi)_{\dot{\beta}}$.

References

[1] A. Bilal, *Introduction to supersymmetry*, Section 3, arXiv:hep-th/0101055.
[2] L. Castellani, R. D'Auria and P. Fre, *Supergravity And Superstrings: A Geometric Perspective. Vol. 1: Mathematical Foundations*, Chapters II.4 and II.5, World Scientific Singapore (1991).
[3] J.D. Lykken, *Introduction to Supersymmetry*, Section 2, TASI 96, arXiv: hep-th/9612114.
[4] M.F. Sohnius, *Introducing Supersymmetry*, Sections 3 and 5.4, *Phys. Rep.* 128 (1985).
[5] S. Weinberg, *The quantum theory of fields. Vol. 1: Foundations*, Chapter 13.1, Cambridge University Press (2000).
[6] M. T. Grisaru and H. N. Pendleton, Soft Spin 3/2 Fermions Require Gravity and Supersymmetry, *Phys. Lett. B* **67** (1977), 323.

Chapter 4

Superspace and Superfields

The usual space-time Lagrangian formulation is not the most convenient one for describing supersymmetric field theories. This is because in ordinary space-time, supersymmetry is not manifest. In fact, an extension of ordinary space-time, known as *superspace*, happens to be the best and most natural framework in which to formulate supersymmetric theories. The basic idea of ($N = 1$) superspace is to enlarge the space-time labelled with coordinates x^μ, associated to the generators P_μ, by adding 2 + 2 anti-commuting Grassmann coordinates $\theta_\alpha, \bar{\theta}_{\dot{\alpha}}$, associated to the supersymmetry generators $Q_\alpha, \overline{Q}_{\dot{\alpha}}$, and obtain an eight coordinate superspace labelled by $(x^\mu, \theta_\alpha, \bar{\theta}_{\dot{\alpha}})$. In such an apparently exotic space, many mysterious (or hidden) properties of supersymmetric field theories become manifest. As we will see, at the price of learning a few new mathematical ingredients, the goal of constructing supersymmetric field theories will be achieved much more easily, and within a framework in which many classical and quantum properties of supersymmetry will be more transparent.

In this chapter we will introduce superspace and superfields. In subsequent chapters we will use this formalism to construct supersymmetric field theories and study their dynamics.

4.1 Superspace as a coset

Let us start recalling the relation between ordinary Minkowski space and the Poincaré group. Minkowski space is a four-dimensional coset space defined as

$$\mathcal{M}_{1,3} = \frac{ISO(1,3)}{SO(1,3)}, \tag{4.1}$$

where $ISO(1,3)$ is the Poincaré group and $SO(1,3)$ the Lorentz group. The Poincaré group $ISO(1,3)$ is nothing but the isometry group of this coset space, which means that any point of $\mathcal{M}_{1,3}$ can be reached from the origin with a Poincaré transformation. This transformation, however, is defined up to Lorentz transformations. Therefore, each coset class ($\equiv$ a point in space-time) has a unique representative being just a translation, which can be parametrized by a coordinate x^μ

$$x^\mu \longleftrightarrow e^{(x^\mu P_\mu)}. \tag{4.2}$$

Superspace can be defined along similar lines. The first thing we need to do is to extend the Poincaré group to the so-called superPoincaré group. In order to do this, given that a group is the exponent of the algebra, we have to rewrite the whole supersymmetry algebra in terms of commutators, namely as a Lie algebra. This is achieved by introducing a set of *constant* Grassmann numbers $\theta_\alpha, \bar{\theta}_{\dot{\alpha}}$, defined such that they anti-commute with everything fermionic and commute with everything bosonic

$$\{\theta^\alpha, \theta^\beta\} = 0, \quad \{\bar{\theta}_{\dot{\alpha}}, \bar{\theta}_{\dot{\beta}}\} = 0, \quad \{\theta^\alpha, \bar{\theta}_{\dot{\beta}}\} = 0. \tag{4.3}$$

This allows to transform anti-commutators of the supersymmetry algebra into commutators, and get

$$\left[\theta Q, \overline{\theta Q}\right] = 2\, \theta\sigma^\mu\bar{\theta}P_\mu, \quad [\theta Q, \theta Q] = \left[\overline{\theta Q}, \overline{\theta Q}\right] = 0, \tag{4.4}$$

where as usual $\theta Q \equiv \theta^\alpha Q_\alpha$, $\overline{\theta Q} \equiv \bar{\theta}_{\dot{\alpha}}\overline{Q}^{\dot{\alpha}}$. This way, one can write the supersymmetry algebra solely in terms of commutators. Exponentiating this algebra, one gets the superPoincaré group. A generic group element can then be written as

$$G(x, \theta, \bar{\theta}, \omega) = \exp\left(ixP + i\theta Q + i\overline{\theta Q} + \frac{1}{2}i\omega M\right), \tag{4.5}$$

where xP is a shorthand notation for $x^\mu P_\mu$ and ωM a shorthand notation for $\omega^{\mu\nu}M_{\mu\nu}$.

The superPoincaré group, mathematically, is $\overline{Osp(4|1)}$. Let us open a brief parenthesis and explain such a notation. Let us define the graded Lie algebra $Osp(2p|N)$ as the grade one Lie algebra $\mathbb{L} = \mathbb{L}_0 \oplus \mathbb{L}_1$ whose generic element can be written as a matrix of complex dimension $(2p+N)\times(2p+N)$

$$\begin{pmatrix} A & B \\ C & D \end{pmatrix} \tag{4.6}$$

where A is a $(2p \times 2p)$ matrix, B a $(2p \times N)$ matrix, C a $(N \times 2p)$ matrix and D a $(N \times N)$ matrix. An element of $\mathbb{L}_0$ and $\mathbb{L}_1$ has entries

$$\begin{pmatrix} A & 0 \\ 0 & D \end{pmatrix}, \quad \text{respectively} \quad \begin{pmatrix} 0 & B \\ C & 0 \end{pmatrix} \tag{4.7}$$

where

$$A^T\Omega_{(2p)} + \Omega_{(2p)}A = 0$$
$$D^T\Omega_{(N)} + \Omega_{(N)}D = 0$$
$$C = \Omega_{(N)}B^T\Omega_{(2p)}$$

and

$$\Omega^2_{(2p)} = -\mathbb{I}, \quad \Omega^T_{(N)} = \Omega_{(N)}, \quad \Omega^T_{(2p)} = -\Omega_{(2p)}. \tag{4.8}$$

This implies that the matrices A span a $Sp(2p,\mathbb{C})$ algebra and the matrices D a $O(N,\mathbb{C})$ algebra. Therefore we have that

$$\mathbb{L}_0 = Sp(2p) \otimes O(N), \tag{4.9}$$

hence the name $Osp(2p|N)$ for the whole superalgebra. A generic element of the superalgebra has the form

$$Q = q^a t_a + q^l t_l, \tag{4.10}$$

where $t_a \in \mathbb{L}_0$ and $t_l \in \mathbb{L}_1$ are a basis of the corresponding vector spaces, and we have introduced complex numbers q^a for $\mathbb{L}_0$ and Grassmann numbers q^l for $\mathbb{L}_1$ (recall why and how we introduced the fermionic parameters $\theta_\alpha, \bar{\theta}_{\dot\alpha}$ before).

Taking $p = 2$, we get the algebra $Osp(4|N)$. This is not yet what we are after. The last step to take, which we do not describe in detail here, amounts to the so-called Inonu-Wigner contraction. Essentially, one has to rescale (almost) all generators by a constant $1/\bar{e}$, rewrite the algebra in terms of the rescaled generators and take the limit $\bar{e} \to 0$. What one ends up with is the N-extended supersymmetry algebra in Minkowski space we all know, dubbed $\overline{Osp(4|N)}$, where in the aforementioned limit one gets the identification

$$A \to P_\mu, M_{\mu\nu} \quad D \to Z^{IJ} \quad B, C \to Q_I, \overline{Q}_I. \tag{4.11}$$

Taking $N = 1$ one finally gets the unextended supersymmetry algebra $\overline{Osp(4|1)}$.

Given the generic group element of the superPoincaré group (4.5), $N = 1$ *superspace* is defined as the $(4+4$ dimensional) group coset

$$\mathcal{M}_{4|1} = \frac{\overline{Osp(4|1)}}{SO(1,3)}. \tag{4.12}$$

A point in superspace (point in a loose sense, of course, given the non-commutative nature of the Grassmann parameters $\theta_\alpha, \bar{\theta}_{\dot\alpha}$) gets identified with the coset representative corresponding to a so-called super-translation through the one-to-one map

$$(x^\mu, \theta_\alpha, \bar{\theta}_{\dot{\alpha}}) \longleftrightarrow e^{(x^\mu P_\mu)}\, e^{(\theta Q + \bar{\theta}\bar{Q})}. \tag{4.13}$$

The $2+2$ anti-commuting Grassmann numbers $\theta_\alpha, \bar{\theta}_{\dot{\alpha}}$ can then be thought of as coordinates in superspace (in four-component notation, they correspond to a Marojana spinor θ). For these Grassmann numbers, all usual spinor identities hold.

Thus far we have introduced what is known as $N=1$ superspace. When discussing extended supersymmetry one should introduce, in principle, a larger superspace. There exist (two, at least) formulations of $N=2$ superspace. However, these formulations present some subtleties and problems that are beyond the scope of this course to discuss. And no formulation is known of $N=4$ superspace. In this course we will use $N=1$ superspace even when discussing extended supersymmetry, as it is typically done in most of the literature.

4.2 Superfields as fields in superspace

Superfields are nothing but fields in superspace: functions of the superspace coordinates $(x^\mu, \theta_\alpha, \bar{\theta}_{\dot{\alpha}})$. Since θ_α and $\bar{\theta}_{\dot{\alpha}}$ anticommute, any product involving more than two θ's or two $\bar{\theta}$'s vanishes: given that $\theta_\alpha\theta_\beta = -\theta_\beta\theta_\alpha$, we have that $\theta_\alpha\theta_\beta = 0$ for $\alpha = \beta$ and therefore $\theta_\alpha\theta_\beta\theta_\gamma = 0$, since at least two indices in this product are the same. Hence, the most general superfield $Y = Y(x, \theta, \bar{\theta})$ has the following simple Taylor-like expansion

$$\begin{aligned} Y(x,\theta,\bar{\theta}) &= f(x) + \theta\psi(x) + \bar{\theta}\,\overline{\chi}(x) + \theta\theta\, m(x) + \bar{\theta}\bar{\theta}\, n(x) + \\ &\quad + \theta\sigma^\mu\bar{\theta}\, v_\mu(x) + \theta\theta\,\bar{\theta}\,\overline{\lambda}(x) + \bar{\theta}\bar{\theta}\,\theta\rho(x) + \theta\theta\,\bar{\theta}\bar{\theta}\, d(x). \end{aligned} \tag{4.14}$$

Each entry above is a field (possibly with some non-trivial tensor structure). In this sense, a superfield is nothing but a finite collection (a multiplet) of ordinary fields.

We aim at constructing supersymmetric Lagrangians out of superfields. In such Lagrangians, superfields get multiplied to one another, and we sometimes act on them with derivatives, etc... Moreover, integration in superspace will be needed, eventually. Therefore, it is necessary to pause a bit and recall how operations of this kind work for Grassmann variables.

Derivation in superspace is defined as follows

$$\partial_\alpha \equiv \frac{\partial}{\partial\theta^\alpha} \quad \text{and} \quad \partial^\alpha = \epsilon^{\alpha\beta}\partial_\beta,\ \ \bar{\partial}_{\dot{\alpha}} \equiv \frac{\partial}{\partial\bar{\theta}^{\dot{\alpha}}} \quad \text{and} \quad \bar{\partial}^{\dot{\alpha}} = \epsilon^{\dot{\alpha}\dot{\beta}}\bar{\partial}_{\dot{\beta}}, \tag{4.15}$$

where

$$\partial_\alpha \theta^\beta = \delta_\alpha^\beta, \quad \bar{\partial}_{\dot{\alpha}} \bar{\theta}^{\dot{\beta}} = \delta_{\dot{\alpha}}^{\dot{\beta}}, \quad \partial_\alpha \bar{\theta}_{\dot{\beta}} = 0, \quad \bar{\partial}^{\dot{\alpha}} \theta^\beta = 0. \tag{4.16}$$

Note that this implies that $\partial^\alpha \theta_\beta = -\delta_\beta^\alpha$ and $\bar{\partial}^{\dot{\alpha}} \bar{\theta}_{\dot{\beta}} = -\delta_{\dot{\beta}}^{\dot{\alpha}}$ (that's life!).

Let us consider a single Grassmann variable θ (either $\theta_1, \theta_2, \bar{\theta}_{\dot{1}}$ or $\bar{\theta}_{\dot{2}}$ in our case). Integration in θ is defined as follows

$$\int d\theta = 0 \quad \int d\theta \; \theta = 1. \tag{4.17}$$

This implies that for a generic function $f(\theta) = f_0 + \theta f_1$, the following results hold

$$\int d\theta \; f(\theta) = f_1, \quad \int d\theta \; \delta(\theta) f(\theta) = f_0 \quad \longrightarrow \int = \partial, \quad \theta = \delta(\theta). \tag{4.18}$$

These relations can be easily generalized to $N = 1$ superspace, provided

$$d^2\theta \equiv \frac{1}{2} d\theta^1 d\theta^2, \quad d^2\bar{\theta} \equiv \frac{1}{2} d\bar{\theta}^{\dot{2}} d\bar{\theta}^{\dot{1}}. \tag{4.19}$$

With these definitions one can prove the following useful identities

$$\int d^2\theta \; \theta\theta = \int d^2\bar{\theta} \; \bar{\theta}\bar{\theta} = 1, \quad \int d^2\theta d^2\bar{\theta} \; \theta\theta \; \bar{\theta}\bar{\theta} = 1$$

$$\int d^2\theta = \frac{1}{4} \epsilon^{\alpha\beta} \partial_\alpha \partial_\beta, \quad \int d^2\bar{\theta} = -\frac{1}{4} \epsilon^{\dot{\alpha}\dot{\beta}} \bar{\partial}_{\dot{\alpha}} \bar{\partial}_{\dot{\beta}}. \tag{4.20}$$

Another crucial question we need to answer is: how does a superfield transform under supersymmetry transformations? In order to answer this question we first need to realize the supersymmetry generators $Q_\alpha, \overline{Q}_{\dot{\alpha}}$ as *differential* operators, in the same way we do for the generators of Poincaré algebra (translations, rotations and boosts).

Let us recall how the story goes in ordinary space-time and consider a translation generated by $\mathcal{P}_\mu$ with infinitesimal parameter a^μ, on a field $\phi(x)$ (for notational convenience, we will momentarily use calligraphic letters for the abstract operator and latin ones for the representation of the same operator as a differential operator in field space). This is defined as

$$\phi(x + a) = e^{-ia\mathcal{P}} \phi(x) e^{ia\mathcal{P}} = \phi(x) - ia^\mu \left[\mathcal{P}_\mu, \phi(x)\right] + \cdots . \tag{4.21}$$

On the other hand, Taylor expanding the left hand side we get

$$\phi(x + a) = \phi(x) + a^\mu \partial_\mu \phi(x) + \cdots . \tag{4.22}$$

Equating the right hand sides of the two equations above we then get

$$[\phi(x), \mathcal{P}_\mu] = -i\partial_\mu \phi(x) \equiv P_\mu \phi(x), \tag{4.23}$$

where $\mathcal{P}_\mu$ is the generator of translations and P_μ is its representation as a differential operator in field space (recall that ∂_μ is an operator and from $(\partial_\mu)^* = \partial_\mu$ one gets that $(\partial_\mu)^\dagger = -\partial_\mu$; hence P_μ is hermitian, as it should be). So, a translation of a field by parameter a^μ induces a change on the field itself as

$$\delta_a \phi \equiv \phi(x+a) - \phi(x) = i a^\mu P_\mu \, \phi. \tag{4.24}$$

Notice that here and below we are using right multiplication, when acting on fields.

We now want to apply the same procedure to a superfield. A translation in superspace (*i.e.* a supersymmetry transformation) on a superfield $Y(x,\theta,\bar\theta)$ by a quantity $(\epsilon_\alpha, \bar\epsilon_{\dot\alpha})$, where $\epsilon_\alpha, \bar\epsilon_{\dot\alpha}$ are spinorial parameters, is defined as

$$Y(x+\delta x, \theta+\delta\theta, \bar\theta+\delta\bar\theta) = e^{-i\left(\epsilon\mathcal{Q}+\bar\epsilon\overline{\mathcal{Q}}\right)} Y(x,\theta,\bar\theta) e^{i\left(\epsilon\mathcal{Q}+\bar\epsilon\overline{\mathcal{Q}}\right)}, \tag{4.25}$$

with

$$\delta_{\epsilon,\bar\epsilon} Y(x,\theta,\bar\theta) \equiv Y(x+\delta x, \theta+\delta\theta, \bar\theta+\delta\bar\theta) - Y(x,\theta,\bar\theta) \tag{4.26}$$

the variation of the superfield under the supersymmetry transformation.

To find the representation of $\mathcal{Q}$ and $\overline{\mathcal{Q}}$ as differential operators, there are two questions we need to answer first. What is the explicit expression for $\delta x, \delta\theta, \delta\bar\theta$? Why are we supposing here $\delta x \neq 0$, given we are not acting with the generator of space-time translations $\mathcal{P}_\mu$, but just with supersymmetry generators?

First notice that eq. (4.25) can be written as

$$\begin{aligned} &Y(x+\delta x, \theta+\delta\theta, \bar\theta+\delta\bar\theta) \\ &\quad = e^{-i\left(\epsilon\mathcal{Q}+\bar\epsilon\overline{\mathcal{Q}}\right)} e^{-i\left(x\mathcal{P}+\theta\mathcal{Q}+\bar\theta\overline{\mathcal{Q}}\right)} Y(0;0,0) e^{i\left(x\mathcal{P}+\theta\mathcal{Q}+\bar\theta\overline{\mathcal{Q}}\right)} e^{i\left(\epsilon\mathcal{Q}+\bar\epsilon\overline{\mathcal{Q}}\right)}. \end{aligned} \tag{4.27}$$

Let us now evaluate the last two exponentials, for which we need to recall the Baker-Campbell-Hausdorff formula for non-commuting objects which says that

$$e^A e^B = e^C \quad \text{where } C = \sum_{n=1}^{\infty} \frac{1}{n!} C_n(A,B) \tag{4.28}$$

with

$$C_1 = A+B, \quad C_2 = [A,B], \quad C_3 = \frac{1}{2}[A,[A,B]] - \frac{1}{2}[B,[B,A]] \quad \ldots . \tag{4.29}$$

We then have

$$\begin{aligned}
&\exp\{i\left(x\mathcal{P}+\theta\mathcal{Q}+\overline{\theta\mathcal{Q}}\right)\}\ \exp\{i\left(\epsilon\mathcal{Q}+\bar{\epsilon}\overline{\mathcal{Q}}\right)\}\\
&\quad=\exp\{ix^\mu\mathcal{P}_\mu+i(\epsilon+\theta)\mathcal{Q}+i(\bar{\epsilon}+\bar{\theta})\overline{\mathcal{Q}}-\frac{1}{2}\left[\overline{\theta\mathcal{Q}},\epsilon\mathcal{Q}\right]-\frac{1}{2}\left[\theta\mathcal{Q},\bar{\epsilon}\overline{\mathcal{Q}}\right]\}\\
&\quad=\exp\{ix^\mu\mathcal{P}_\mu+i(\epsilon+\theta)\mathcal{Q}+i(\bar{\epsilon}+\bar{\theta})\overline{\mathcal{Q}}+\epsilon\sigma^\mu\bar{\theta}\,\mathcal{P}_\mu-\theta\sigma^\mu\bar{\epsilon}\,\mathcal{P}_\mu\}\\
&\quad=\exp\{i(x^\mu+i\theta\sigma^\mu\bar{\epsilon}-i\epsilon\sigma^\mu\bar{\theta})\mathcal{P}_\mu+i(\epsilon+\theta)\mathcal{Q}+i(\bar{\epsilon}+\bar{\theta})\overline{\mathcal{Q}}\}
\end{aligned}\tag{4.30}$$

which means that

$$\begin{aligned}
\delta x^\mu&=i\theta\sigma^\mu\bar{\epsilon}-i\epsilon\sigma^\mu\bar{\theta}\\
\delta\theta^\alpha&=\epsilon^\alpha\\
\delta\bar{\theta}^{\dot{\alpha}}&=\bar{\epsilon}^{\dot{\alpha}}
\end{aligned}\tag{4.31}$$

This answers the first question. Notice, further, the expression for δx^μ, which is non-vanishing. This is needed, in order to be consistent with the supersymmetry algebra, $\{Q_\alpha,\overline{Q}_{\dot{\alpha}}\}\sim P_\mu$: two subsequent supersymmetry transformations generate a space-time translation. This answers the second question.

We can now find the representation of the supersymmetry generators $\mathcal{Q}_\alpha$ and $\overline{\mathcal{Q}}_{\dot{\alpha}}$ as differential operators. Let us take eq. (4.26) and, recalling eqs. (4.32), let us Taylor expand the right hand side which becomes

$$\begin{aligned}
\delta_{\epsilon,\bar{\epsilon}}Y(x,\theta,\bar{\theta})&=Y(x,\theta,\bar{\theta})+i\left(\theta\sigma^\mu\bar{\epsilon}-\epsilon\sigma^\mu\bar{\theta}\right)\partial_\mu Y(x,\theta,\bar{\theta})+\\
&\quad+\epsilon^\alpha\partial_\alpha Y(x,\theta,\bar{\theta})+\bar{\epsilon}^{\dot{\alpha}}\bar{\partial}_{\dot{\alpha}}Y(x,\theta,\bar{\theta})+\cdots-Y(x,\theta,\bar{\theta})\\
&=\left[\epsilon^\alpha\partial_\alpha+\bar{\epsilon}^{\dot{\alpha}}\bar{\partial}_{\dot{\alpha}}+i\left(\theta\sigma^\mu\bar{\epsilon}-\epsilon\sigma^\mu\bar{\theta}\right)\partial_\mu+\cdots\right]Y(x,\theta,\bar{\theta})
\end{aligned}\tag{4.32}$$

On the other hand, from eq. (4.25) we get

$$\begin{aligned}
&\delta_{\epsilon,\bar{\epsilon}}Y(x,\theta,\bar{\theta})\\
&\quad=\left(1-i\epsilon\mathcal{Q}-i\bar{\epsilon}\overline{\mathcal{Q}}+\cdots\right)Y(x,\theta,\bar{\theta})\left(1+i\epsilon\mathcal{Q}+i\bar{\epsilon}\overline{\mathcal{Q}}+\cdots\right)-Y(x,\theta,\bar{\theta})\\
&\quad=-i\epsilon^\alpha\left[\mathcal{Q}_\alpha,Y(x,\theta,\bar{\theta})\right]+i\bar{\epsilon}^{\dot{\alpha}}\left[\overline{\mathcal{Q}}_{\dot{\alpha}},Y(x,\theta,\bar{\theta})\right]+\cdots,
\end{aligned}\tag{4.33}$$

(recall that $i\bar{\epsilon}\overline{Q}\equiv i\bar{\epsilon}_{\dot{\alpha}}\overline{Q}^{\dot{\alpha}}=-i\bar{\epsilon}^{\dot{\alpha}}\overline{Q}_{\dot{\alpha}}$). Defining

$$[Y,\mathcal{Q}_\alpha]\equiv Q_\alpha Y,\quad \left[Y,\overline{\mathcal{Q}}_{\dot{\alpha}}\right]\equiv\overline{Q}_{\dot{\alpha}}Y,\tag{4.34}$$

the previous result implies that the supersymmetry variation of a superfield by parameters $\epsilon,\bar{\epsilon}$ is represented as

$$\delta_{\epsilon,\bar{\epsilon}}Y=\left(i\epsilon Q+i\bar{\epsilon}\overline{Q}\right)\,Y.\tag{4.35}$$

Comparing with eq. (4.32) we get the following expression for the differential operators $Q_\alpha, \overline{Q}_{\dot\alpha}$

$$\begin{cases} Q_\alpha = -i\partial_\alpha - \sigma^\mu_{\alpha\dot\beta}\,\bar\theta^{\dot\beta}\partial_\mu \\ \overline{Q}_{\dot\alpha} = +i\bar\partial_{\dot\alpha} + \theta^\beta\sigma^\mu_{\beta\dot\alpha}\partial_\mu \end{cases} \tag{4.36}$$

Notice that, consistently, $Q^\dagger_\alpha = \overline{Q}_{\dot\alpha}$ (to prove this, recall that $(\sigma^\mu_{\alpha\dot\beta})^\dagger = \sigma^\mu_{\beta\dot\alpha}$).

One can check the validity of the expressions (4.36) by showing that the two differential operators close the supersymmetry algebra, namely that

$$\{Q_\alpha, Q_\beta\} = \{\overline{Q}_{\dot\alpha}, \overline{Q}_{\dot\beta}\} = 0, \quad \{Q_\alpha, \overline{Q}_{\dot\beta}\} = 2\sigma^\mu_{\alpha\dot\beta}P_\mu. \tag{4.37}$$

We can now give a more precise definition for what a superfield actually is: a superfield is *a field in superspace which transforms under a super-translation according to eq. (4.25)*. This implies, in particular, that a product of superfields is still a superfield.

4.3 Supersymmetric invariant actions — general philosophy

Having seen that a supersymmetry transformation is simply a translation in superspace, it is now easy to construct supersymmetric invariant actions. In order for an action to be invariant under superPoincaré transformations it is enough that the Lagrangian is Poincaré invariant (actually, it should transform as a scalar density) and that its supersymmetry variation is a total space-time derivative.

Here is where the formalism we have introduced starts to manifest its powerfulness. The basic point is that the integral in superspace of *any* arbitrary superfield is a supersymmetric invariant quantity. In other words, the following integral

$$\int d^4x\, d^2\theta\, d^2\bar\theta\, Y(x,\theta,\bar\theta) \tag{4.38}$$

is manifestly supersymmetric invariant, if Y is a superfield. This can be proven as follows. The integration measure is translationally invariant by construction since

$$\int d\theta\theta = \int d(\theta+\xi)(\theta+\xi) = 1 \tag{4.39}$$

This implies that

$$\delta_{\epsilon,\bar\epsilon}\int d^4x\, d^2\theta\, d^2\bar\theta\, Y(x,\theta,\bar\theta) = \int d^4x\, d^2\theta\, d^2\bar\theta\;\delta_{\epsilon,\bar\epsilon}Y(x,\theta,\bar\theta). \tag{4.40}$$

Now, using eqs. (4.35) and (4.36) we get

$$\delta_{\epsilon,\bar{\epsilon}} Y = \epsilon^{\alpha}\partial_{\alpha}Y + \bar{\epsilon}_{\dot{\alpha}}\bar{\partial}^{\dot{\alpha}}Y + \partial_{\mu}\left[-i\left(\epsilon\sigma\bar{\theta} - \theta\sigma\bar{\epsilon}\right)Y\right]. \tag{4.41}$$

Integration in $d^2\theta d^2\bar{\theta}$ kills the first two terms since they do not have enough θ's or $\bar{\theta}$'s to compensate for the measure, and leaves *only* the last term, which is a total derivative. In other words, under supersymmetry transformations the integrand in eq. (4.40) transforms as a total space-time derivative plus terms which get killed by integration in superspace. Hence the full integral is supersymmetric invariant

$$\delta_{\epsilon,\bar{\epsilon}} \int d^4x\, d^2\theta\, d^2\bar{\theta}\, Y(x,\theta,\bar{\theta}) = 0. \tag{4.42}$$

Supersymmetric invariant actions are constructed this way, *i.e.* by integrating a suitably defined superfield in superspace. Such superfield, call it $\mathcal{A}$, should not be generic, of course. It should have the right structure to give rise, upon integration over Grassman coordinates, to a Lagrangian density, which is a real, dimension-four operator, transforming as a scalar density under Poincaré transformations. The end result will be a supersymmetric invariant action $\mathcal{S}$

$$\mathcal{S} = \int d^4x\, d^2\theta\, d^2\bar{\theta}\ \mathcal{A}(x;\theta,\bar{\theta}) = \int d^4x\ \mathcal{L}\left(\phi(x), \psi(x), A_{\mu}(x), \ldots\right). \tag{4.43}$$

Let us emphasize again: one does not need to prove $\mathcal{S}$ to be invariant under supersymmetry transformations. If it comes from an integral of a superfield in superspace, this is just automatic: by construction, the Lagrangian $\mathcal{L}$ on the r.h.s. of eq. (4.43), an apparently innocent-looking function of ordinary fields, is guaranteed to be supersymmetric invariant, up to total space-time derivatives.

The superfield $\mathcal{A}$ will be in general a product of superfields (recall that a product of superfields is still a superfield). However, the general superfield (4.14) cannot be the basic object of this construction: it contains too many field components to correspond to an irreducible representation of the supersymmetry algebra. We have to put (supersymmetric invariant) constraints on Y and restrict its form to contain only a subset of fields. Being the supersymmetric invariant constraint, this reduced set of fields will still be a superfield, and hence it will carry a representation of the supersymmetry algebra. In what follows, we will start discussing a few such constraints, the so-called chiral and real constraints. These will be the relevant ones for our purposes, as they will lead to chiral and vector superfields, the right superfields to accommodate matter and radiation, respectively, and to linear superfields, where conserved currents sit.

4.4 Chiral superfields

One can construct *covariant derivatives* $D_\alpha, \overline{D}_{\dot{\alpha}}$ defined as

$$\begin{cases} D_\alpha = \partial_\alpha + i\, \sigma^\mu_{\alpha\dot{\beta}}\, \bar{\theta}^{\dot{\beta}} \partial_\mu \\ \overline{D}_{\dot{\alpha}} = \bar{\partial}_{\dot{\alpha}} + i\, \theta^\beta \sigma^\mu_{\beta\dot{\alpha}} \partial_\mu \end{cases} \tag{4.44}$$

and which anticommute with the supersymmetry generators $Q_\alpha, \overline{Q}_{\dot{\alpha}}$. More precisely we have

$$\{D_\alpha, \overline{D}_{\dot{\beta}}\} = 2i\, \sigma^\mu_{\alpha\dot{\beta}}\, \partial_\mu = -2\sigma^\mu_{\alpha\dot{\beta}}\, P_\mu, \tag{4.45}$$

$$\{D_\alpha, D_\beta \text{ or } Q_\beta \text{ or } \overline{Q}_{\dot{\beta}}\} = 0 \quad (\text{similarly for } \overline{D}_{\dot{\alpha}}). \tag{4.46}$$

This implies that

$$\delta_{\epsilon,\bar{\epsilon}}\,(D_\alpha Y) = D_\alpha\,(\delta_{\epsilon,\bar{\epsilon}} Y), \tag{4.47}$$

since D_α commutes both with ϵQ and $\bar{\epsilon}\overline{Q}$. Therefore, if Y is a superfield, that is a field in superspace transforming as dictated by eq. (4.25) under a supersymmetry transformation, so is $D_\alpha Y$. This means that $D_\alpha Y = 0$ is a supersymmetric invariant constraint we can impose on a superfield Y to reduce the number of its components, while still having the field carrying a representation of the supersymmetry algebra (the same holds for the constraint $\overline{D}_{\dot{\alpha}} Y = 0$).

Recall the generic expression (4.14) for Y and consider $\bar{\partial}_{\dot{\alpha}} Y$: this has fewer components with respect to Y itself, since, for instance, there is no $\theta\theta\bar{\theta}\bar{\theta}$ term. However

$$\left[\bar{\partial}_{\dot{\alpha}}, \epsilon Q\right] = \epsilon^\beta \sigma^\mu_{\beta\dot{\alpha}}\, \partial_\mu. \tag{4.48}$$

This implies that a supersymmetry transformation on $\overline{X}_{\dot{\alpha}} = \bar{\partial}_{\dot{\alpha}} Y$ would generate a $\theta\theta\bar{\theta}\bar{\theta}$ term which $\overline{X}_{\dot{\alpha}}$ did not contain. Hence $\bar{\partial}_{\dot{\alpha}} Y$ is not a superfield, in the sense of providing a basis for a representation of supersymmetry. On the other hand, the covariant derivatives defined in (4.44) anticommute with Q_α and $\overline{Q}_{\dot{\alpha}}$. Hence, if Y is a superfield, $D_\alpha Y, \overline{D}_{\dot{\alpha}} Y$ are also superfields (and so is $\partial_\mu Y$, since P_μ commutes with Q_α and $\overline{Q}_{\dot{\alpha}}$).

A *chiral* superfield Φ is a superfield such that

$$\overline{D}_{\dot{\alpha}} \Phi = 0. \tag{4.49}$$

Seemingly, an *anti-chiral* superfield Ψ is a superfield such that

$$D_\alpha \Psi = 0. \tag{4.50}$$

Notice that if Φ is chiral, its hermitian conjugate is anti-chiral. This implies that a chiral superfield cannot be real. Indeed, in this case it is easy to show that it should be a constant. Taking the hermitian conjugate of eq. (4.49) one would conclude that the field would also be anti-chiral. Acting now on it with the anticommutator in eq. (4.45) one would get $\partial_\mu \Phi = 0$. This is the superfield analogue of what we have seen in the previous chapter, when we constructed the chiral multiplet and we learned that the scalar field ϕ had to be complex.

We would like to find the most general expression for a chiral superfield in terms of ordinary fields, as we did for the general superfield (4.14). In other words, we have to integrate the constraint (4.49). To this aim, it is useful to define new coordinates

$$y^\mu = x^\mu + i\theta\sigma^\mu\bar{\theta}, \quad \bar{y}^\mu = x^\mu - i\theta\sigma^\mu\bar{\theta}. \tag{4.51}$$

It easily follows that

$$\overline{D}_{\dot{\alpha}}\theta_\beta = \overline{D}_{\dot{\alpha}} y^\mu = 0, \quad D_\alpha \bar{\theta}_{\dot{\beta}} = D_\alpha \bar{y}^\mu = 0. \tag{4.52}$$

Recalling the definition (4.49) this implies that Φ depends only on (y^μ, θ_α) explicitly, but not on $\bar{\theta}_{\dot{\alpha}}$ (the $\bar{\theta}$-dependence is hidden inside y^μ). In this (super)coordinate system the chiral constraint is easily solved by

$$\Phi(y,\theta) = \phi(y) + \sqrt{2}\theta\psi(y) - \theta\theta F(y). \tag{4.53}$$

Taylor-expanding the above expression around x we get for the actual $\Phi(x,\theta,\bar{\theta})$

$$\begin{aligned}\Phi(x,\theta,\bar{\theta}) = \phi(x) &+ \sqrt{2}\theta\psi(x) + i\theta\sigma^\mu\bar{\theta}\partial_\mu\phi(x) - \theta\theta F(x) + \\ &- + \frac{i}{\sqrt{2}}\theta\theta\partial_\mu\psi(x)\sigma^\mu\bar{\theta} - \frac{1}{4}\theta\theta\bar{\theta}\bar{\theta}\Box\phi(x),\end{aligned} \tag{4.54}$$

which can also be conveniently recast as $\Phi(x,\theta,\bar{\theta}) = e^{i\theta\sigma^\mu\partial_\mu\bar{\theta}}\Phi(x,\theta)$. We see that, as expected, this superfield has less components than the general superfield Y, and some of them are related to each other.

The chiral superfield (4.54) is worth its name, since it is a superfield which encodes precisely the degrees of freedom of the chiral multiplet of fields we have previously constructed. On-shell, it corresponds to a $N = 1$ multiplet of states, hence carrying an irreducible representation of the $N = 1$ supersymmetry algebra.

A similar story holds for the anti-chiral superfield $\overline{\Phi}$ (throughout this book, unless otherwise stated, we will use the symbol — instead of † to mean hermitian conjugation, to adapt to the two-component spinor notation for which $\psi^\dagger = \overline{\psi}_{\dot\alpha}$). We would get

$$\overline{\Phi}(x, \theta, \bar\theta) = \overline{\phi}(\bar y) + \sqrt{2}\bar\theta\overline{\psi}(\bar y) - \bar\theta\bar\theta\overline{F}(\bar y) \tag{4.55}$$

and, in turn

$$\begin{aligned}\overline{\Phi}(x, \theta, \bar\theta) = \overline{\phi}(x) &+ \sqrt{2}\bar\theta\overline{\psi}(x) - i\theta\sigma^\mu\bar\theta\partial_\mu\overline{\phi}(x) - \bar\theta\bar\theta\overline{F}(x) + \\ &+ \frac{i}{\sqrt{2}}\bar\theta\bar\theta\theta\sigma^\mu\partial_\mu\overline{\psi}(x) - \frac{1}{4}\theta\theta\bar\theta\bar\theta\Box\overline{\phi}(x). \end{aligned}\tag{4.56}$$

Let us now try and see how does a chiral (or anti-chiral) superfield transform under supersymmetry transformations. This amounts to computing

$$\delta_{\epsilon,\bar\epsilon}\Phi(y;\theta) = \left(i\epsilon Q + i\bar\epsilon\overline{Q}\right)\Phi(y;\theta) \tag{4.57}$$

(and similarly for $\overline{\Phi}$). To compute eq. (4.57) it is convenient to write the differential operators $Q_\alpha, \overline{Q}_{\dot\alpha}$ in the $(y^\mu, \theta_\alpha, \bar\theta_{\dot\alpha})$ coordinate system. This amounts to trading the partial derivatives taken with respect to $(x^\mu, \theta_\alpha, \bar\theta_{\dot\alpha})$ for those taken with respect to the new system $(y^\mu, \theta_\alpha, \bar\theta_{\dot\alpha})$ and plugging this into eq. (4.36). The final result reads

$$\begin{cases} Q_\alpha^{new} &= -i\partial_\alpha \\ \overline{Q}_{\dot\alpha}^{new} &= +i\bar\partial_{\dot\alpha} + 2\theta^\alpha\sigma^\mu_{\alpha\dot\alpha}\frac{\partial}{\partial y^\mu} \end{cases}. \tag{4.58}$$

Plugging these expressions into eq. (4.57) one gets

$$\begin{aligned}\delta_{\epsilon,\bar\epsilon}\Phi(y;\theta) &= \left(\epsilon^\alpha\partial_\alpha + 2i\theta^\alpha\sigma^\mu_{\alpha\dot\beta}\bar\epsilon^{\dot\beta}\frac{\partial}{\partial y^\mu}\right)\Phi(y;\theta) \\ &= \sqrt{2}\epsilon\psi - 2\epsilon\theta F + 2i\theta\sigma^\mu\bar\epsilon\left(\frac{\partial}{\partial y^\mu}\phi + \sqrt{2}\theta\frac{\partial}{\partial y^\mu}\psi\right) \\ &= \sqrt{2}\epsilon\psi + \sqrt{2}\theta\left(-\sqrt{2}\epsilon F + \sqrt{2}i\sigma^\mu\bar\epsilon\frac{\partial}{\partial y^\mu}\phi\right) - \theta\theta\left(-i\sqrt{2}\bar\epsilon\bar\sigma^\mu\frac{\partial}{\partial y^\mu}\psi\right). \end{aligned}\tag{4.59}$$

Therefore, the final expression for the supersymmetry variation of the different field components of the chiral superfield Φ reads

$$\begin{cases}\delta\phi = \sqrt{2}\epsilon\psi \\ \delta\psi_\alpha = \sqrt{2}i(\sigma^\mu\bar\epsilon)_\alpha\partial_\mu\phi - \sqrt{2}\epsilon_\alpha F \\ \delta F = i\sqrt{2}\partial_\mu\psi\sigma^\mu\bar\epsilon\end{cases}. \tag{4.60}$$

It is left to the reader to derive the corresponding expressions for an anti-chiral superfield. In this case, one should write the generators $Q_\alpha, \overline{Q}_{\dot\alpha}$ in the $(\bar{y}^\mu, \theta_\alpha, \bar\theta_{\dot\alpha})$ coordinate system.

4.5 Real (aka Vector) superfields

In order to have gauge interactions we clearly need to find a supersymmetric invariant projection which saves the vector field v^μ in the general expression (4.14) and makes it real (this was not the case for the constraint (4.49), for which the vector component is $\sim \partial_\mu \phi$). The right thing to do is to impose a *reality* condition on the general superfield Y. Indeed, under hermitian conjugation, $Y \to \overline{Y}$, one has $v^\mu \to \overline{v}^\mu$. So, by imposing a reality condition, the vector component not only survives as an independent degree of freedom, but becomes real.

A real (aka *vector*) superfield V is a superfield such that

$$V = \overline{V}. \tag{4.61}$$

Looking at the general expression (4.14) this leads to the following expansion for V

$$\begin{aligned} V(x,\theta,\bar\theta) = {} & C(x) + i\,\theta\chi(x) - i\,\bar\theta\overline{\chi}(x) + \theta\sigma^\mu\bar\theta v_\mu + \frac{i}{2}\,\theta\theta\,(M(x)+iN(x)) + \\ & -\frac{i}{2}\,\bar\theta\bar\theta\,(M(x)-iN(x)) + i\,\theta\theta\bar\theta\left(\overline{\lambda}(x) + \frac{i}{2}\,\bar\sigma^\mu\partial_\mu\chi(x)\right) + \\ & -i\,\bar\theta\bar\theta\theta\left(\lambda(x) + \frac{i}{2}\,\sigma^\mu\partial_\mu\overline{\chi}(x)\right) + \frac{1}{2}\,\theta\theta\bar\theta\bar\theta\left(D(x) - \frac{1}{2}\,\partial^2 C(x)\right). \end{aligned} \tag{4.62}$$

Notice that, as such, this superfield has $8_B{+}8_F$ degrees of freedom. The next step is to introduce the supersymmetric version of gauge transformations. As we shall see, after gauge fixing, this will reduce the number of off-shell degrees of freedom to $4_B{+}4_F$, which become $2_B{+}2_F$ on-shell (for a massless representation), as it should be the case for a massless vector multiplet of states; see eq. (3.15).

First notice that if Φ is a chiral superfield, then $\Phi + \overline{\Phi}$ is a (very special) vector superfield. So, under the following transformation on V

$$V \to V + \Phi + \overline{\Phi} \tag{4.63}$$

one gets a superfield V which is still real. Under the shift (4.63) the vector component v_μ in (4.62) transforms as $v_\mu \to v_\mu {-} \partial_\mu\,(2\ \mathrm{Im}\phi)$. This is precisely how an ordinary (abelian) gauge transformation acts on a vector field. Therefore, eq. (4.63) is a natural definition for the supersymmetric version of a gauge transformation. Under eq. (4.63) the component fields of V transform as

$$\begin{cases} C & \to & C + 2\,\mathrm{Re}\phi \\ \chi & \to & \chi - i\sqrt{2}\psi \\ M & \to & M - 2\,\mathrm{Im}F \\ N & \to & N + 2\,\mathrm{Re}F \\ D & \to & D \\ \lambda & \to & \lambda \\ v^{\mu} & \to & v^{\mu} - 2\,\partial_{\mu}\mathrm{Im}\phi \end{cases} \tag{4.64}$$

where the components of Φ have been dubbed (ϕ, ψ, F). From the transformations above, by properly choosing Φ, namely choosing

$$\mathrm{Re}\phi = -\frac{C}{2}, \quad \psi = -\frac{i}{\sqrt{2}}\chi, \quad \mathrm{Re}F = -\frac{N}{2}, \quad \mathrm{Im}F = \frac{M}{2}, \tag{4.65}$$

one can gauge away (namely put to zero) C, M, N, χ. The choice above is called Wess-Zumino gauge. In this gauge, a vector superfield can be written as

$$V_{WZ} = \theta\sigma^{\mu}\bar{\theta}\, v_{\mu}(x) + i\theta\theta\,\bar{\theta}\bar{\lambda}(x) - i\bar{\theta}\bar{\theta}\,\theta\lambda(x) + \frac{1}{2}\theta\theta\,\bar{\theta}\bar{\theta}D(x). \tag{4.66}$$

Therefore, taking into account gauge invariance (that is, the redundancy of one of the vector degrees of freedom, the one associated to the transformation $v_{\mu} \to v_{\mu} - \partial_{\mu}\,(2\ \mathrm{Im}\phi)$, which the WZ gauge does not fix), we end up with $4_B + 4_F$ degrees of freedom off-shell. As we shall see later, D will turn out to be an auxiliary field; therefore, by imposing the equations of motion for D, the spinor λ and the vector v^{μ}, one will end up with $2_B + 2_F$ degrees of freedom on-shell. Since we like to formulate gauge theories keeping gauge invariance manifest off-shell, the WZ gauge is defined as a gauge where $C = M = N = \chi = 0$, but no restrictions on v^{μ}. This way, while remaining in the WZ gauge, we still have the freedom to do ordinary gauge transformations. In other words, once in the WZ gauge, we can still perform a supersymmetric gauge transformation (4.63) with parameters $\phi = -\overline{\phi}, \psi = 0, F = 0$.

Let us end this section with two important comments. First, notice that in the WZ gauge, each term in the expansion of V_{WZ} contains at least one θ.

Therefore

$$V_{WZ}^2 = \frac{1}{2}\theta\theta\bar{\theta}\bar{\theta}v_\mu v^\mu, \quad V_{WZ}^n = 0 \;\; n \geq 3. \tag{4.67}$$

These identities will simplify things a lot when it comes to constructing supersymmetric gauge actions. Second, it should be remarked that the WZ gauge does not commute with supersymmetry. Acting with a supersymmetry transformation on a vector superfield in the WZ gauge, one obtains a new superfield which is not in the WZ gauge. Hence, when working in this gauge, after a supersymmetry transformation, one has to do a compensating supersymmetric gauge transformation (4.63), with a properly chosen Φ, to come back to the WZ gauge.

4.6 (Super)Current superfields

The two superfields described above are what we need to describe matter and radiation in a supersymmetric theory, if we are not interested in gravitational interactions. However, in a supersymmetric theory, composite operators should also sit in superfields. There are at least two other types of superfields which accommodate important composite operators, *i.e.* conserved currents and the supersymmetry current (supercurrent for short), the latter being ubiquitous in a supersymmetric QFT, as this is the current associated to the supersymmetry charge itself. Both these superfields turn out to be real superfields, as the superfield described in the previous section, but current conservation implies extra supersymmetric invariant conditions they should satisfy which make them a particular class of real superfields. In what follows, we will briefly describe both of them.

4.6.1 *Internal symmetry current superfields*

Because of Nöether theorem, any continuous symmetry in a local QFT is associated to a conserved current j_μ satisfying $\partial^\mu j_\mu = 0$, and to the corresponding conserved charge Q defined as $Q = \int d^3x\, j^0$. Here we are referring to non-R symmetries. R-symmetry will be discussed later.

As any other operator, a conserved current in a supersymmetric theory should sit in a superfield. It turns out that this is a real scalar superfield $\mathcal{J}$ satisfying the following extra constraint

$$D^2\mathcal{J} = \overline{D}^2\mathcal{J} = 0. \tag{4.68}$$

A real superfield satisfying the constraint above is called *linear* superfield. Working a little bit one can show that a real superfield subject to the conditions (4.68) has the following component expression

$$\mathcal{J} = J(x) + i\theta j(x) - i\bar{\theta}\bar{j}(x) + \theta\sigma^{\mu}\bar{\theta}j_{\mu}(x) + \\ + \frac{1}{2}\theta^{2}\bar{\theta}\bar{\sigma}^{\mu}\partial_{\mu}j(x) - \frac{1}{2}\bar{\theta}^{2}\theta\sigma^{\mu}\partial_{\mu}\bar{j}(x) + \frac{1}{4}\theta^{2}\bar{\theta}^{2}\Box J(x), \qquad (4.69)$$

where J is a real scalar and j_{α} a spinor. By imposing eq. (4.68) on the above expression one easily sees that the current j_{μ} satisfies $\partial^{\mu}j_{\mu} = 0$, *i.e.* is a conserved current. So the constraint (4.68) is indeed the correct supersymmetric generalization of current conservation. Note that while the condition (4.68) is compatible with supersymmetry (both D^2 and $\overline{D}^2$ commute with supersymmetry transformations), it stands on a slightly different footing with respect to the conditions (4.49), (4.50) and (4.61). The latter constrain the dependence of a superfield as a function of the fermionic coordinates $(\theta_{\alpha}, \bar{\theta}_{\dot{\alpha}})$, but they do not say anything about space-time dependence. On the contrary, eq. (4.68) constrains also the space-time dependence of some of the fields imposing differential equations in x-space, one obvious example being the conservation equation $\partial^{\mu}j_{\mu} = 0$. In this sense, (4.68) is an on-shell constraint.

A few comments are in order. First notice that, as compared to a general real superfield (4.62), a linear superfield has less independent components. This is due to the extra condition (4.68) a linear superfield has to satisfy. Another comment regards the spin content of $\mathcal{J}$. One condition that $\mathcal{J}$ should (and does) satisfy is that it should not contain fields with spin higher than one. If this were the case, one could not gauge the current j^{μ} without introducing higher-spin gauge fields, something which is expected not to be consistent in a local interacting QFT with rigid supersymmetry (recall our discussion in the previous chapter). This implies that $\mathcal{J}$ should be a real *scalar* superfield, namely its lowest component J should be a scalar. Finally, it is worth noticing that the detailed structure of $\mathcal{J}$ is not uniquely fixed, but in fact defined up to Schwinger terms entering the current algebra.

This can be understood as follows. Because the conserved charge Q is a non-R symmetry charge, it commutes with supersymmetry generators, $[Q_{\alpha}, Q] = 0$. This implies that the operator on the right hand side of the current algebra

$$[Q_{\alpha}, j_{\mu}] = \mathcal{O}_{\alpha\mu}, \qquad (4.70)$$

should be an operator which vanishes when acting with ∂_μ, as does j_μ, and it should also be a total space-time derivative for $\mu = 0$, say $\mathcal{O}_{\alpha 0} = \partial^\nu A_{\alpha\nu}$, so that it integrates to zero, because it happens likewise to the left hand side, given that

$$\int d^4x\, [Q_\alpha, j_0] = \int dt\, [Q_\alpha, Q] = 0. \tag{4.71}$$

An operator of this kind is known as Schwinger term. Different Schwinger terms provide different versions of the superfield $\mathcal{J}$, which is hence not univocally defined. The superfield defined in eq. (4.69) is one such possible completion, for which $\mathcal{O}_{\alpha\mu} = -2i(\sigma_{\mu\nu})_\alpha^{\ \beta}\,\partial^\nu j_\beta$. This can be easily checked using eqs. (4.34)–(4.35). This said, the first terms in the expansion of $\mathcal{J}$, namely $J + i\theta j - i\overline{\theta}\overline{j} + \theta\sigma^\mu\bar{\theta} j_\mu(x)$, are universal in the sense that they turn out not to depend on the specific Schwinger term appearing in eq. (4.70).

4.6.2 *Supercurrent superfields*

While currents associated to internal symmetries might or might not be there, in any supersymmetric theory there always exists, by definition, a conserved current — the supersymmetry current $S_{\alpha\mu}$ — associated to the conservation of the supercharge Q_α. In terms of the supercurrent, the supersymmetry charge is $Q_\alpha = \int d^3x\, S_\alpha^{\ 0}$. Such supercurrent should sit in a superfield.

An equation analogous to eq. (4.70) is imposed by the supersymmetry algebra, which reads

$$\{\overline{Q}_{\dot\alpha}, S_{\alpha\nu}\} = 2\sigma^\mu_{\alpha\dot\alpha} T_{\mu\nu} + \mathcal{O}_{\alpha\dot\alpha\nu}, \tag{4.72}$$

where $T_{\mu\nu}$ is the (conserved) energy-momentum tensor and $\mathcal{O}_{\alpha\dot\alpha\nu}$ is again a Schwinger term. Note that, unlike eq. (4.70), the $\nu = 0$ component of the left hand side of eq. (4.72) does not integrate to zero now but is proportional to $\int dt\, P_\mu$ by the supersymmetry algebra, namely to $\int d^4x\, T_\mu^{\ 0}$. This is why, on top of a Schwinger term, the energy-momentum tensor appears on the right hand side of eq. (4.72). This also shows that the supercurrent and the energy-momentum tensor sit in the same superfield, $T_{\mu\nu}$ being the highest spin field of the representation (otherwise, it would be problematic to couple supersymmetry with gravity avoiding higher spin currents). This is the current operators counterpart of the fact that the graviton and the gravitino sit in the same multiplet. These properties are enough to fix the universal structure of the supercurrent superfield which reads

$$\mathcal{J}_\mu = j_\mu + \theta\,(S_\mu + \cdots) + \bar\theta\,(\overline{S}_\mu + \cdots) + \theta\sigma^\nu\bar\theta\,(2T_{\mu\nu} + \cdots) + \cdots\,. \tag{4.73}$$

The arbitrariness of the Schwinger term gives rise to different possible completions of $\mathcal{J}_\mu$. The most known of such completions is due to Ferrara and Zumino. The FZ multiplet can be described by a pair of superfields $(\mathcal{J}_\mu, X)$ satisfying the relation

$$2\,\overline{D}^{\dot\alpha}\sigma^\mu_{\alpha\dot\alpha}\,\mathcal{J}_\mu = D_\alpha X, \tag{4.74}$$

with $\mathcal{J}_\mu$ being a real vector superfield, and X a chiral superfield, $\overline{D}_{\dot\alpha}X = 0$. The same comment we made on the on-shell nature of the condition (4.68) holds also in this case. From the defining equation above one can work out the component expression of these two superfields. They read

$$\begin{aligned}\mathcal{J}_\mu = j_\mu &+ \theta\left(S_\mu - \frac{1}{3}\sigma_\mu\overline{S}\right) + \bar\theta\left(\overline{S}_\mu + \frac{1}{3}\bar\sigma_\mu S\right) + \frac{i}{2}\theta^2\partial_\mu x^* - \frac{i}{2}\bar\theta^2\partial_\mu x\, + \\ &+ \theta\sigma^\nu\bar\theta\left(2T_{\mu\nu} - \frac{2}{3}\eta_{\mu\nu}T + \frac{1}{2}\varepsilon_{\mu\nu\rho\sigma}\partial^\rho j^\sigma\right) + \cdots\end{aligned} \tag{4.75}$$

and

$$X = x + \frac{2}{3}\theta S + \theta^2\left(\frac{2}{3}T + i\,\partial^\mu j_\mu\right) + \cdots\,, \tag{4.76}$$

where ... stand for the supersymmetric completion and we have defined the trace operators $T \equiv T^\mu{}_\mu$ and $S_\alpha \equiv \sigma^\mu_{\alpha\dot\alpha}\overline{S}^{\dot\alpha}_\mu$. All in all, the FZ superfield contains a (in general non-conserved) R-current j_μ, a symmetric and conserved $T_{\mu\nu}$, a conserved $S_{\alpha\mu}$, and a complex scalar x. That j_μ is a R-current follows from the fact that it sits in the same superfield where $S_{\alpha\mu}$ sits and so it does not commute with supersymmetry. From the above expression one can also see that whenever X vanishes, the current j_μ becomes conserved and all trace operators vanish. In this case, the theory is conformal and j_μ becomes the always present (and conserved) superconformal R-current.

We have seen that the FZ multiplet contains a non-conserved R-current. What if a theory admits a $U(1)_R$ R-symmetry? For theories with an R-symmetry there exists an alternative supermultiplet accommodating the energy-momentum tensor and the supercurrent, the so-called $\mathcal{R}$ multiplet.

It turns out this is again defined in terms of a pair of superfields $(\mathcal{R}_\mu, \chi_\alpha)$ which now satisfy a different on-shell condition

$$2\,\overline{D}^{\dot\alpha}\sigma^\mu_{\alpha\dot\alpha}\,\mathcal{R}_\mu = \chi_\alpha, \tag{4.77}$$

where $\mathcal{R}_\mu$ is a real vector superfield and χ_α a chiral superfield which, besides $\overline{D}_{\dot\alpha}\chi_\alpha = 0$, also satisfies the identity $\overline{D}_{\dot\alpha}\overline{\chi}^{\dot\alpha} - D^\alpha\chi_\alpha = 0$. This implies that $\partial^\mu\mathcal{R}_\mu = 0$, from which it follows that the lowest component of $\mathcal{R}_\mu$ is now a conserved current, the R-current j^R_μ. The component expression of the superfields making up the $\mathcal{R}$ multiplet reads

$$\mathcal{R}_\mu = j^R_\mu + \theta S_\mu + \bar\theta\,\overline{S}_\mu + \theta\sigma^\nu\bar\theta\left(2T_{\mu\nu} + \frac{1}{2}\varepsilon_{\mu\nu\rho\sigma}(\partial^\rho j^\sigma + C^{\rho\sigma})\right) + \cdots \tag{4.78}$$

and

$$\chi_\alpha = -2S_\alpha - \left(4\delta_\alpha^\beta T + 2i\,(\sigma^\rho\bar\sigma^\tau)_\alpha^\beta C_{\rho\tau}\right)\theta_\beta + 2i\,\theta^2\sigma^\nu_{\alpha\dot\alpha}\partial_\nu\overline{S}^{\dot\alpha} + \cdots \tag{4.79}$$

where again ... stand for the supersymmetric completion, and $C_{\mu\nu}$ is a closed two-form. That j^R_μ is an R-current can be easily seen by noticing that the current algebra now reads $\left[Q_\alpha, j^R_\mu\right] = S_{\alpha\mu}$. Taking the time-component and integrating, this implies that $\int dt\,\left[Q_\alpha, Q^R\right] = \int dt\,Q_\alpha$, which is what is expected for a R-symmetry; recall eq. (2.80). Notice, finally, that when $X = 0$, the FZ multiplet (4.75) becomes a (special instance of an) R-multiplet, one for which $\chi_\alpha = 0$. Indeed, its lowest component j_μ now becomes the conserved superconformal R-current.

The FZ and $\mathcal{R}$ multiplets are the more common supercurrent multiplets. However, there are instances in which a theory does not admit a R-symmetry (and hence the $\mathcal{R}$ multiplet cannot be defined) and the FZ multiplet is not a well-defined operator, *e.g.* it is not gauge invariant. In these cases, one should consider yet another multiplet where the supercurrent can sit, the so-called $\mathcal{S}$ multiplet, which is bigger than the two above. We will not discuss the $\mathcal{S}$ multiplet here, and refer to the references given at the end of this chapter. On the contrary, there exist theories in which both the FZ and the $\mathcal{R}$ multiplets can be defined. In such cases it turns out that the two are related by a so-called shift transformation defined as

$$\mathcal{R}_\mu = \mathcal{J}_\mu + \frac{1}{4}\bar\sigma^{\dot\alpha\alpha}_\mu\left[D_\alpha, \overline{D}_{\dot\alpha}\right]U, \quad X = -\frac{1}{2}\overline{D}^2U, \quad \chi_\alpha = \frac{3}{2}\overline{D}^2D_\alpha U, \tag{4.80}$$

where U is a real superfield associated to a non-conserved (and non-R) current.

4.7 Exercises

Exercise 4.1. Prove identities (4.20).

Exercise 4.2. Check that the differential operators Q_α and $\overline{Q}_{\dot\alpha}$ (4.36) close the supersymmetry algebra (4.37).

Hint: recall that all θ's and $\bar\theta$'s anti-commute between themselves, and that

$$\{a_i, a_j\} = 0 \quad \longrightarrow \quad \frac{\partial}{\partial a_i} a_j = \frac{\partial a_j}{\partial a_i} - a_j \frac{\partial}{\partial a_i} \,, \tag{4.81}$$

which implies that, e.g.

$$\{\partial_\alpha, \bar\theta^{\dot\gamma}\} = 0, \quad \{\partial_\alpha, \theta^\beta\} = \delta^\beta_\alpha, \quad \{\bar\partial_{\dot\alpha}, \bar\theta^{\dot\beta}\} = \delta^{\dot\beta}_{\dot\alpha}. \tag{4.82}$$

Exercise 4.3. Check that the covariant derivatives D_α and $\overline{D}_{\dot\alpha}$ (4.44) anticommute between themselves and with the supercharge operators (4.36).

Exercise 4.4. Compute how the field components of an anti-chiral superfield Ψ transform under supersymmetry transformations. Show that if $\Psi = \overline{\Phi}$ one gets the hermitian conjugate of the transformations (4.60).

Exercise 4.5. Compute the supersymmetric variation of a vector superfield in the WZ gauge, and find the explicit form of the chiral superfield Φ which, via a compensating supersymmetric gauge transformation, brings the vector superfield back to the WZ gauge.

References

[1] A. Bilal, *Introduction to supersymmetry*, Section 4, arXiv:hep-th/0101055.
[2] L. Castellani, R. D'Auria and P. Frè, *Supergravity And Superstrings: A Geometric Perspective. Vol. 1: Mathematical Foundations*, Chapters II.2.5 and II.2.6, World Scientific Singapore (1991).
[3] J. D. Lykken, *Introduction to Supersymmetry*, Sections 2 and 3, arXiv:hep-th/9612114.
[4] A. Salam and J. A. Strathdee, Supergauge Transformations, *Nucl. Phys. B* **76** (1974), 477.
A. Salam and J. A. Strathdee, On Superfields and Fermi-Bose Symmetry, *Phys. Rev. D* **11** (1975), 1521.

[5] Z. Komargodski and N. Seiberg, *Comments on Supercurrent Multiplets, Supersymmetric Field Theories and Supergravity*, Section 1, arXiv:1002.2228 [hep-th].
[6] T. T. Dumitrescu and Z. Komargodski, Aspects of supersymmetry and its breaking, Sections 2 and 4, *Nucl. Phys. Proc. Suppl.* **216** (2011), 44.
[7] S. Ferrara and B. Zumino, Transformation Properties of the Supercurrent, *Nucl. Phys. B* **87** (1975), 207.

Chapter 5

Supersymmetric Actions: Minimal Supersymmetry

In the previous chapter we have introduced the basic superfields one needs to construct $N = 1$ supersymmetric theories, if one is not interested in describing gravitational interactions. We are now ready to look for supersymmetric actions describing the dynamics of these superfields. We will first concentrate on matter actions and construct the most general supersymmetric action describing the interaction of a set of chiral superfields. Then we will introduce SuperYang-Mills theory which is nothing but the supersymmetric version of Yang-Mills. Finally, we will couple the two sectors with the final goal of deriving the most general $N = 1$ supersymmetric action describing the interaction of radiation with matter. In all these cases, we will consider both renormalizable as well as non-renormalizable theories, the latter being relevant to describe effective low energy theories.

In the following we will deal with gauge theories, and hence gauge groups, like $SU(N)$ and alike. In order to avoid confusion, in the remainder of this book we will use calligraphic N when referring to the number of supersymmetry, $\mathcal{N} = 1, 2$ or 4 respectively.

5.1 $\mathcal{N} = 1$ matter actions

Following the general strategy outlined in Section 4.3 we want to construct a supersymmetric invariant action describing the interaction of a (set of) chiral superfield(s). Let us first notice that a product of chiral superfields is still a chiral superfield and a product of anti-chiral superfields is an anti-chiral superfield. Conversely, the product of a chiral superfield with its hermitian conjugate (which is anti-chiral) is a (very special, in fact) real superfield.

Let us start analyzing the theory of a single chiral superfield Φ. Consider the following integral

$$\int d^2\theta\, d^2\bar{\theta}\, \overline{\Phi}\Phi. \tag{5.1}$$

This integral satisfies all necessary conditions to be a supersymmetric Lagrangian. First, it is supersymmetric invariant (up to total space-time derivative) since it is the integral in superspace of a superfield. Second, it is real and a scalar object. Indeed, the first component of $\overline{\Phi}\Phi$ is $\overline{\phi}\phi$ which is real and a scalar. Now, the $\theta^2\bar{\theta}^2$ component of a superfield, which is the only term contributing to the above integral, has the same tensorial structure as its first component since $\theta^2\bar{\theta}^2$ does not have any free space-time indices and is real, that is $(\theta^2\bar{\theta}^2)^\dagger = \theta^2\bar{\theta}^2$. Finally, the above integral has also the right physical dimensions for being a Lagrangian, *i.e.* $[M]^4$. Indeed, from the expansion of a chiral superfield, one can see that θ and $\bar{\theta}$ have both dimension $[M]^{-1/2}$ (compare the first two components of a chiral superfield, $\phi(x)$ and $\theta\psi(x)$, and recall that a spinor in four dimensions has physical dimension $[M]^{3/2}$). This means that the $\theta^2\bar{\theta}^2$ component of a superfield Y has dimension $[Y]+2$ if $[Y]$ is the dimension of the superfield (which is that of its lowest component). Since the dimension of $\overline{\Phi}\Phi$ is 2, it follows that its $\theta^2\bar{\theta}^2$ component has dimension 4 (notice that, consistently, $\int d^2\theta d^2\bar{\theta}\;\theta^2\bar{\theta}^2$ is dimensionless since $d\theta$ ($d\bar{\theta}$) has opposite dimensions with respect to θ ($\bar{\theta}$), given that the differential is equivalent to a derivative, for Grassman variables). Summarizing, eq. (5.1) is an object of dimension 4, is real, and transforms as a total space-time derivative under SuperPoincaré transformations.

To perform the integration in superspace one can start from the expression of Φ and $\overline{\Phi}$ in the y (resp. $\bar{y}$) coordinate system, take the product of $\overline{\Phi}(\bar{y},\bar{\theta})\Phi(y,\theta)$, expand the result in the $(x,\theta,\bar{\theta})$ space, and finally pick up the $\theta^2\bar{\theta}^2$ component, only. The end result is

$$\mathcal{L}_{kin} = \int d^2\theta\, d^2\bar{\theta}\, \overline{\Phi}\Phi = \partial_\mu\overline{\phi}\,\partial^\mu\phi + \frac{i}{2}\left(\partial_\mu\psi\;\sigma^\mu\overline{\psi} - \psi\;\sigma^\mu\partial_\mu\overline{\psi}\right) + \overline{F}F, \tag{5.2}$$

up to total space-time derivatives, which we do not care about since they will anyhow being killed by the integral in ordinary space-time. What we get is precisely the kinetic term describing the degrees of freedom of a free chiral superfield! In doing so we also see that, as anticipated, the F field is an auxiliary field, namely a non-propagating degree of freedom. Integrating it out (which is trivial in this case since its equation of motion is simply $F = 0$) one gets a (supersymmetric) Lagrangian describing physical degrees of freedom, only. Notice that after integrating F out, supersymmetry is realized on-shell, namely upon imposing the equation of motions on the propagating degrees of freedom.

The equations of motion for ϕ, ψ and F following from the Lagrangian (5.2) can be easily derived using superfield formalism readily from the expression in superspace. This might not look obvious at first sight since varying the action (5.1) with respect to $\overline{\Phi}$ would get $\Phi = 0$, which does not provide the equations of motion we would expect, as it can be easily inferred expanding it in components. The point is that the integral in eq. (5.1) is a constrained one, since Φ is a chiral superfield and hence subject to the constraint $\overline{D}_{\dot{\alpha}}\Phi = 0$. One can rewrite the above integral as an unconstrained one noticing that

$$\int d^2\theta\, d^2\bar{\theta}\ \overline{\Phi}\Phi = -\frac{1}{4}\int d^2\bar{\theta}\ \overline{\Phi}D^2\Phi. \tag{5.3}$$

In getting the right hand side we have used the fact that $\int d\theta_\alpha = D_\alpha$, up to total space-time derivative, and that Φ is a chiral superfield (hence $D_\alpha\overline{\Phi} = 0$). Now, varying with respect to $\overline{\Phi}$ we get

$$D^2\Phi = 0, \tag{5.4}$$

which, upon expansion in $(x, \theta, \bar{\theta})$, does correspond to the equations of motion for ϕ, ψ and F one would obtain from the Lagrangian (5.2).

That's great. However, we want to describe interactions, not just a set of freely propagating fields. Can we have a more general Lagrangian than just (5.1)? Let us try to consider a more generic function of Φ and $\overline{\Phi}$, call it $K(\Phi, \overline{\Phi})$, and consider the integral

$$\int d^2\theta\, d^2\bar{\theta}\ K(\Phi, \overline{\Phi}). \tag{5.5}$$

In order for the integral (5.5) to be a promising object to describe a supersymmetric Lagrangian, the function K should satisfy a number of properties. First, it should be a superfield. This ensures supersymmetric invariance. Second, it should be a real and scalar function. As before, this is needed since a Lagrangian should have these properties and the $\theta^2\bar{\theta}^2$ component of K, which is the only one contributing to the above integral, is a real scalar object, if so is the superfield K. Third, K should have mass dimension 2, since then its $\theta^2\bar{\theta}^2$ component will have dimension 4, as a Lagrangian should have. Finally, K should be a function of Φ and $\overline{\Phi}$ but not of $D_\alpha\Phi$ and $\overline{D}_{\dot{\alpha}}\overline{\Phi}$. The reason is that, as it can be easily checked, covariant derivatives would provide $\theta\theta\bar{\theta}\bar{\theta}$-term contributions giving a higher derivative theory (third order and higher), which cannot be accepted for a local field theory. It is not difficult to be convinced that the most general expression for K which is compatible with all these properties is

$$K(\Phi, \overline{\Phi}) = \sum_{m,n=1}^{\infty} c_{mn} \overline{\Phi}^{m} \Phi^{n} \quad \text{where} \quad c_{mn} = c_{nm}^{*}, \tag{5.6}$$

where the reality condition on K is ensured by the relation $c_{mn} = c_{nm}^{*}$. Note that all coefficients c_{mn} with either m or n greater than one have negative mass dimension, while c_{11} is dimensionless. This means that, in general, a contribution as that in eq. (5.5) will describe a supersymmetric but non-renormalizable theory, typically defined below some cut-off scale Λ. Indeed, the coefficients c_{mn} will be of the form

$$c_{mn} \sim \Lambda^{2-(m+n)} \tag{5.7}$$

with the constant of proportionality being a pure number. The function K is called *Kähler potential.* The reason for such fancy name will become clear later (see Section 5.1.1).

If renormalizability is an issue, the lowest component of K should not contain operators of dimension bigger than 2, given that the $\theta^2\bar{\theta}^2$ component has dimension $[K] + 2$. In this case all c_{mn} but c_{11} should vanish and the Kähler potential would just be equal to $\overline{\Phi}\Phi$, the object we already considered before and which leads to the renormalizable (but free) Lagrangian (5.2).

In passing, notice that the combination $\Phi + \overline{\Phi}$ respects all the physical requirements discussed above. However, a term like that would not give any contribution since its $\theta^2\bar{\theta}^2$ component is a total derivative. This means that two Kähler potentials K and K' related as

$$K(\Phi, \overline{\Phi})' = K(\Phi, \overline{\Phi}) + \Phi + \overline{\Phi}, \tag{5.8}$$

are different, but their integrals in full superspace, which is all that matters for us, are the same

$$\int d^4x\, d^2\theta\, d^2\bar{\theta}\; K(\Phi, \overline{\Phi})' = \int d^4x\, d^2\theta\, d^2\bar{\theta}\; K(\Phi, \overline{\Phi}). \tag{5.9}$$

This is the reason why we did not consider $m = 0$ or $n = 0$ in the expansion (5.6) (recall that products of (anti)chiral superfields is still a (anti)chiral superfield so (5.8) holds with arbitrary powers of Φ and $\overline{\Phi}$).

Thus far, we have not been able to describe any renormalizable interaction, like non-derivative scalar interactions and Yukawa interactions.

How to describe them? As we have just seen, the simplest possible integral in superspace fulfilling the minimal and necessary physical requirements, $\overline{\Phi}\Phi$, already gives two-derivative contributions; see eq. (5.2). What can we do then?

When dealing with chiral superfields, there is yet another possibility to construct supersymmetric invariant superspace integrals. Let us consider a generic chiral superfield Σ (which can be obtained from products of Φ's, in our case). Integrating it in full superspace would give

$$\int d^4x\, d^2\theta\, d^2\bar{\theta}\ \Sigma = 0, \tag{5.10}$$

since its $\theta^2\bar{\theta}^2$ component is a total derivative. Consider instead integrating Σ in half superspace

$$\int d^4x\, d^2\theta\ \Sigma. \tag{5.11}$$

Differently from the previous one, this integral does not vanish, since now it is the θ^2 component which contributes, and this is not a total derivative for a chiral superfield. Note that in computing (5.11) one can work in the $(y, \theta, \bar{\theta})$ coordinate space, taking $\Sigma = \Sigma(y, \theta)$, and then evaluate the result at $y^\mu = x^\mu$. The terms one is missing would just provide total space-time derivatives, which do not contribute to $\int d^4x$. Another way to reach the same conclusion is to notice that in $(x^\mu, \theta_\alpha, \bar{\theta}_{\dot{\alpha}})$ coordinate, the chiral superfield Σ reads $\Sigma(x, \theta, \bar{\theta}) = \exp(i\theta\sigma^\mu\bar{\theta}\partial_\mu)\Sigma(x, \theta)$.

Besides being non-vanishing, (5.11) is also supersymmetric invariant, since the θ^2 component of a chiral superfield transforms as a total derivative under supersymmetry transformations, as can be seen from eq. (4.59).

An integral like (5.11) is more general than an integral like (5.5). The reason is the following. Any integral in full superspace can be re-written as an integral in half superspace. Indeed, for any superfield Y

$$\int d^4x\, d^2\theta\, d^2\bar{\theta}\ Y = -\frac{1}{4}\int d^4x\, d^2\theta\, \overline{D}^2 Y. \tag{5.12}$$

(In passing, let us notice that for any arbitrary Y, $\overline{D}^2 Y$ is manifestly chiral, since $\overline{D}^3 = 0$ identically). This is because when going from $d\bar{\theta}$ to $\overline{D}$ the difference is just a total space-time derivative, which does not contribute to the above integral. On the other hand, the converse is not true in general. Consider a term like

$$\int d^4x\, d^2\theta\, \Phi^n, \tag{5.13}$$

where Φ is a chiral superfield. This integral cannot be converted into an integral in full superspace, essentially because there are no covariant derivatives to play with. Integrals like (5.13), which cannot be converted into integral in full superspace, are called F-terms. All others, like (5.12), are called D-terms.

Coming back to our problem, it is clear that since the simplest non-vanishing integral in full superspace, eq. (5.1), already contains field derivatives, we must turn to F-terms. First notice that any holomorphic function of Φ, namely a function $W(\Phi)$ such that $\partial W/\partial\overline{\Phi} = 0$, is a chiral superfield, if Φ is. Indeed

$$\overline{D}_{\dot{\alpha}} W(\Phi) = \frac{\partial W}{\partial \Phi}\overline{D}_{\dot{\alpha}}\Phi + \frac{\partial W}{\partial \overline{\Phi}}\overline{D}_{\dot{\alpha}}\overline{\Phi} = 0. \tag{5.14}$$

The proposed term for describing interactions in a theory of a chiral superfield is

$$\mathcal{L}_{int} = \int d^2\theta\, W(\Phi) + \int d^2\bar{\theta}\, \overline{W}(\overline{\Phi}), \tag{5.15}$$

where W is a holomorphic function of Φ and the hermitian conjugate has been added to make the whole thing real. The function W is called *superpotential.* Which properties should W satisfy? First, as already noticed, W should be a holomorphic function of Φ. This ensures it to be a chiral superfield and hence (5.15) to be a supersymmetric invariant quantity (modulo total space-time derivatives). Second, W should not contain covariant derivatives since $D_\alpha\Phi$ is not a chiral superfield, given that D_α and $\overline{D}_{\dot{\alpha}}$ do not (anti)commute. Finally, $[W] = 3$, to make the expression (5.15) have dimension 4. The upshot is that the superpotential should have an expression like

$$W(\Phi) = \sum_{n=1}^{\infty} a_n\, \Phi^n. \tag{5.16}$$

If renormalizability is an issue, the lowest component of W should not contain operators of dimensionality bigger than 3, given that the θ^2 component has dimensionality $[W] + 1$. Since Φ has dimension one, it follows that to avoid non-renormalizable operators the highest power in the expansion (5.16) should be $n = 3$, so that the θ^2 component will have operators of dimension 4, at most. In other words, a renormalizable superpotential should be at most cubic.

The superpotential is also constrained by R-symmetry. Given a chiral superfield Φ, if the $U(1)_R$ charge of its lowest component ϕ is r, then that of ψ is $r-1$ and that of F is $r-2$. This follows from the commutation relations (2.78). Therefore, we have

$$R[\theta]=1,\quad R[\bar{\theta}]=-1,\quad R[d\theta]=-1,\quad R[d\bar{\theta}]=1 \tag{5.17}$$

(recall that $d\theta=\partial/\partial\theta$, and similarly for $\bar{\theta}$). In theories with a R-symmetry, it follows that the superpotential should have R-charge equal to 2

$$R[W]=2, \tag{5.18}$$

in order for the Lagrangian (5.15) to have R-charge 0 and hence be R-symmetry invariant. Does R-symmetry constraint also the Kähler potential? Let us first notice that the integral measure in full superspace has R-charge 0, because of eqs. (5.17). This implies that for theories with a R-symmetry, the Kähler potential should have itself R-charge 0. This is trivially the case for a canonical Kähler potential, since $\overline{\Phi}\Phi$ has R-charge 0. If one allows for non-canonical Kähler potential, then besides the reality condition, one should also impose that $c_{nm}=0$ whenever $n\neq m$, see eq. (5.6).

The integration in superspace of the Lagrangian (5.15) is easily done recalling the expansion of the superpotential in powers of θ. We have

$$W(\Phi)=W(\phi)+\sqrt{2}\frac{\partial W}{\partial\phi}\theta\psi-\theta\theta\left(\frac{\partial W}{\partial\phi}F+\frac{1}{2}\frac{\partial^2 W}{\partial\phi\partial\phi}\psi\psi\right), \tag{5.19}$$

where

$$\frac{\partial^n W}{\partial\phi^n}\equiv\left.\frac{\partial^n W}{\partial\Phi^n}\right|_{\Phi=\phi}. \tag{5.20}$$

So, modulo total space-time derivatives, we get

$$\mathcal{L}_{int}=\int d^2\theta\, W(\Phi)+\int d^2\bar{\theta}\,\overline{W}(\overline{\Phi})=-\frac{\partial W}{\partial\phi}F-\frac{1}{2}\frac{\partial^2 W}{\partial\phi\partial\phi}\psi\psi+h.c., \tag{5.21}$$

where, again, the r.h.s is evaluated at $y^\mu=x^\mu$.

Summing up the two contributions (5.5) and (5.15) we can now write down the most generic $\mathcal{N}=1$ supersymmetric Lagrangian describing the dynamics of a chiral superfield Φ, which reads

$$\mathcal{L}=\int d^2\theta\, d^2\bar{\theta}\, K(\Phi,\overline{\Phi})+\int d^2\theta\, W(\Phi)+\int d^2\bar{\theta}\,\overline{W}(\overline{\Phi}). \tag{5.22}$$

As already emphasized, renormalizability restricts the structure of the Kähler potential and of the superpotential as

$$K(\Phi, \overline{\Phi}) = \overline{\Phi}\Phi, \; W(\Phi) = \sum_{1}^{3} a_n \, \Phi^n. \tag{5.23}$$

In this case, upon integrating in superspace we get

$$\begin{aligned}\mathcal{L} &= \int d^2\theta \, d^2\bar{\theta} \, \overline{\Phi}\Phi + \int d^2\theta \, W(\Phi) + \int d^2\bar{\theta} \, \overline{W}(\overline{\Phi}) \\ &= \partial_\mu \bar{\phi} \, \partial^\mu \phi + \frac{i}{2} \left(\partial_\mu \psi \, \sigma^\mu \overline{\psi} - \psi \, \sigma^\mu \partial_\mu \overline{\psi} \right) + \overline{F} F - \frac{\partial W}{\partial \phi} F + \\ &\quad - \frac{1}{2} \frac{\partial^2 W}{\partial \phi \partial \phi} \psi\psi + h.c.,\end{aligned} \tag{5.24}$$

where it is understood that W is at most cubic. We can now integrate out the auxiliary fields F and $\overline{F}$ by substituting in the Lagrangian their equations of motion which read

$$\overline{F} = \frac{\partial W}{\partial \phi}, \quad F = \frac{\partial \overline{W}}{\partial \bar{\phi}}. \tag{5.25}$$

Doing so, we get a Lagrangian where only physical fields enter, that is

$$\begin{aligned}\mathcal{L}_{on-shell} &= \partial_\mu \bar{\phi} \, \partial^\mu \phi + \frac{i}{2} \left(\partial_\mu \psi \, \sigma^\mu \overline{\psi} - \psi \, \sigma^\mu \partial_\mu \overline{\psi} \right) + \\ &\quad - \left| \frac{\partial W}{\partial \phi} \right|^2 - \frac{1}{2} \frac{\partial^2 W}{\partial \phi \partial \phi} \psi\psi - \frac{1}{2} \frac{\partial^2 \overline{W}}{\partial \bar{\phi} \partial \bar{\phi}} \overline{\psi}\overline{\psi}.\end{aligned} \tag{5.26}$$

From this we can read the scalar potential which is

$$V(\phi, \bar{\phi}) = \left| \frac{\partial W}{\partial \phi} \right|^2 = \overline{F} F, \tag{5.27}$$

where the last equality holds on-shell, namely upon use of eqs. (5.25).

All that we said so far can be easily generalized to a set of chiral superfields Φ^i where $i = 1, 2, \ldots, n$. In this case the most general Lagrangian reads

$$\mathcal{L} = \int d^2\theta \, d^2\bar{\theta} \, K(\Phi^i, \overline{\Phi}_i) + \int d^2\theta \, W(\Phi^i) + \int d^2\bar{\theta} \, \overline{W}(\overline{\Phi}_i). \tag{5.28}$$

For renormalizable theories we have

$$K(\Phi^i, \overline{\Phi}_i) = \overline{\Phi}_i \Phi^i \quad \text{and} \quad W(\Phi^i) = a_i \Phi^i + \frac{1}{2} m_{ij} \Phi^i \Phi^j + \frac{1}{3} g_{ijk} \Phi^i \Phi^j \Phi^k, \tag{5.29}$$

where summation over dummy indices is understood (notice that a quadratic Kähler potential can always be brought to such diagonal form by means of a $GL(n,\mathbb{C})$ transformation on the most general term $K^i_j\,\overline{\Phi}_i\Phi^j$, where K^i_j is a constant hermitian matrix). In this case the scalar potential reads

$$V(\phi^i,\overline{\phi}_i)=\sum_{i=1}^{n}\left|\frac{\partial W}{\partial\phi^i}\right|^2=\overline{F}_iF^i, \tag{5.30}$$

where

$$\overline{F}_i=\frac{\partial W}{\partial\phi^i}\,,\quad F^i=\frac{\partial\overline{W}}{\partial\overline{\phi}_i}. \tag{5.31}$$

5.1.1 *Non-linear sigma model I*

The possibility to deal with non-renormalizable supersymmetric field theories we alluded to previously, is not just academic. In fact, one often has to deal with effective field theories at low energy. The Standard Model itself, though renormalizable, is best thought of as an effective field theory, valid up to a scale of order the TeV scale or slightly higher. Not to mention other effective field theories which are relevant beyond the realm of particle physics. In this section we would like to say something more about the Lagrangian (5.28) once the most general Kähler potential and superpotential are allowed, and show that what one ends-up with in this case is a supersymmetric version of a non-linear σ-model. Though a bit heavy notation-wise, the effort we are going to do here will be very instructive as it will show the deep relation between supersymmetry and geometry.

Since we do not care about renormalizability here, the superpotential is no more restricted to be cubic and the Kähler potential is no more restricted to be quadratic (though it must still be real and with no covariant derivatives acting on the chiral superfields Φ^i). For later purposes it is convenient to define the following quantities

$$K_i=\frac{\partial}{\partial\phi^i}K(\phi,\overline{\phi}),\quad K^i=\frac{\partial}{\partial\overline{\phi}_i}K(\phi,\overline{\phi}),\quad K^j_i=\frac{\partial^2}{\partial\phi^i\partial\overline{\phi}_j}K(\phi,\overline{\phi})$$

$$W_i=\frac{\partial}{\partial\phi^i}W(\phi),\quad W^i=\overline{W}_i,\quad W_{ij}=\frac{\partial^2}{\partial\phi^i\partial\phi^j}W(\phi),\quad W^{ij}=\overline{W}_{ij},$$

where in the above fromulæ both the Kähler potential and the superpotential are meant as their restriction to the scalar component of the chiral superfields, while ϕ stands for the full n-dimensional vector made out of the n scalar fields ϕ^i (similarly for $\overline{\phi}$).

Extracting the F-term contribution in terms of the above quantities is pretty simple. The superpotential can be written as

$$W(\Phi) = W(\phi) + W_i \Delta^i + \frac{1}{2} W_{ij} \Delta^i \Delta^j , \tag{5.32}$$

where we have defined

$$\Delta^i(y) = \Phi^i - \phi^i(y) = \sqrt{2}\theta\psi^i(y) - \theta\theta F^i(y), \tag{5.33}$$

and we get for the F-term

$$\int d^2\theta\, W(\Phi^i) + \int d^2\bar{\theta}\, \overline{W}(\overline{\Phi}_i) = \left(-W_i F^i - \frac{1}{2} W_{ij}\psi^i\psi^j\right) + h.c., \tag{5.34}$$

where we remind that all quantities on the r.h.s. are evaluated at $y^\mu = x^\mu$.

Extracting the D-term contribution is more tricky (but much more instructive). Let us first define

$$\Delta^i(x) = \Phi^i - \phi^i(x), \quad \overline{\Delta}_i(x) = \overline{\Phi}_i - \overline{\phi}_i(x) \tag{5.35}$$

which read

$$\begin{aligned} \Delta^j(x) &= \sqrt{2}\theta\psi^j(x) + i\theta\sigma^\mu\bar{\theta}\partial_\mu\phi^j(x) - \theta\theta F^j(x) + \\ &\quad - \frac{i}{\sqrt{2}}\theta\theta\partial_\mu\psi^j(x)\sigma^\mu\bar{\theta} - \frac{1}{4}\theta\theta\,\bar{\theta}\bar{\theta}\Box\phi^j(x) \end{aligned} \tag{5.36}$$

$$\begin{aligned} \overline{\Delta}_j(x) &= \sqrt{2}\bar{\theta}\overline{\psi}_j(x) - i\theta\sigma^\mu\bar{\theta}\partial_\mu\overline{\phi}_j(x) - \bar{\theta}\bar{\theta}\overline{F}_j(x) + \\ &\quad + \frac{i}{\sqrt{2}}\bar{\theta}\bar{\theta}\theta\sigma^\mu\partial_\mu\overline{\psi}_j(x) - \frac{1}{4}\theta\theta\,\bar{\theta}\bar{\theta}\Box\overline{\phi}_j(x). \end{aligned} \tag{5.37}$$

Note that $\Delta^i\Delta^j\Delta^k = \overline{\Delta}_i\overline{\Delta}_j\overline{\Delta}_k = 0$. With these definitions the Kähler potential can be written as follows

$$\begin{aligned} K(\Phi,\overline{\Phi}) &= K(\phi,\overline{\phi}) + K_i\Delta^i + K^i\overline{\Delta}_i + \frac{1}{2}K_{ij}\Delta^i\Delta^j + \frac{1}{2}K^{ij}\overline{\Delta}_i\overline{\Delta}_j + \\ &\quad + K_i^j\Delta^i\overline{\Delta}_j + \frac{1}{2}K_{ij}^k\Delta^i\Delta^j\overline{\Delta}_k + \frac{1}{2}K_k^{ij}\overline{\Delta}_i\overline{\Delta}_j\Delta^k + \\ &\quad + \frac{1}{4}K_{ij}^{kl}\Delta^i\Delta^j\overline{\Delta}_k\overline{\Delta}_l. \end{aligned} \tag{5.38}$$

We can now compute the D-term contribution to the Lagrangian. We get

$$\begin{aligned}&\int d^2\theta\, d^2\bar{\theta}\; K(\Phi,\overline{\Phi})\\&= -\frac{1}{4}K_i\Box\phi^i - \frac{1}{4}K^i\Box\overline{\phi}_i - \frac{1}{4}K_{ij}\partial_\mu\phi^i\partial^\mu\phi^j - \frac{1}{4}K^{ij}\partial_\mu\overline{\phi}_i\partial^\mu\overline{\phi}_j +\\&+K_i^j\left(F^i\overline{F}_j + \frac{1}{2}\partial_\mu\phi^i\partial^\mu\overline{\phi}_j - \frac{i}{2}\psi^i\sigma^\mu\partial_\mu\overline{\psi}_j + \frac{i}{2}\partial_\mu\psi^i\sigma^\mu\overline{\psi}_j\right)+\\&+\frac{i}{4}K_{ij}^k\left(\psi^i\sigma^\mu\overline{\psi}_k\partial_\mu\phi^j + \psi^j\sigma^\mu\overline{\psi}_k\partial_\mu\phi^i - 2i\psi^i\psi^j\overline{F}_k\right) + h.c. +\\&+\frac{1}{4}K_{ij}^{kl}\psi^i\psi^j\overline{\psi}_k\overline{\psi}_l, \qquad (5.39)\end{aligned}$$

up to total derivatives. Notice now that

$$\begin{aligned}\Box K(\phi,\overline{\phi}) = K_i\Box\phi^i + K^i\Box\overline{\phi}_i + 2K_i^j\partial_\mu\overline{\phi}_j\partial^\mu\phi^i + K_{ij}\partial_\mu\phi^i\partial^\mu\phi^j +\\ + K^{ij}\partial_\mu\overline{\phi}_i\partial^\mu\overline{\phi}_j. \qquad (5.40)\end{aligned}$$

Using this identity we can eliminate K_{ij} and K^{ij}, and rewrite eq. (5.39) as

$$\begin{aligned}&\int d^2\theta\, d^2\bar{\theta}\; K(\Phi,\overline{\Phi})\\&= K_i^j\left(F^i\overline{F}_j + \partial_\mu\phi^i\partial^\mu\overline{\phi}_j - \frac{i}{2}\psi^i\sigma^\mu\partial_\mu\overline{\psi}_j + \frac{i}{2}\partial_\mu\psi^i\sigma^\mu\overline{\psi}_j\right)+\\&+\frac{i}{4}K_{ij}^k\left(\psi^i\sigma^\mu\overline{\psi}_k\partial_\mu\phi^j + \psi^j\sigma^\mu\overline{\psi}_k\partial_\mu\phi^i - 2i\psi^i\psi^j\overline{F}_k\right) + h.c. +\\&+\frac{1}{4}K_{ij}^{kl}\psi^i\psi^j\overline{\psi}_k\overline{\psi}_l. \qquad (5.41)\end{aligned}$$

A few important comments are in order. As just emphasized, independently whether the fully holomorphic and fully anti-holomorphic Kähler potential components, K_{ij} and K^{ij} respectively, are or are not vanishing, they do not enter the final result (5.41). In other words, from a practical view point it is as if they are not there. The only two-derivative contribution entering the effective Lagrangian is K_i^j. This means that given any holomorphic function of ϕ, $\Lambda(\phi)$, the transformation

$$K(\phi,\overline{\phi}) \to K(\phi,\overline{\phi}) + \Lambda(\phi) + \overline{\Lambda}(\overline{\phi}), \qquad (5.42)$$

known as Kähler transformation, is a symmetry of the theory (in fact, such symmetry applies to the full Kähler potential, as we have already observed). This is important for our second comment.

The function K_i^j which normalizes the kinetic term of all fields in eq. (5.41), is hermitian, *i.e.* $K_i^j = K_j^{i\,\dagger}$, since $K(\phi,\overline{\phi})$ is a real function. Moreover, it is positive definite and non-singular, because of the correct sign for the kinetic terms of all non-auxiliary fields. That is to say K_i^j has all necessary properties to be interpreted as a metric of a manifold $\mathcal{M}$ of complex dimension n whose coordinates are the scalar fields ϕ^i themselves. The metric K_i^j is in fact the second derivative of a (real) scalar function K, since

$$K_i^j = \frac{\partial^2}{\partial \phi^i \partial \overline{\phi}_j} K(\phi,\overline{\phi}). \tag{5.43}$$

In this case we speak of a Kähler metric and the manifold $\mathcal{M}$ is mathematically known as Kähler manifold. The scalar fields are maps from space-time to this Riemanniann manifold, which supersymmetry dictates to be Kähler. This is the (supersymmetric) σ-model. Actually, in order to prove that the Lagrangian is a σ-model, with target space the Kähler manifold $\mathcal{M}$, we should prove that not only the kinetic term but any other term in the Lagrangian can be written in terms of geometric quantities defined on $\mathcal{M}$, e.g. the affine connection and the curvature tensor. With some work one can compute both of them out of the Kähler metric K_i^j and, using the auxiliary field equations of motion

$$F^i = (K^{-1})^i_k W^k - \frac{1}{2}(K^{-1})^i_k K^k_{lm} \psi^l \psi^m \tag{5.44}$$

(remark: the above equation shows that when the Kähler potential is non-canonical, the auxiliary fields can depend also on fermion fields!), get for the Lagrangian

$$\begin{aligned}\mathcal{L} = {}& K_i^j \left(\partial_\mu \phi^i \partial^\mu \overline{\phi}_j + \frac{i}{2} D_\mu \psi^i \sigma^\mu \overline{\psi}_j - \frac{i}{2} \psi^i \sigma^\mu D_\mu \overline{\psi}_j \right) - (K^{-1})^i_j W_i W^j + \\ & -\frac{1}{2}\left(W_{ij} - \Gamma^k_{ij} W_k\right)\psi^i\psi^j - \frac{1}{2}\left(W^{ij} - \Gamma^{ij}_k W^k\right)\overline{\psi}_i\overline{\psi}_j + \frac{1}{4} R^{kl}_{ij}\psi^i\psi^j\overline{\psi}_k\overline{\psi}_l,\end{aligned} \tag{5.45}$$

where

$$V(\phi,\overline{\phi}) = (K^{-1})^i_j W_i W^j \tag{5.46}$$

is the scalar potential, and the covariant derivatives appearing in the fermion kinetic terms are

$$\begin{aligned} D_\mu \psi^i &= \partial_\mu \psi^i + \Gamma^i_{jk} \partial_\mu \phi^i \psi^k \\ D_\mu \overline{\psi}_i &= \partial_\mu \overline{\psi}_i + \Gamma^{kj}_i \partial_\mu \overline{\phi}_k \overline{\psi}_j . \end{aligned}$$

With our conventions on indices, $\Gamma^i_{jk} = (K^{-1})^i_m K^m_{jk}$ while $\Gamma^{kj}_i = (K^{-1})^m_i K^{kj}_m$ and $R^{kl}_{ij} = K^{kl}_{ij} - K^m_{ij}(K^{-1})^n_m K^{kl}_n$.

The upshot is that a complicated component field Lagrangian is uniquely characterized by the geometry of the target space. Once a Kähler potential is specified, anything in the Lagrangian (masses and couplings) depends geometrically on this potential (and on W, of course). This shows the strong connection between supersymmetry and geometry. There are of course infinitely many Kähler metrics and therefore infinitely many $\mathcal{N} = 1$ supersymmetric σ-models. The normalizable case, $K^j_i = \delta^j_i$, is just the simplest such instances.

The more supersymmetry the more constraints, hence one could imagine that there should be more restrictions on the geometric structure of the σ-model for theories with extended supersymmetry. This is indeed the case, as we will see explicitly when discussing the $\mathcal{N} = 2$ version of the supersymmetric σ-model. In this case, the scalar manifold is further restricted to be a special class of Kähler manifolds, known as special-Kähler manifolds. For $\mathcal{N} = 4$ constraints are even sharper. In fact, in this case the Lagrangian turns out to be unique, the only possible scalar manifold being the trivial one, $\mathcal{M} = \mathbb{R}^{6n}$, if n is the number of $\mathcal{N} = 4$ vector multiplets (recall that a $\mathcal{N} = 4$ vector multiplet contains six scalars). So, for $\mathcal{N} = 4$ supersymmetry the only allowed Kähler potential is the canonical one! As we will discuss later, this has drastic consequences on the quantum behavior of theories with $\mathcal{N} = 4$ supersymmetry.

5.2 $\mathcal{N} = 1$ supersymmetric Yang-Mills

We would now like to find a supersymmetric invariant action describing the dynamics of vector superfields. In other words, we want to write down the supersymmetric version of Yang-Mills theory (SYM for short). Let us start considering an abelian theory, with gauge group $G = U(1)$. The basic object we should play with is the vector superfield V, which is the supersymmetric extension of a spin one field. Notice, however, that the vector v^μ appears explicitly in V so the first thing to do is to find a suitable supersymmetric generalization of the field strength, which is the gauge invariant object which should enter the action. Let us define the following superfield.

$$W_\alpha = -\frac{1}{4}\overline{DD}D_\alpha V, \quad \overline{W}_{\dot\alpha} = -\frac{1}{4}DD\overline{D}_{\dot\alpha}V, \tag{5.47}$$

and see if this can do the job. First, W_α is obviously a superfield, since V is a superfield and both $\overline{D}_{\dot{\alpha}}$ and D_α commute with supersymmetry transformations. In fact, W_α is a chiral superfield, since $\overline{D}^3 = 0$ identically. The chiral superfield W_α is invariant under the gauge transformation (4.63). Indeed

$$W_\alpha \to W_\alpha - \frac{1}{4}\overline{DD}D_\alpha\left(\Phi + \overline{\Phi}\right) = W_\alpha + \frac{1}{4}\overline{D}^{\dot{\beta}}\overline{D}_{\dot{\beta}}D_\alpha\Phi$$
$$= W_\alpha + \frac{1}{4}\overline{D}^{\dot{\beta}}\{\overline{D}_{\dot{\beta}}, D_\alpha\}\Phi = W_\alpha + \frac{i}{2}\sigma^\mu_{\alpha\dot{\beta}}\partial_\mu\overline{D}^{\dot{\beta}}\Phi = W_\alpha. \quad (5.48)$$

This also means that, as anticipated, as far as we deal with W_α, we can stick to the WZ-gauge without bothering about compensating gauge transformations or anything.

In order to find the component expression for W_α it is useful to use the $(y, \theta, \bar{\theta})$ coordinate system, momentarily. In the WZ gauge the vector superfield reads

$$V_{WZ} = \theta\sigma^\mu\bar{\theta}\, v_\mu(y) + i\theta\theta\,\bar{\theta}\overline{\lambda}(y) - i\bar{\theta}\bar{\theta}\,\theta\lambda(y) + \frac{1}{2}\theta\theta\,\bar{\theta}\bar{\theta}\,(D(y) - i\partial_\mu v^\mu(y)). \quad (5.49)$$

It is a simple exercise we leave to the reader to prove that expanding in $(x, \theta, \bar{\theta})$ coordinate system, the above expression reduces to eq. (4.66). Acting with D_α written in the $(y, \theta, \bar{\theta})$ coordinate system, we get

$$D_\alpha V_{WZ} = \sigma^\mu_{\alpha\dot{\beta}}\bar{\theta}^{\dot{\beta}} v_\mu + 2i\theta_\alpha\bar{\theta}\overline{\lambda} - i\bar{\theta}\bar{\theta}\lambda_\alpha + \theta_\alpha\bar{\theta}\bar{\theta}D +$$
$$+ 2i(\sigma^{\mu\nu})_\alpha^{\ \beta}\,\theta_\beta\bar{\theta}\bar{\theta}\partial_\mu v_\nu + \theta\theta\bar{\theta}\bar{\theta}\sigma^\mu_{\alpha\dot{\beta}}\partial_\mu\overline{\lambda}^{\dot{\beta}} \quad (5.50)$$

and finally, using the identity $\overline{DD}\bar{\theta}\bar{\theta} = -4$

$$W_\alpha = -i\lambda_\alpha + \theta_\alpha D + i\left(\sigma^{\mu\nu}\theta\right)_\alpha F_{\mu\nu} + \theta\theta\left(\sigma^\mu\partial_\mu\overline{\lambda}\right)_\alpha, \quad (5.51)$$

where $F_{\mu\nu} = \partial_\mu v_\nu - \partial_\nu v_\mu$ is the usual gauge field strength and y-dependence of all fields is understood. Since it contains the field strength $F_{\mu\nu}$ this seems to be the right superfield we were searching for! W_α is the so-called supersymmetric field strength and it is an instance of a chiral superfield whose lowest component is not a scalar field, as we have been used to, but in fact a Weyl fermion, λ_α, the gaugino. For this reason, W_α is also called gaugino superfield.

Given that W_α is a chiral superfield, a putative supersymmetric Lagrangian could be constructed out of the following integral in chiral superspace

$$\int d^2\theta \, W^\alpha W_\alpha . \tag{5.52}$$

Indeed, this integral is supersymmetric invariant (modulo total space-time derivatives) since it is an integral in half-superspace of a chiral superfield (the product of chiral superfields is still a chiral superfield), gauge invariant, since so is W_α and of dimension 4 since $W^\alpha W_\alpha$ has dimension 3. Plugging eq. (5.51) into the expression above and computing the superspace integral one gets after some simple algebra

$$\int d^2\theta \, W^\alpha W_\alpha = -\frac{1}{2} F_{\mu\nu} F^{\mu\nu} - 2i\lambda\sigma^\mu \partial_\mu \overline{\lambda} + D^2 + \frac{i}{4} \epsilon^{\mu\nu\rho\sigma} F_{\mu\nu} F_{\rho\sigma} . \tag{5.53}$$

One can get a real object by adding the hermitian conjugate to (5.53), having finally

$$\mathcal{L}_{gauge} = \int d^2\theta \, W^\alpha W_\alpha + \int d^2\bar{\theta} \, \overline{W}_{\dot{\alpha}} \overline{W}^{\dot{\alpha}} = -F_{\mu\nu} F^{\mu\nu} - 4i\lambda\sigma^\mu \partial_\mu \overline{\lambda} + 2D^2 . \tag{5.54}$$

This is the supersymmetric version of the abelian gauge Lagrangian (up to an overall normalization to be fixed later). As anticipated, D is an auxiliary (real) field.

The Lagrangian (5.54) has been written as an integral over chiral superspace, so one might be tempted to say it is a F-term. This is wrong since (5.54) is *not* a true F-term. Indeed, it can be re-written as an integral in full superspace (while F-terms cannot)

$$\int d^2\theta \, W^\alpha W_\alpha = \int d^2\theta \, d^2\bar{\theta} \; D^\alpha V \cdot W_\alpha , \tag{5.55}$$

and so it is in fact a D-term. As we will see later, this fact has important consequences at the quantum level, when discussing renormalization properties of supersymmetric Lagrangians.

All what we said, so far, has to do with abelian interactions. What changes if we consider a non-abelian gauge group G? First, we have to promote the vector superfield to

$$V = V_a T^a \;\; a = 1, \ldots, \dim G , \tag{5.56}$$

where T^a are hermitian generators and V_a are $n = \dim G$ vector superfields. Second, it is useful to define the finite version of the gauge transformation

(4.63) which can be written as

$$e^V \to e^{i\overline{\Lambda}} e^V e^{-i\Lambda}. \tag{5.57}$$

One can easily check that at leading order in Λ this indeed reduces to (4.63), upon the identification $\Phi = -i\Lambda$. Again, it is straightforward to set the WZ gauge for which

$$e^V = 1 + V + \frac{1}{2}V^2. \tag{5.58}$$

In what follows this gauge choice is always understood. The gaugino superfield is generalized as follows

$$W_\alpha = -\frac{1}{4}\overline{DD}\left(e^{-V} D_\alpha e^V\right), \quad \overline{W}_{\dot\alpha} = -\frac{1}{4}DD\left(e^V \overline{D}_{\dot\alpha} e^{-V}\right) \tag{5.59}$$

which again reduces to the expression (5.47) to first order in V. Let us look at eq. (5.59) more closely. Under the gauge transformation (5.57) W_α transforms as

$$\begin{aligned} W_\alpha \to & -\frac{1}{4}\overline{DD}\left[e^{i\Lambda} e^{-V} e^{-i\overline{\Lambda}} D_\alpha \left(e^{i\overline{\Lambda}} e^V e^{-i\Lambda}\right)\right] \\ = & -\frac{1}{4}\overline{DD}\left[e^{i\Lambda} e^{-V}\left(\left(D_\alpha e^V\right) e^{-i\Lambda} + e^V D_\alpha e^{-i\Lambda}\right)\right] \\ = & -\frac{1}{4} e^{i\Lambda}\overline{DD}\left(e^{-V} D_\alpha e^V\right) e^{-i\Lambda} = e^{i\Lambda} W_\alpha e^{-i\Lambda}, \end{aligned} \tag{5.60}$$

where we used the fact that, given that Λ (and products thereof) is a chiral superfield, $\overline{D}_{\dot\alpha} e^{-i\Lambda} = 0$, $D_\alpha e^{i\overline{\Lambda}} = 0$ and also $\overline{DD}D_\alpha e^{-i\Lambda} = 0$. The end result is that W_α transforms covariantly under a finite gauge transformation, as a field strength should whenever one is dealing with non-abelian groups. Similarly, one can prove that

$$\overline{W}_{\dot\alpha} \to e^{i\overline{\Lambda}} \overline{W}_{\dot\alpha} e^{-i\overline{\Lambda}}. \tag{5.61}$$

Let us now expand W_α in component fields. We would expect the non-abelian generalization of eq. (5.51). We have

$$\begin{aligned} W_\alpha &= -\frac{1}{4}\overline{DD}\left[\left(1 - V + \frac{1}{2}V^2\right) D_\alpha \left(1 + V + \frac{1}{2}V^2\right)\right] \\ &= -\frac{1}{4}\overline{DD}D_\alpha V - \frac{1}{8}\overline{DD}D_\alpha V^2 + \frac{1}{4}\overline{DD}V D_\alpha V \end{aligned}$$

$$= -\frac{1}{4}\overline{DD}D_\alpha V - \frac{1}{8}\overline{DD}VD_\alpha V - \frac{1}{8}\overline{DD}D_\alpha V \cdot V + \frac{1}{4}\overline{DD}VD_\alpha V$$
$$= -\frac{1}{4}\overline{DD}D_\alpha V + \frac{1}{8}\overline{DD}\,[V, D_\alpha V].$$

The first term is the same as the one we already computed in the abelian case. As for the second term we get

$$\frac{1}{8}\overline{DD}\,[V, D_\alpha V] = \frac{1}{2}(\sigma^{\mu\nu}\theta)_\alpha\,[v_\mu, v_\nu] - \frac{i}{2}\theta\theta\sigma^\mu_{\ \alpha\dot\beta}\left[v_\mu, \overline{\lambda}^{\dot\beta}\right]. \tag{5.62}$$

Adding everything up simply amounts to turn ordinary derivatives into covariant ones, finally obtaining

$$W_\alpha = -i\lambda_\alpha(y) + \theta_\alpha D(y) + i\,(\sigma^{\mu\nu}\theta)_\alpha F_{\mu\nu} + \theta\theta\left(\sigma^\mu D_\mu\overline{\lambda}(y)\right)_\alpha \tag{5.63}$$

with

$$F_{\mu\nu} = \partial_\mu v_\nu - \partial_\nu v_\mu - \frac{i}{2}[v_\mu, v_\nu], \quad D_\mu = \partial_\mu \; - \frac{i}{2}[v_\mu,], \tag{5.64}$$

which provide the correct non-abelian generalization for the field strength and the (covariant) derivatives.

In view of coupling the pure SYM Lagrangian with matter, it is convenient to introduce the coupling constant g explicitly, making the redefinition

$$V \to 2gV \;\Leftrightarrow\; v_\mu \to 2gv_\mu, \quad \lambda \to 2g\lambda, \quad D \to 2gD, \tag{5.65}$$

which implies the following changes in the final Lagrangian. First, we have now

$$F_{\mu\nu} = \partial_\mu v_\nu - \partial_\nu v_\mu - ig\,[v_\mu, v_\nu]\,, \quad D_\mu = \partial_\mu \; - ig\,[v_\mu,]\,. \tag{5.66}$$

Moreover, the gaugino superfield (5.63) should be multiplied by $2g$ and the (non-abelian version of the) Lagrangian (5.54) by $1/4g^2$. The end result for the SYM Lagrangian is

$$\begin{aligned}\mathcal{L}_{\rm SYM} &= \frac{1}{32\pi}{\rm Im}\left(\tau\int d^2\theta\,{\rm Tr}W^\alpha W_\alpha\right)\\ &= {\rm Tr}\Big[-\frac{1}{4}F_{\mu\nu}F^{\mu\nu} - i\lambda\sigma^\mu D_\mu\overline{\lambda} + \frac{1}{2}D^2\Big] + \frac{\theta_{\rm YM}}{32\pi^2}\,g^2{\rm Tr}F_{\mu\nu}\widetilde{F}^{\mu\nu},\end{aligned} \tag{5.67}$$

where we have introduced the complexified gauge coupling

$$\tau = \frac{\theta_{\rm YM}}{2\pi} + \frac{4\pi i}{g^2} \tag{5.68}$$

and the dual field strength

$$\widetilde{F}^{\mu\nu} = \frac{1}{2}\epsilon^{\mu\nu\rho\sigma}F_{\rho\sigma}, \tag{5.69}$$

while gauge group generators are normalized as $\mathrm{Tr}\, T^a T^b = \delta^{ab}$.

5.3 $\mathcal{N} = 1$ gauge-matter actions

We want now to couple radiation with matter in a supersymmetric consistent way. To this end, let us consider a chiral superfield Φ transforming in some representation R of the gauge group G, $T^a \to (T^a_{\rm R})^i_j$ where $i, j = 1, 2, \ldots, \mathrm{dim}R$. Under the gauge transformation (5.57) we expect Φ to transform as

$$\Phi \to \Phi' = e^{i\Lambda}\Phi, \quad \Lambda = \Lambda_a T^a_R. \tag{5.70}$$

Note that since Λ is a chiral superfield, Φ' is still a chiral superfield. This looks promising but in this way it turns out that the chiral superfield kinetic action we have derived previously would not be gauge invariant since

$$\overline{\Phi}\Phi \to \overline{\Phi}e^{-i\overline{\Lambda}}e^{i\Lambda}\Phi \neq \overline{\Phi}\Phi. \tag{5.71}$$

This means that we have to change the kinetic action. As we are going to show in the following, the correct expression for the kinetic term happens to be

$$\overline{\Phi}e^V\Phi, \tag{5.72}$$

which can be easily shown to be a supersymmetric invariant quantity (modulo total space-time derivatives) and also gauge invariant, when integrated in superspace. With this modification, the complete Lagrangian for charged matter hence reads

$$\mathcal{L}_{\rm matter} = \int d^2\theta\, d^2\bar{\theta}\, \overline{\Phi}e^V\Phi + \int d^2\theta\, W(\Phi) + \int d^2\bar{\theta}\, \overline{W}(\overline{\Phi}). \tag{5.73}$$

Obviously the superpotential should be compatible with the gauge symmetry, *i.e.* it should be gauge invariant itself. This means that a term like

$$a_{i_1 i_2 \ldots i_n}\Phi^{i_1}\Phi^{i_2}\ldots\Phi^{i_n} \tag{5.74}$$

is allowed only if $a_{i_1 i_2 \ldots i_n}$ is an invariant tensor of the gauge group and if $R \times R \times \cdots \times R$ n times contains the singlet representation of the gauge group G.

As an explicit example, take the gauge group of strong interactions, $G = SU(3)$, and consider quarks as matter field. In this case R is the fundamental representation, $R = \underline{3}$. Since $\underline{3} \times \underline{3} \times \underline{3} = \underline{1} + \ldots$ and ϵ_{ijk} is an invariant tensor of $SU(3)$, while $\underline{1} \not\subset \underline{3} \times \underline{3}$ it follows that a supersymmetric and gauge invariant cubic term is allowed, but a mass term is not. In order to have mass terms for quarks, one needs $R = \underline{3} + \underline{\bar{3}}$ corresponding to a chiral superfield Φ in the $\underline{3}$ (quark) and a chiral superfield $\widetilde{\Phi}$ in the $\underline{\bar{3}}$ (anti-quark). In this case $\widetilde{\Phi}\Phi$ is gauge invariant and does correspond to a mass term. This is consistent with the fact that a chiral superfield contains a Weyl fermion only and quarks are described by Dirac fermions. The lesson is that to construct supersymmetric actions with colour charged matter, one needs to introduce two sets of chiral superfields which transform in conjugate representations of the gauge group. This is just the supersymmetric version of what happens in ordinary QCD or in any non-abelian gauge theory with fermions transforming in complex representations ($G = SU(2)$ is an exception because $\underline{2} \simeq \underline{\bar{2}}$).

Let us now compute the D-term of the Lagrangian (5.73). We have (as usual we work in the WZ gauge)

$$\overline{\Phi} e^V \Phi = \overline{\Phi}\Phi + \overline{\Phi} V \Phi + \frac{1}{2} \overline{\Phi} V^2 \Phi. \tag{5.75}$$

The first term is the one we have already calculated, so let us focus on the D-term contribution of the other two. After some algebra we get

$$\begin{aligned} \overline{\Phi} V \Phi|_{\theta\theta\bar{\theta}\bar{\theta}} &= \frac{i}{2} \overline{\phi} v^\mu \partial_\mu \phi - \frac{i}{2} \partial_\mu \overline{\phi} v^\mu \phi - \frac{1}{2} \overline{\psi} \bar{\sigma}^\mu v_\mu \psi + \\ &\quad + \frac{i}{\sqrt{2}} \overline{\phi} \lambda \psi - \frac{i}{\sqrt{2}} \overline{\psi\lambda} \phi + \frac{1}{2} \overline{\phi} D \phi \\ \overline{\Phi} V^2 \Phi|_{\theta\theta\bar{\theta}\bar{\theta}} &= \frac{1}{2} \overline{\phi} v^\mu v_\mu \phi. \end{aligned}$$

Putting everything together we finally get (up to total derivatives)

$$\overline{\Phi} e^V \Phi|_{\theta\theta\bar{\theta}\bar{\theta}} = \left(\overline{D_\mu \phi}\right) D^\mu \phi - i \overline{\psi} \bar{\sigma}^\mu D_\mu \psi + \overline{F} F + \frac{i}{\sqrt{2}} \overline{\phi} \lambda \psi - \frac{i}{\sqrt{2}} \overline{\psi \lambda} \phi + \frac{1}{2} \overline{\phi} D \phi, \tag{5.76}$$

where $D_\mu = \partial_\mu - \frac{i}{2} v_\mu^a T_{\mathrm{R}}^a$.

Performing the rescaling $V \to 2gV$ and rewriting $\overline{\psi} \bar{\sigma}^\mu D_\mu \psi = \psi \sigma^\mu D_\mu \overline{\psi}$ (recall the spinor identity $\chi \sigma^\mu \overline{\psi} = -\overline{\psi} \bar{\sigma}^\mu \chi$) we get finally

$$\begin{aligned} \overline{\Phi} e^{2gV} \Phi|_{\theta\theta\bar{\theta}\bar{\theta}} &= \left(\overline{D_\mu \phi}\right) D^\mu \phi - i \psi \sigma^\mu D_\mu \overline{\psi} + \overline{F} F + i\sqrt{2} g \overline{\phi} \lambda \psi - i \sqrt{2} g \overline{\psi \lambda} \phi + \\ &\quad + g \overline{\phi} D \phi, \end{aligned} \tag{5.77}$$

where now $D_\mu = \partial_\mu - igv_\mu^a T_{\rm R}^a$. The above result shows that the D-term in the Lagrangian (5.73) not only provides matter kinetic terms but also interaction terms between matter fields ϕ, ψ and gauginos λ, where it is understood that

$$\overline{\phi}\lambda\psi = \overline{\phi}_i (T_{\rm R}^a)^i_j \lambda_a \psi^j , \qquad (5.78)$$

and similarly for the other couplings.

To get the most general action there is one term still missing: the so called Fayet-Iliopulos term. Suppose that the gauge group is not semi-simple, *i.e.* it contains $U(1)$ factors. Let V^A be the vector superfields corresponding to the abelian factors, $A = 1, 2, \ldots, m$, where m is the number of abelian factors. The D-term of V^A transforms as a total derivative under supersymmetric gauge transformations, since

$$V^A \to V^A - i\Lambda + i\overline{\Lambda} \ : \ D^A \to D^A + \partial_\mu \partial^\mu (\ldots) \, . \qquad (5.79)$$

Therefore a Lagrangian of this type

$$\mathcal{L}_{FI} = \sum_A \xi_A \int d^2\theta d^2\bar{\theta} \, V^A = \frac{1}{2} \sum_A \xi_A D^A \qquad (5.80)$$

is supersymmetric invariant (since V^A are superfields) and gauge invariant, modulo total space-time derivatives.

We can now assemble all ingredients and write down the most general $\mathcal{N} = 1$ supersymmetric Lagrangian (with canonical Kähler potential, hence renormalizable, if the superpotential is at most cubic) which reads

$$\begin{aligned}
\mathcal{L} &= \mathcal{L}_{\rm SYM} + \mathcal{L}_{\rm matter} + \mathcal{L}_{FI} \\
&= \frac{1}{32\pi} {\rm Im} \left(\tau \int d^2\theta \, {\rm Tr} W^\alpha W_\alpha \right) + 2g \sum_A \xi_A \int d^2\theta d^2\bar{\theta} \, V^A + \\
&\quad + \int d^2\theta d^2\bar{\theta} \, \overline{\Phi} e^{2gV} \Phi + \int d^2\theta \, W(\Phi) + \int d^2\bar{\theta} \, \overline{W}(\overline{\Phi}) \\
&= {\rm Tr} \Big[-\frac{1}{4} F_{\mu\nu} F^{\mu\nu} - i\lambda\sigma^\mu D_\mu \overline{\lambda} + \frac{1}{2} D^2 \Big] + \frac{\theta_{\rm YM}}{32\pi^2} \, g^2 {\rm Tr} F_{\mu\nu} \widetilde{F}^{\mu\nu} + \\
&\quad + g \sum_A \xi_A D^A + (\overline{D_\mu \phi}) D^\mu \phi - i\psi\sigma^\mu D_\mu \overline{\psi} + \overline{F} F + \\
&\quad + i\sqrt{2} g \overline{\phi} \lambda \psi - i\sqrt{2} g \overline{\psi} \overline{\lambda} \phi + \\
&\quad + g\overline{\phi} D \phi - \frac{\partial W}{\partial \phi^i} F^i - \frac{\partial \overline{W}}{\partial \overline{\phi}_i} \overline{F}_i - \frac{1}{2} \frac{\partial^2 W}{\partial \phi^i \partial \phi^j} \psi^i \psi^j - \frac{1}{2} \frac{\partial^2 \overline{W}}{\partial \overline{\phi}_i \partial \overline{\phi}_j} \overline{\psi}_i \overline{\psi}_j .
\end{aligned} \qquad (5.81)$$

Both D^a and F^i are auxiliary fields and can be integrated out. Their equations of motion read

$$\overline{F}_i = \frac{\partial W}{\partial \phi^i}, \quad D^a = -g\overline{\phi}T^a\phi - g\xi^a \quad (\xi^a = 0 \quad \text{if} \quad a \neq A). \tag{5.82}$$

These can plugged back into (5.81) leading to the following on-shell Lagrangian

$$\begin{aligned}\mathcal{L} = \mathrm{Tr}\Big[-\frac{1}{4}F_{\mu\nu}F^{\mu\nu} - i\lambda\sigma^\mu D_\mu\overline{\lambda}\Big] + \frac{\theta_{\mathrm{YM}}}{32\pi^2}\, g^2\mathrm{Tr}F_{\mu\nu}\widetilde{F}^{\mu\nu} + \\ + \overline{D_\mu\phi}D^\mu\phi - i\psi\sigma^\mu D_\mu\overline{\psi} + i\sqrt{2}g\overline{\phi}\lambda\psi - i\sqrt{2}g\overline{\psi}\overline{\lambda}\phi + \\ - \frac{1}{2}\frac{\partial^2 W}{\partial\phi^i\partial\phi^j}\psi^i\psi^j - \frac{1}{2}\frac{\partial^2\overline{W}}{\partial\overline{\phi}_i\partial\overline{\phi}_j}\overline{\psi}_i\overline{\psi}_j - V(\phi,\overline{\phi}),\end{aligned} \tag{5.83}$$

where the scalar potential $V(\phi,\overline{\phi})$ is

$$\begin{aligned}V\left(\phi,\overline{\phi}\right) &= \frac{\partial W}{\partial\phi^i}\frac{\partial\overline{W}}{\partial\overline{\phi}_i} + \frac{g^2}{2}\sum_a |\overline{\phi}_i(T^a)^i_j\phi^j + \xi^a|^2 \\ &= \overline{F}F + \frac{1}{2}D^2\Big|_{\text{on the solution}} \geq 0.\end{aligned} \tag{5.84}$$

where in the last step we have used eqs. (5.82). The above equation shows that the potential is a semi-positive definite quantity in supersymmetric theories.

Expressing the potential in terms of auxiliary fields, as above, provides a very direct way to understand whether a supersymmetric theory admits supersymmetric vacua and also suggests how to parametrize such vacua.

First recall that a vacuum is a Lorentz invariant state configuration. This means that all field derivatives and all fields but scalar ones should vanish in a vacuum state. Hence, the only non trivial thing of the Hamiltonian which can be different from zero in the vacuum is the non-derivative scalar part, which, by definition, is the scalar potential. Therefore, the vacua of a theory, which are the minimal energy states, are in one-to-one correspondence with the (global or local) minima of the scalar potential.

As we have already seen, in a supersymmetric theory the energy of any state is semi-positive definite. This holds also for vacua. For a vacuum Ω we have

$$\langle\Omega|P^0|\Omega\rangle \sim \sum_\alpha \left(||Q_\alpha|\Omega\rangle||^2 + ||Q^\dagger_\alpha|\Omega\rangle||^2\right) \geq 0. \tag{5.85}$$

This means that the vacuum energy is 0 if and only if it is a supersymmetric state, that is $Q_\alpha|\Omega\rangle = 0, \overline{Q}_{\dot{\alpha}}|\Omega\rangle = 0$ with $\alpha, \dot{\alpha} = 1, 2$. Conversely, supersymmetry is broken (in the perturbative theory based on this vacuum) if and only if the vacuum energy is positive. This implies that *supersymmetric vacua* are in one-to-one correspondence with the zero's of the scalar potential. From eq. (5.84) we see that, if they exist, supersymmetric vacua are described by the set of scalar field VEVs which solve simultaneously the so-called D-term and F-term equations

$$\overline{F}_i(\phi) = 0, \quad D^a(\phi, \overline{\phi}) = 0. \tag{5.86}$$

To find such zero's, the most convenient thing to do is to look first for the space of scalar field VEVs such that

$$D^a(\phi, \overline{\phi}) = 0, \tag{5.87}$$

which is called the space of *D-flat directions.* If a superpotential is present, one should then consider the F-term equations, which may put further constraints on the subset of scalar field VEVs already satisfying the D-term equations (5.87). The subspace of the space of D-flat directions which is also F-flat, *i.e.* which also satisfies the equations

$$\overline{F}_i(\phi) = 0, \tag{5.88}$$

is called (classical) *moduli space* and represents the space of (classical) supersymmetric vacua, $\mathcal{M}_{\rm cl}$. Note that in solving for the D-term equations, one should mod out by gauge transformations, since solutions which are related by gauge transformations are physically equivalent and describe the same vacuum state.

The space of flat directions is the space of fields the potential does not depend on and is called moduli space because each flat direction has a massless particle associated to it, a *modulus.* The moduli represent the lightest degrees of freedom of the low energy effective theory (think about the supersymmetric σ-model we discussed in Section 5.1.1). As one moves along the moduli space one spans physically inequivalent (supersymmetric) vacua, since the mass spectrum changes from point to point as, generically, particle masses will depend on scalar field VEVs.

Let us anticipate an important and far-reaching fact that we will derive later. While in a non-supersymmetric theory (or in a supersymmetry breaking vacuum of a supersymmetric theory) the space of classical flat directions, if any, is generically lifted by radiative corrections (which can be computed at leading order by e.g. the Coleman-Weinberg potential), in supersymmetric theories this does not happen. If the ground state energy is zero at tree level, it remains so at all orders in perturbation theory. This is because perturbations around a supersymmetric vacuum are themselves described by a supersymmetric Lagrangian and quantum corrections are protected by cancellations between fermionic and bosonic loops. This means that the only way to lift a classical supersymmetric vacuum, namely to break supersymmetry, if not at tree level by some cleverly chosen superpotential, are non-perturbative corrections. We will have much more to say about this issue in later chapters.

5.3.1 *Classical moduli space: Examples*

To make concrete the previous discussion on moduli space, in what follows we would like to consider two examples explicitly. Before we do that, however, we want to rephrase our definition of moduli space presenting an alternative (but equivalent) way to describe it.

Suppose we are considering a theory without superpotential. For such a theory the space of D-flat directions coincides with the moduli space. The space of D-flat directions is defined as the set of scalar field VEVs satisfying the D-flat conditions

$$\mathcal{M}_{\rm cl} = \{\langle\phi_i\rangle / D^a = 0 \ \ \forall a\}/\text{gauge transformations.} \tag{5.89}$$

Generically it is not at all easy to solve the above constraints and find a simple parametrization of $\mathcal{M}_{\rm cl}$. An equivalent, though less transparent definition of the space of D-flat directions can help in this respect. It turns out that the same space can be defined as the space spanned by all (single trace) gauge invariant operator VEVs made out of scalar fields, modulo classical relations between them

$$\mathcal{M}_{\rm cl} = \{\langle\text{Gauge invariant operators } \equiv X_r(\phi)\rangle\}/\text{classical relations.} \tag{5.90}$$

The latter parametrization is very convenient since, up to classical relations, the construction of the moduli space is unconstrained. In other words, the gauge invariant operators provide a direct parametrization of the space of scalar field VEVs satisfying the D-flat conditions (5.87).

Notice that if a superpotential is present, this is not the end of the story: F-equations will put extra constraints on the $X_r(\phi)$'s and may lift part of (or even all) the moduli space of supersymmetric vacua. In later chapters we will discuss some such instances in detail. Here, in order clarify the equivalence between definitions (5.89) and (5.90), we will instead consider two models with no superpotential term.

Massless SQED. The first example we want to consider is (massless) SQED, the supersymmetric version of quantum electrodynamics. This is a supersymmetric gauge theory theory with gauge group $U(1)$, F (couples of) chiral superfields $(Q_i, \widetilde{Q}_i)$ having opposite charge with respect to the gauge group (we will set for definiteness the charges to be ± 1) and no superpotential, $W = 0$. The vanishing of the superpotential implies that for this system the space of D-flat directions coincides with the moduli space of supersymmetric vacua. The Lagrangian is an instance of the general one we derived before and reads

$$\mathcal{L}_{SQED} = \frac{1}{32\pi}\mathrm{Im}\left(\tau \int d^2\theta\; W^\alpha W_\alpha\right) + \int d^2\theta d^2\bar{\theta}\;\left(Q_i^\dagger\, e^{2V} Q^i + \widetilde{Q}_i^\dagger\, e^{-2V}\widetilde{Q}^i\right) \tag{5.91}$$

(in order to ease the notation, we have come back to the most common notation to indicate hermitian conjugation).

The (only one) D-equation reads

$$D = Q_i^\dagger Q^i - \widetilde{Q}_i^\dagger \widetilde{Q}^i = 0, \tag{5.92}$$

where here and in the following a $\langle\ \rangle$ is understood whenever Q_i or $\widetilde{Q}_i$ appear.

What is the moduli space? Let us first use the definition (5.89). The number of putative complex scalar fields parametrizing the moduli space is $2F$. We have one D-term equation only, which provides one real condition, plus gauge invariance

$$Q^i \to e^{i\alpha} Q^i, \quad \widetilde{Q}^i \to e^{-i\alpha}\widetilde{Q}^i, \tag{5.93}$$

which provides another real condition. Therefore, the complex dimension of the moduli space is

$$\dim_{\mathbb{C}}\mathcal{M}_{\mathrm{cl}} = 2F - \frac{1}{2} - \frac{1}{2} = 2F - 1. \tag{5.94}$$

This is what was expected. At a generic point of the moduli space the gauge group $U(1)$ is broken. The -1 above corresponds to the complex scalar field which, together with its fermionic superpartner, gets eaten by the vector superfield to give a massive vector superfield. One component of the complex scalar field provides the third polarization to the otherwise massless photon; the other real component provides the real physical scalar field a massive vector superfield has. Finally, the Weyl fermion provides the extra degrees of freedom to let the photino become massive. This is nothing but the supersymmetric version of the Higgs mechanism. As anticipated, the vacua are physically inequivalent, generically, since e.g. the mass of the photon depends on the VEV of the scalar fields.

Let us now repeat the above analysis using the definition (5.90). The only gauge invariants we can construct are

$$M^i_j = Q^i \widetilde{Q}_j , \tag{5.95}$$

the so-called mesons. They look as F^2 degrees of freedom but there are classical relations between them, that we now want to find. The matrix (5.95) is a symmetric $F \times F$ complex matrix with rank one since so is the rank of Q and $\widetilde{Q}$ (Q and $\widetilde{Q}$ are vectors of length F, since the gauge group is abelian). This implies that the meson matrix has only one non-vanishing eigenvalue which means

$$\det (M - \lambda \mathbf{1}) = \lambda^{F-1}(\lambda - \lambda_0)(-1)^F . \tag{5.96}$$

Recalling that for a matrix A

$$\det A \epsilon_{j_1 j_2 \ldots j_F} = \epsilon_{i_1 i_2 \ldots i_F} A^{i_1}_{j_1} A^{i_2}_{j_2} \ldots A^{i_F}_{j_F} , \tag{5.97}$$

with $\epsilon_{i_1 i_2 \ldots i_F}$ the fully antisymmetric tensor with F indices, we have

$$\epsilon_{i_1 i_2 \ldots i_F} (M^{i_1}_{j_1} - \lambda \delta^{i_1}_{j_1}) \ldots (M^{i_F}_{j_F} - \lambda \delta^{i_F}_{j_F}) = \lambda^{F-1}(\lambda - \lambda_0)(-1)^F \epsilon_{j_1 j_2 \ldots j_F} , \tag{5.98}$$

which means that from the left hand side only the coefficients of the terms λ^F and λ^{F-1} survive. The next contribution, proportional to λ^{F-2}, should vanish, that is

$$\epsilon_{i_1 i_2 \ldots i_F} M^{i_1}_{j_1} M^{i_2}_{j_2} \epsilon^{j_1 j_2 \ldots j_F} = 0. \tag{5.99}$$

One can show that of the above complex equations, only $(F-1)^2$ give independent conditions, while the contributions proportional to lower powers of λ, that is $\lambda^{F-3}, \lambda^{F-4}, \ldots$ do not give new constraints. So we finally get that

$$\dim_{\mathbb{C}} \mathcal{M}_{\rm cl} = F^2 - (F-1)^2 = 2F - 1. \tag{5.100}$$

which coincides with what we have found before, eq. (5.94)! A quick way to get the same result looking at the meson matrix is to observe that being of rank one, the matirx M^i_j is fully determined by the first row and the first column, namely by $2F-1$ elements.

The parametrization in terms of (single trace) gauge invariant operators is very useful if one wants to find the low energy effective theory around the supersymmetric vacua. Indeed, up to classical relations, these gauge invariant operators (in fact, their fluctuations) directly parametrize the massless degrees of freedom of the perturbation theory constructed upon these same vacua.

Using that $Q, \widetilde{Q}$ and that the meson matrix M have rank one, one can show that on the moduli space (5.92)

$$\mathrm{Tr}\, Q^\dagger Q = \mathrm{Tr}\, \widetilde{Q}^\dagger \widetilde{Q} = \mathrm{Tr}\, \sqrt{M^\dagger M}. \tag{5.101}$$

Therefore, the Kähler potential, which is canonical in terms of the microscopic UV degrees of freedom Q and $\widetilde{Q}$, once projected on the moduli space reads

$$K = \mathrm{Tr}\, \left[Q^\dagger Q + \widetilde{Q}^\dagger \widetilde{Q}\right] = 2\mathrm{Tr}\, \sqrt{M^\dagger M}. \tag{5.102}$$

The Kähler metric of the non-linear σ-model hence reads

$$ds^2 = K_{MM^\dagger} dM dM^\dagger = \frac{1}{2} \frac{1}{\sqrt{M^\dagger M}} dM dM^\dagger, \tag{5.103}$$

which is manifestly non-canonical. Notice that the (scalar) kinetic term

$$\frac{1}{2} \int d^4x \, \frac{1}{\sqrt{M^\dagger M}} \partial_\mu M \partial^\mu M^\dagger \tag{5.104}$$

is singular at the origin, since the Kähler metric diverges there. This has a clear physical interpretation: at the origin the theory is unhiggsed, the photon becomes massless and the correct low energy effective theory should include it in the description. This is a generic feature in this all business: singularities showing-up at specific points of the moduli space are a signal of extra-massless degrees of freedom that, for a reason or another, show up at those specific points

$$\text{Singularities} \longleftrightarrow \text{New massless d.o.f..} \tag{5.105}$$

The correct low energy, singularity-free, effective description of the theory should include them. The singular behavior of $K_{MM^\dagger}$ at the origin is simply telling us that.

Massless SQCD. Let us now consider the non-abelian version of the previous theory. We have now a non-abelian gauge group — which we take for definiteness to be $SU(N)$ — F flavors and again no superpotential, $W = 0$. The quarks superfields Q and $\widetilde{Q}$ are $F \times N$ complex matrices. Looking at the Lagrangian, which is the obvious generalization of (5.91), we see there are two independent flavor symmetries, one associated to Q and one to $\widetilde{Q}$, $SU(F)_{\rm L}$ and $SU(F)_{\rm R}$ respectively. To make the different (flavor and gauge) symmetries manifest we split matter indexes as (i, a) with i an index in the (anti)fundamental of $SU(F)_{L,R}$ and a in the (anti)fundamental of $SU(N)$. Matter quantum numbers are summarized below

	$SU(N)$	$SU(F)_{\rm R}$	$SU(F)_{\rm R}$
Q^i_a	N	F	1
$\widetilde{Q}^b_j$	$\overline{N}$	1	$\overline{F}$

where $i, j = 1, 2, \ldots, F$ and $a, b = 1, 2, \ldots, N$. The convention for gauge indices is that lower indices are for an object transforming in the fundamental representation and upper indices for an object transforming in the anti-fundamental. The convention for flavor indices is chosen to be the opposite one. Given these conventions, the D-term equations read

$$\begin{aligned} D^A &= {Q^\dagger}^b_i \, (T^A_N)^c_b \, Q^i_c + \widetilde{Q^\dagger}^i_b \, (T^a_{\bar N})^b_c \, \widetilde{Q}^c_i = {Q^\dagger}_i^{\ b} \, (T^A)^c_b \, Q^i_c - \widetilde{Q^\dagger}^i_{\ b} \, (T^A)^b_c \, \widetilde{Q}^c_{\ i} \\ &= {Q^\dagger}^b_i \, (T^A)^c_b \, Q^i_c - \widetilde{Q}_i^{\ b} \, (T^A)^c_b \, \widetilde{Q^\dagger}^i_{\ c} = 0, \end{aligned} \tag{5.106}$$

where $A = 1, 2, \ldots, N^2 - 1$ is an index in the adjoint of $SU(N)$, and we used the fact that $(T^A_N)^a_b = -(T^A_{\bar N})^a_b \equiv (T^A)^a_b$.

Let us first focus on the case $F < N$. Using the two $SU(F)$ flavor symmetries and the (global part of the) gauge symmetry $SU(N)$, one can show that on the moduli space (5.106) the matrices Q and $\widetilde{Q}$ can be put, at most, in the following form (recall that the maximal rank of Q and $\widetilde{Q}$ is F in this case, since $F < N$)

$$Q = \begin{pmatrix} v_1 & 0 & \ldots & 0 & \ldots \\ 0 & v_2 & \ldots & 0 & \ldots \\ \ldots & \ldots & \ldots & \ldots & \ldots \\ 0 & 0 & \ldots & v_F & \ldots \end{pmatrix} = \widetilde{Q}^T \tag{5.107}$$

This means that at a generic point of the moduli space the gauge group is broken to $SU(N - F)$. So, the complex dimension of the classical moduli space is

$$\dim_{\mathbb{C}} \mathcal{M}_{\rm cl} = 2FN - \left\{ N^2 - 1 - \left[(N - F)^2 - 1 \right] \right\} = F^2. \tag{5.108}$$

Let us now use the parametrization in terms of gauge invariant single trace operators (5.90). In this case we have

$$M^i_{\ j} = Q^i_a \widetilde{Q}^a_j \tag{5.109}$$

(notice the contraction on the N gauge indices). The meson matrix has now maximal rank, since $F < N$, so there are no classical constraints it has to satisfy: its F^2 entries are all independent. In terms of the meson matrix the classical moduli space dimension is then (trivially) F^2, in agreement with eq. (5.108). Again, playing with global symmetries, the meson matrix can be diagonalized in terms of F complex eigenvalues V_i, which, not suprprisingly, turn out to be the square of the ones in (5.107), $V_i = v_i^2$.

Also in this case one can write down the (classical) effective action. On the moduli space we have $Q^{\dagger}{}_i^a Q_b^i = \widetilde{Q}_i^a \widetilde{Q^{\dagger}}{}^i{}_b$. Using this identity we get

$$\left(M^{\dagger}M\right)^i_j = \widetilde{Q^{\dagger}}{}^i{}_a Q^{\dagger}{}^a_k\, Q^k_b \widetilde{Q}^b_j = \widetilde{Q^{\dagger}}{}^i{}_a \widetilde{Q}^a_k\, \widetilde{Q^{\dagger}}{}^k{}_b \widetilde{Q}^b_j, \tag{5.110}$$

which implies $\widetilde{Q}^{\dagger}\widetilde{Q} = \sqrt{M^{\dagger}M}$ as a matrix equation. So the Kähler potential is

$$K = 2\mathrm{Tr}\,\sqrt{M^{\dagger}M}. \tag{5.111}$$

The Kähler metric is singular whenever the meson matrix M is not invertible. This does not only happen at the origin of field space as for SQED, but actually on the subspace where some of the $N^2 - 1 - \left[(N-F)^2 - 1\right] = (2N-F)F$ massive gauge bosons parametrizing the coset $SU(N)/SU(N-F)$ become massless (and they need to be included in the low energy effective description).

Let us now consider the case $F \geq N$. Following a similar procedure as the one before, the matrices Q and $\widetilde{Q}$ can be brought to the following form on the moduli space

$$Q = \begin{pmatrix} v_1 & 0 & \dots & 0 \\ 0 & v_2 & \dots & 0 \\ \dots & \dots & \dots & \dots \\ 0 & 0 & \dots & v_N \\ 0 & 0 & 0 & 0 \\ \dots & \dots & \dots & \dots \\ 0 & 0 & 0 & 0 \end{pmatrix}, \quad \widetilde{Q}^T = \begin{pmatrix} \tilde{v}_1 & 0 & \dots & 0 \\ 0 & \tilde{v}_2 & \dots & 0 \\ \dots & \dots & \dots & \dots \\ 0 & 0 & \dots & \tilde{v}_N \\ 0 & 0 & 0 & 0 \\ \dots & \dots & \dots & \dots \\ 0 & 0 & 0 & 0 \end{pmatrix} \tag{5.112}$$

where $|v_i|^2 - |\tilde{v}_i|^2 = a$, with a a i-independent number. Since $F \geq N$, at a generic point on the moduli space the gauge group is now completely higgsed. Therefore, the dimension of the classical moduli space is

$$\dim_{\mathbb{C}} \mathcal{M}_{\rm cl} = 2NF - (N^2 - 1). \tag{5.113}$$

The parametrization in terms of gauge invariant operators is slightly more involved, in this case. The mesons are still there, and defined as in eq. (5.109). However, there are non-trivial classical constraints one should take into account, since the rank of the meson matrix, which is at most N, is now smaller than its dimension, F, like in the SQED example. Moreover, besides the mesons, there are now new gauge invariant single trace operators one can build, the baryons, which are operators made out of N fields Q respectively N fields $\widetilde{Q}$, with fully anti-symmetrized indices.

As an explicit example of this richer structure, let us apply the above rationale to the case $N = F$. According to eq. (5.113), in this case $\dim_{\mathbb{C}} \mathcal{M}_{\text{cl}} = F^2 + 1$. The gauge invariant operators are the meson matrix plus two baryons, B and $\widetilde{B}$, defined as

$$B = \epsilon^{a_1 a_2 \ldots a_N} \, Q^1_{a_1} Q^2_{a_2} \ldots Q^N_{a_N}$$
$$\widetilde{B} = \epsilon_{a_1 a_2 \ldots a_N} \, \widetilde{Q}^{a_1}_1 \widetilde{Q}^{a_2}_2 \ldots \widetilde{Q}^{a_N}_N$$

Notice that since $F = N$ the anti-symmetrization on the flavor indices is automatically taken care of, once anti-symmetrization on the gauge indices is imposed. All in all we have, naively, $F^2 + 2$ complex moduli space directions. There is however one classical constraints between them which reads

$$\det M - B\widetilde{B} = 0, \tag{5.114}$$

as can be easily checked from the definition of the meson matrix (5.109) and that of the baryons above. Hence, the actual dimension of the moduli space is $F^2 + 2 - 1 = F^2 + 1$, as expected. As for the case $F < N$, there is a subspace in the moduli space, which includes the origin, where some fields become massless and the low energy effective analysis should be modified to include them.

All we said, so far, is true classically. As we will see when discussing the quantum dynamics of SQCD, quantum corrections sensibly change this picture and the exact structure of the moduli space differs in many respects from the classical one. For instance, focusing again on $F = N$ SQCD it turns out that the classical constraint (5.114) is modified at the quantum level. This has the effect of excising the origin of field space from the actual quantum moduli space removing all singular subspace and the corresponding extra massless degrees of freedom, which are then just an artifact of the classical analysis, in this case.

We will have much more to say about SQCD and its classical and quantum properties in later chapters.

5.3.2 *Non-linear sigma model II*

In section 5.1.1 we discussed the supersymmetric non-linear σ-model for matter fields, which is relevant to describe supersymmetric low energy effective theories. Though it is not often the case, it may happen to face effective theories with some left over propagating gauge degrees of freedom at low energy. Therefore, in this section we will generalize the σ-model of section 5.1.1 to such a situation: a supersymmetric but non-renormalizable effective theory coupled to gauge fields. Note that the choice of the gauge group cannot be arbitrary here. In order to preserve the structure of the non-linear σ-model one can gauge only a subgroup G of the isometry group of the scalar manifold.

Following previous strategy, one gets easily convinced that the pure SYM part changes simply by promoting the (complexified) gauge coupling τ to a holomorphic function of the chiral superfields, getting

$$\tau \int d^2\theta \, \mathrm{Tr} W^\alpha W_\alpha \quad \longrightarrow \quad \int d^2\theta \, \mathcal{F}_{ab}(\Phi) \, W^{\alpha a} \, W^b_\alpha, \tag{5.115}$$

where the chiral superfield $\mathcal{F}_{ab}(\Phi)$ should transform in the $Adj \times Adj$ of the gauge group G in order for the whole action to be G-invariant. Notice that for $\mathcal{F}_{ab} = \tau \, \mathrm{Tr} T_a T_b$ one gets back the usual result (recall that we have normalized the gauge group generators as $\mathrm{Tr}\, T_a T_b = \delta_{ab}$). For this reason the function $\mathcal{F}_{ab}$ (actually its restriction to the scalar fields) is dubbed generalized complexified gauge coupling.

As for the matter Lagrangian, given what we have already seen, namely that whenever one has to deal with charged matter fields the gauge invariant combination is $(\overline{\Phi} e^{2gV})_i \Phi^i$, one should simply observe that the same holds for any real G-invariant function of Φ and $\overline{\Phi}$. In other words, the σ-model Lagrangian for charged chiral superfields is obtained from the one we derived in section 5.1.1 upon the substitution

$$K(\Phi^i, \overline{\Phi}_i) \quad \longrightarrow \quad K\left(\Phi^i, (\overline{\Phi} e^{2gV})_i\right). \tag{5.116}$$

The end result is then

$$\begin{aligned} \mathcal{L} = \frac{1}{32\pi} \, \mathrm{Im} & \left[\int d^2\theta \, \mathcal{F}_{ab}(\Phi) \, W^{\alpha a} \, W^b_\alpha \right] + \\ & + \int d^2\theta d^2\bar\theta K(\Phi^i, (\overline{\Phi} e^{2gV})_i) + \int d^2\theta \, W(\Phi^i) + \int d^2\bar\theta \, \overline{W}(\overline{\Phi}_i). \end{aligned} \tag{5.117}$$

By expanding and integrating in superspace one gets the final result. The derivation is a bit lengthy and we omit it here. Let us just mention some important differences with respect to our previous results. The gauge part has the imaginary part of $\mathcal{F}_{ab}$ multiplying the kinetic term (the generalized gauge coupling) and the real part multiplying the instanton term (generalized θ-angle). Moreover, there are higher order couplings between fields belonging to vector and scalar multiplets which are proportional to derivatives of $\mathcal{F}_{ab}$ with respect to the scalar fields, and which are obviously absent for the renormalizable Lagrangian (5.81). As for the matter part, one important difference with respect to the σ-model Lagrangian (5.45) is that all derivatives are (also) gauge covariantized. More precisely we have

$$\widetilde{D}_\mu\psi^i = \partial_\mu\psi^i - igv_\mu^a T_{\mathrm{R}}^a\psi^i + \Gamma^i_{jk}\partial_\mu\phi^i\psi^k$$
$$\widetilde{D}_\mu\overline{\psi}_j = \partial_\mu\overline{\psi}_j - igv_\mu^a T_{\mathrm{R}}^a\overline{\psi}_j + \Gamma^{ki}_j\partial_\mu\overline{\phi}_k\overline{\psi}_i,$$

which are covariant both with respect to the σ-model metric and the gauge connection. As compared to the Lagrangian (5.81) the Yukawa-like couplings have the Kähler metric inserted, that is

$$\overline{\phi}\lambda\psi \quad\longrightarrow\quad K^i_j\overline{\phi}_i\lambda\psi^j = K^i_j\overline{\phi}_i T^a_{\mathrm{R}}\lambda_a\psi^j = K^i_j(\overline{\phi}_i)_M(T^a_{\mathrm{R}})^M_N\lambda_a(\psi^j)^N, \quad (5.118)$$

where M, N are gauge indices. Moreover, the term $g\overline{\phi}D\phi$ is also modified into $g\overline{\phi}_iDK^i$, where as usual $K^i = \partial K/\partial\overline{\phi}_i$.

All these changes are important to keep in mind. However, it is worth noticing that in $\mathcal{N}=1$ supersymmetry vectors belong to different multiplets with respect to those where scalars sit. Hence, any geometric operation on the scalar manifold $\mathcal{M}$ will not have much effect on the vectors, and viceversa. In other words, the structure of the $\mathcal{N}=1$ non-linear σ-model is essentially unchanged by gauging some of the isometries of the scalar manifold. This is very different from what happens in models with extended supersymmetry, as we will see in the next chapter.

After solving for the auxiliary fields which read (with obvious notation)

$$F^i = (K^{-1})^i_j W^j - \frac{1}{2}\Gamma^i_{jk}\psi^j\psi^k - i\frac{g^2}{16\pi}(K^{-1})^i_j(\mathcal{F}_{ab,j})^\dagger\overline{\lambda}^a\overline{\lambda}^b \quad (5.119)$$
$$D^a = -\frac{4\pi}{g^2}\left(\mathrm{Im}\,\mathcal{F}\right)^{-1}_{ab}\left(g\overline{\phi}_i T^b K^i + g^2\frac{1}{8\pi\sqrt{2}}\left[(\mathcal{F}_{bc,i})\psi^i\lambda^c + h.c.\right]\right) \quad (5.120)$$

one finds for the potential

$$V(\phi,\overline{\phi}) = (K^{-1})^i_j\, W_i W^j + 2\pi\, (\mathrm{Im}\,\mathcal{F})^{-1}_{ab}\, (\overline{\phi}_i T^a K^i)\, (\overline{\phi}_j T^b K^j), \qquad (5.121)$$

which is the σ-model version of the potential (5.84).

As far as the potential, we cannot resist making a comment which will be relevant later, when we will discuss supersymmetry breaking. Whenever the effective theory one is dealing with does not have any propagating gauge degrees of freedom (due to Higgs mechanism, confinement or alike) the scalar potential (5.121) gets contributions from the first term, only. In this case the zero's of the potential, which correspond to the supersymmetric vacua of the theory, are described just by

$$W_i = 0, \qquad (5.122)$$

as in cases where the Kähler potential is canonical, since K is a positive definite matrix (provided the integrating out procedure has been done correctly along the whole moduli space). This means that it is possible to see whether supersymmetry is broken/unbroken independently of any knowledge of the Kähler potential! This is very different from non-supersymmetric σ models and it is related to what was mentioned earlier, namely that in a supersymmetric theory if supersymmetry is unbroken classically it cannot be broken at any order in perturbation theory but only non-perturbatively. This said, other important features which characterize the vacua, as field VEVs, the exact value of the vacuum energy (if not zero), the mass and the interactions of the lightest excitations, etc... do depend on K. With an abuse of notation, eqs. (5.122) are usually referred to as F-term equations, even though for a theory with non-canonical Kähler potential the correct F-term equations are eqs. (5.119).

5.4 Exercises

Exercise 5.1. Derive eq. (5.2). Using eqs. (4.60), show that the resulting expression transforms as a total space-time derivative under supersymmetry transformations.

Exercise 5.2. Compute $D^2\Phi = 0$ and show that the different components provide the equations of motion for a free massless WZ multiplet.

Exercise 5.3. Consider a theory of a chiral superfield Φ with canonical Kähler potential, $K = \overline{\Phi}\Phi$ and superpotential $W(\Phi) = \frac{1}{2}m\Phi^2 + \frac{1}{3}g\Phi^3$. This is the renowned Wess-Zumino model. Derive the equation of motion in superfield formalism and the corresponding ones in component fields. Compute the off-shell and on-shell space-time Lagrangians. Show that once auxiliary fields are integrated out supersymmetry closes only on-shell, namely that the algebra closes only upon use of (some of) the equations of motion.

Exercise 5.4. Consider the theory of a single chiral superfield Φ and Kähler potential $K = \Lambda^2 \ln(1 + \overline{\Phi}\Phi/\Lambda^2)$, where Λ is a given mass scale. Compute the off-shell and on-shell Lagrangians and study the geometry of the one-dimensional supersymmetric non-linear σ model.

Exercise 5.5. Using all possible available symmetries, show that in $SU(N)$ SQCD with $F < N$ flavors, the complex scalar field matrices parametrizing the moduli space can be put in the form (5.107). Using the same procedure, show that the structure (5.112) holds for $F \geq N$.

Exercise 5.6. Consider the following matter theories

$$1.\ K = \overline{Q}Q,\ W = \frac{1}{2}mQ^2$$

$$2.\ K = \overline{X}X + \overline{Y}Y,\ W = (X - m)\,Y^2$$

$$3.\ K = \overline{X}X + \overline{Y}Y + \overline{Z}Z,\ W = gXYZ$$

$$4.\ K = \Lambda\sqrt{\overline{X}X},\ W = \lambda X.$$

Determine whether there are supersymmetric vacua and, if they exist, compute the mass spectrum of the theory around them.

Exercise 5.7. Consider SQED with F flavors and focus on a point of the moduli space where all scalar field VEVs are the same, that is $v_i = v$ for any i. Starting from the SQED Lagrangian (5.91), derive the supersymmetric Lagrangian describing the dynamics of the system around such vacuum. [Hint: the spectrum should correspond to $2F - 1$ massless chiral multiplets and one massive vector multiplet.]

References

[1] A. Bilal, Introduction to supersymmetry, Sections 4, 5 and 7, arXiv:hep-th/0101055.

[2] M. F. Sohnius, Introducing Supersymmetry, Sections 5, 7, 8, 9 and 10, *Phys. Rept.* **128** (1985), 39.

[3] J. Wess and B. Zumino, Supergauge Transformations in Four-Dimensions, *Nucl. Phys. B* **70** (1974), 39.

[4] J. Wess and B. Zumino, Supergauge Invariant Extension of Quantum Electrodynamics, *Nucl. Phys. B* **78** (1974), 1.
S. Ferrara and B. Zumino, Supergauge Invariant Yang-Mills Theories, *Nucl. Phys. B* **79** (1974), 413.

[5] J. Terning, *Modern supersymmetry: Dynamics and duality*, Chapter 3.4, Oxford University Press (2006).

[6] M. J. Strassler, *An unorthodox introduction to supersymmetric gauge theory*, Sections 4.5 and 5.1, TASI 2001, arXiv:hep-th/0309149.

[7] M. A. Luty and W. Taylor, Varieties of vacua in classical supersymmetric gauge theories, *Phys. Rev. D* **53** (1996), 3399 [hep-th/9506098].

Chapter 6

Supersymmetric Actions: Extended Supersymmetry

So far we have dealt with theories with $\mathcal{N} = 1$ supersymmetry. In this chapter we will discuss the structure of theories with extended supersymmetry. This will also let us emphasize the basic differences which arise at the quantum level between theories with different number of supersymmetries, thanks to powerful non-renormalization theorems.

6.1 $\mathcal{N} = 2$ supersymmetric actions

We would like to construct the most general $\mathcal{N} = 2$ supersymmetric action in four dimensions. We will follow the same logic of the previous chapter, but we will not develop the corresponding $\mathcal{N} = 2$ superspace approach, whose formulation is beyond our scope. Rather, we will use the (by now familiar) $\mathcal{N} = 1$ superspace formalism and see which specific properties does more supersymmetry impose on an otherwise generic $\mathcal{N} = 1$ Lagrangian.

We have two kinds of $\mathcal{N} = 2$ multiplets we have to deal with, vector multiplets and hypermutiplets. What we noticed at the level of representations of the supersymmetry algebra on states, Chapter 3, holds also at the field level. In particular, in $\mathcal{N} = 1$ language a $\mathcal{N} = 2$ vector superfield can be seen as the direct sum of a vector superfield V and a chiral superfield Φ (having the same internal quantum numbers, e.g. they both transform in the adjoint representation of the gauge group). Similarly, in terms of degrees of freedom a hypermultiplet can be constructed out of two $\mathcal{N} = 1$ chiral superfields, H_1 and H_2. Schematically, we have

$$[\mathcal{N} = 2 \text{ vector multiplet}] : V = (\lambda_\alpha, v_\mu, D) \;\oplus\; \Phi = (\phi, \psi_\alpha, F)$$

$$[\mathcal{N} = 2 \text{ hypermultiplet}] : H_1 = (H_1, \psi_{1\alpha}, F_1) \;\oplus\; \overline{H}_2 = \left(\overline{H}_2, \overline{\psi}_{2\dot\alpha}, \overline{F}_2\right)$$

(notice that H_1 and $\overline{H}_2$ transform in the same representations of internal symmetries since they belong to the same supersymmetry representation, while H_2 transforms in complex conjugate representations).

Let us start considering pure SYM. In $\mathcal{N} = 1$ language this is a Lagrangian of the type (5.81) on which, however, two specific requirements should be imposed. First, as already stressed, the chiral superfield Φ should transform in the adjoint representation of the gauge group, as the vector superfield V does. Second, notice that we have now a larger R-symmetry group, whose compact component, $SU(2)_R$, should be a symmetry of the Lagrangian. All bosonic degrees of freedom v_μ, D, F and ϕ are singlets under $SU(2)_R$, but $(\lambda_\alpha, \psi_\alpha)$ transform as a doublet. This is because (Q^1_α, Q^2_α) transform under the fundamental representation of $SU(2)_R$, and the same should hold for λ_α and ψ_α (recall from (3.19) that they are obtained acting with the two supersymmetry generators on the Clifford vacuum $|j=0\rangle$).

The Lagrangian reads

$$\begin{aligned}\mathcal{L}^{\mathcal{N}=2}_{\text{SYM}} &= \frac{1}{32\pi}\,\text{Im}\left(\tau \int d^2\theta\, \text{Tr}\, W^\alpha W_\alpha\right) + \int d^2\theta d^2\bar\theta\, \text{Tr}\,\overline{\Phi} e^{2gV}\Phi \\ &= \text{Tr}\left\{-\frac{1}{4}F_{\mu\nu}F^{\mu\nu} - i\lambda\sigma^\mu D_\mu\overline{\lambda} - i\psi\sigma^\mu D_\mu\overline{\psi} + \overline{D_\mu\phi}D^\mu\phi + \right.\\ &\quad + \frac{\theta_{\text{YM}}}{32\pi^2}g^2 F_{\mu\nu}\widetilde{F}^{\mu\nu} + \frac{1}{2}D^2 + \overline{F}F + i\sqrt{2}g\overline{\phi}\{\lambda,\psi\} + \\ &\quad \left. - i\sqrt{2}g\{\overline{\psi},\overline{\lambda}\}\phi + gD\left[\phi,\overline{\phi}\right]\right\},\end{aligned} \tag{6.1}$$

where

$$\begin{aligned}&\phi = \phi^a T_a, \quad \psi_\alpha = \psi^a_\alpha T_a, \quad F = F^a T_a;\\ &\lambda_\alpha = \lambda^a_\alpha T_a, \quad v_\mu = v^a_\mu T_a, \quad D = D^a T_a\end{aligned}$$

with $a = 1, 2, \ldots, \dim G$. The reason why commutators and anti-commutators appear in the Lagrangian (6.1) as opposed to the Lagrangian (5.81), is just because all fields transform in the adjoint representation of G. For example, given that $(T^a_{adj})_{bc} = -if_{abc}$, the term $\overline{\phi}\lambda\psi$ can be written as

$$\begin{aligned}\overline{\phi}\lambda\psi &= \overline{\phi}^b\lambda^a (T^a_{adj})_{bc}\psi^c = -i\overline{\phi}^b\lambda^a f_{abc}\psi^c = i\overline{\phi}^b\lambda^a f_{bac}\psi^c \\ &= \overline{\phi}^b\lambda^a\psi^c\, \text{Tr}\, T_b\left[T_a, T_c\right] = \text{Tr}\,\overline{\phi}\{\lambda,\psi\},\end{aligned} \tag{6.2}$$

and similarly for all other contributions in eq. (6.1).

As compared to a $\mathcal{N}=1$ Lagrangian describing matter coupled SYM theory, the above Lagrangian is special in many respects.

First and foremost, despite $\mathcal{N}=2$ supersymmetry is not manifest using $\mathcal{N}=1$ superspace language, the above Lagrangian is invariant under two independent supersymmetries, as it should. This can be seen using the $SU(2)_R$ symmetry the Lagrangian enjoys, actually just its center $\mathbb{Z}_2$, under which the two generators Q^1_α, Q^2_α are exchanged. Eq. (6.1) is written in terms of two $\mathcal{N}=1$ superfields and, correspondingly, it is obviously $\mathcal{N}=1$ invariant. Acting now with a $\mathbb{Z}_2$ R-symmetry rotation which acts as $\psi_\alpha \to \lambda_\alpha$ and $\lambda_\alpha \to -\psi_\alpha$, while leaving the bosonic fields invariant, one sees that the same Lagrangian shows an invariance under an independent $\mathcal{N}=1$ supersymmetry acting on two different superfields with entries $(\phi, \lambda_\alpha, F)$ and $(v_\mu, -\psi_\alpha, D)$. So we conclude that the Lagrangian is indeed $\mathcal{N}=2$ supersymmetric invariant.

The existence of a manifest $SU(2)_R$ symmetry has also other related consequences on the structure of the Lagrangian (6.1). The kinetic terms for λ and ψ have the same normalization. Moreover, and more importantly, the Lagrangian has no superpotential, $W=0$. Indeed, a superpotential would give ψ interactions and/or mass terms, that are absent for λ. This is clearly forbidden by the $SU(2)_R$ symmetry. While there is no superpotential, there is a potential, which comes from D-terms. Indeed, the auxiliary fields equations of motion are in this case

$$F^a = 0, \quad D^a = -g\left[\phi, \overline{\phi}\right]^a \tag{6.3}$$

(the auxiliary fields F^a appear only in the non-dynamical kinetic term $\overline{F}_a F^a$ and therefore are trivial). The potential hence reads

$$V(\phi, \overline{\phi}) = \frac{1}{2}D^2 = \frac{1}{2}g^2 \,\mathrm{Tr}\left[\phi, \overline{\phi}\right]^2 . \tag{6.4}$$

The above expression shows that pure $\mathcal{N}=2$ SYM enjoys a huge moduli space of supersymmetric vacua. Indeed, the potential vanishes whenever the fields ϕ belong to the Cartan subalgebra of the gauge group G. At a generic point of the moduli space the scalar field matrix can be diagonalized and the gauge group is broken to $U(1)^r$, with r the rank of G. The low energy effective dynamics is that of r massless vector multiplets and $\dim G - r$ massive vector multiplets whose masses depend on the scalar fields VEVs. The theory is said to be in a Coulomb phase, since charged external sources will feel a Coulomb-like potential. The (classical) moduli space is a r-dimensional complex manifolds, parametrized by r massless complex scalars. Singularities arise whenever some VEVs become degenerate and the theory gets partially unhiggsed (in particular, at the origin of the moduli space one recovers the full gauge symmetry G). This is classical analysis and, as for $\mathcal{N}=1$ theories, in later chapters we will see if and how this description gets modified once (non-perturbative) quantum corrections are taken into account.

Let us now consider the addition of hypermultiplets. In this case the scalar fields, H_1 and $\overline{H}_2$ form a $SU(2)_R$ doublet (again, recall how hypermultiplets were constructed from the Clifford vacuum state of the corresponding $\mathcal{N}=2$ supersymmetry representation). Hypermultiplets cannot interact between themselves since no cubic $SU(2)$ invariant is possible. Therefore, for renormalizable theories a superpotential is not allowed and interactions turn out to be all gauge interactions.

Let us now suppose that matter transforms under some non-trivial representation of the gauge group. We get for the corresponding $\mathcal{N}=2$ hypermultiplet Lagrangian

$$\begin{aligned}\mathcal{L}^{\mathcal{N}=2}_{\text{matter}} &= \int d^2\theta d^2\bar{\theta}\,(\overline{H}_1 e^{2gV_R}H_1 + \overline{H}_2 e^{-2gV_R}H_2) + \\ &\quad + \int d^2\theta\sqrt{2}gH_1\Phi H_2 + h.c.,\end{aligned} \tag{6.5}$$

where the suffix R on the vector superfield V refers to the representation of the gauge group G carried by the hypermultiplets. The F-term coupling the hypermultiplets with the chiral multiplet Φ belonging to the $\mathcal{N}=2$ vector multiplet is there because of $\mathcal{N}=2$ supersymmetry (it is, say, the supersymmetric partner of the kinetic terms which couple the hypermultiplets to V). So we see that eventually a cubic interaction does arise, but it is a gauge interaction, in the sense that it vanishes once the gauge coupling g is switched off.

Eliminating the auxiliary fields F_1 and F_2, the scalar potential for the hypermultiplets can be recast as a D-term contribution only and reads

$$V(H_1, H_2) = \frac{1}{2}D^2 = \frac{1}{2}g^2|\overline{H}_1 T^a_R H_1 - \overline{H}_2 T^a_R H_2|^2. \tag{6.6}$$

Notice finally that a mass term can be present and has the form

$$mH_1H_2. \tag{6.7}$$

However, a term of this sort can be there only for BPS hypermultiplets (which, as discussed in Chapter 3, are short enough to close the algebra within maximal spin 1/2 particle states).

6.1.1 *Non-linear sigma model III*

In the previous section we have constructed the most general renormalizable $\mathcal{N}=2$ supersymmetric Lagrangian. Like for $\mathcal{N}=1$ supersymmetry, one can relax renormalizability and get a $\mathcal{N}=2$ σ-model.

Let us start with pure SYM. Differently from the $\mathcal{N}=1$ case, this is a meaningful thing to do, since scalar fields are present in a $\mathcal{N}=2$ vector multiplet and a σ-model can exist. To write it down it is sufficient to take the $\mathcal{N}=1$ σ-model Lagrangian (5.117), set the superpotential to zero, and take into account that the chiral superfield Φ transforms in the adjoint representation of the gauge group. On general grounds, one would expect the Kähler potential K to be related, in a $\mathcal{N}=2$ consistent way, to the generalized complexified gauge coupling $\mathcal{F}_{ab}$, since the scalars spanning the manifold $\mathcal{M}$ sit in the same multiplets where the vectors sit (in particular, one would expect that an isometry transformation on $\mathcal{M}$ should have effects on the vectors, too). Equivalently, one can notice from (5.117) that the imaginary part of the generalized complexified gauge coupling multiplies the gaugino kinetic term while the Kähler metric that of the matter fermion fields. These should transform as a doublet under $SU(2)_R$ and then one would expect $\mathcal{F}_{ab}$ and K_{ab} to be exchanged under a $\mathbb{Z}_2$ R-symmetry rotation

$$\mathrm{Im}\,\mathcal{F}_{ab}\times\lambda^a\sigma^\mu D_\mu\overline{\lambda}^b \quad\longleftrightarrow\quad K_a^b\times\psi^a\sigma^\mu D_\mu\overline{\psi}_b. \tag{6.8}$$

What one actually finds is that $\mathcal{F}_{ab}$ and K can be written in terms of one and the same holomorphic function $\mathcal{F}(\Phi)$, dubbed *prepotential* and read

$$\mathcal{F}_{ab}(\Phi) = \frac{\partial^2\mathcal{F}(\Phi)}{\partial\Phi^a\partial\Phi^b} \tag{6.9}$$

$$\begin{aligned} K(\Phi,\overline{\Phi}) &= -\frac{i}{32\pi}\overline{\Phi}_a\frac{\partial\mathcal{F}(\Phi)}{\partial\Phi^a} + h.c. \\ &= -\frac{i}{32\pi}\overline{\Phi}_a\mathcal{F}^a(\Phi) + \frac{i}{32\pi}\overline{\mathcal{F}_a(\Phi)}\Phi^a. \end{aligned} \tag{6.10}$$

This is the very non-trivial statement that the σ-model action is uniquely determined by a single holomorphic function, the prepotential $\mathcal{F}(\Phi)$.

The end result reads

$$
\begin{aligned}
\mathcal{L}_{eff}^{\mathcal{N}=2} &= \frac{1}{64\pi i}\int d^2\theta \mathcal{F}_{ab}(\Phi)W^{\alpha a}W_{\alpha}^{b} + \\
&+ \frac{1}{32\pi i}\int d^2\theta d^2\bar{\theta}\left(\overline{\Phi}e^{2gV}\right)^a \mathcal{F}_a(\Phi) + h.c. \\
&= \frac{1}{32\pi}\,\mathrm{Im}\left[\int d^2\theta \mathcal{F}_{ab}(\Phi)W^{\alpha a}W_{\alpha}^{b} + 2\int d^2\theta d^2\bar{\theta}\left(\overline{\Phi}e^{2gV}\right)^a \mathcal{F}_a(\Phi)\right].
\end{aligned}
\tag{6.11}
$$

Using eqs. (6.9)–(6.10) one can compute the Kähler metric and see its relation with the complexified gauge coupling which is

$$
K_a^b(\phi,\overline{\phi}) = \frac{\partial^2 K(\phi,\overline{\phi})}{\partial\phi^a\partial\overline{\phi}_b} = -\frac{i}{32\pi}\left(\frac{\partial^2\mathcal{F}(\phi)}{\partial\phi^a\partial\phi^b} - \frac{\partial^2\overline{\mathcal{F}(\phi)}}{\partial\overline{\phi}_a\partial\overline{\phi}_b}\right) = \frac{1}{16\pi}\,\mathrm{Im}\mathcal{F}_{ab}(\phi).
\tag{6.12}
$$

Therefore, we finally get for the potential (recall that $W = 0$)

$$
V(\phi,\overline{\phi}) = -\frac{1}{2\pi}\left(\mathrm{Im}\ \mathcal{F}_{ab}(\phi)\right)^{-1}\left[\overline{\phi},\mathcal{F}_c(\phi)T^c\right]^a\left[\overline{\phi},\mathcal{F}_d(\phi)T^d\right]^b. \tag{6.13}
$$

A Kähler manifold where the Kähler potential can be written in terms of a holomorphic function as in eq. (6.10) is called *special* Kähler manifold. From a geometric point of view this corresponds to a Kähler manifold endowed with a symplectic structure (a $2n_v$ symplectic bundle, where n_v is the number of vector multiplets). The renormalizable Lagrangian (6.1) is recovered just taking $\mathcal{F}(\Phi) = \frac{1}{2}\tau\,\mathrm{Tr}\Phi^2$.

This is not the end of the story. To the σ-model action we have constructed one can add hypermultiplets. This will be a (very) special version of the Lagrangian (5.45). We refrain to present its precise structure here and just make two comments. Hypermultiplets contain two complex scalars. What one finds is that the corresponding σ-model is defined on a quaternionic manifold, known as HyperKähler manifold, which is, essentially, the quaternionic extension of a Kähler manifold (in particular, there are three rather than just one complex structures).

To sum-up, in $\mathcal{N} = 2$ supersymmetry, due to the existence of two sets of scalars, those belonging to matter multiplets and those belonging to gauge multiplets, the most general scalar manifold is (classically) of the form

$$
\mathcal{M} = \mathcal{M}^V \times \mathcal{M}^H, \tag{6.14}
$$

where $\mathcal{M}^V$ is a special Kähler manifold and $\mathcal{M}^H$ a HyperKähler manifold. Notice that, once renormalizability is relaxed, quartic superpotential couplings (and higher, if $SU(2)_R$ singlets) are possible. We will have much more to say about $\mathcal{N}=2$ σ-models in later chapters.

6.2 $\mathcal{N}=4$ supersymmetric actions

Let us now discuss the structure of the $\mathcal{N}=4$ Lagrangian. In this case there is only one kind of multiplet, the vector multiplet. So, from a $\mathcal{N}=4$ perspective, we speak of pure SYM theories. The decomposition of the $\mathcal{N}=4$ vector superfield in terms of $\mathcal{N}=1$ representations is as follows

$$[\mathcal{N}=4 \text{ vector multiplet}] : V=(\lambda_\alpha, v_\mu, D) \;\oplus\; \Phi_M=\left(\phi^M, \psi_\alpha^M, F^M\right),$$

where $M=1,2,3$. The propagating degrees of freedom are a vector field, six real scalars (two for each complex scalar ϕ_A) and four gauginos. The Lagrangian is a special instance of the Lagrangian (5.81) and very much constrained by $\mathcal{N}=4$ supersymmetry. First, the chiral superfields Φ_M should transform in the adjoint representation of the gauge group, since internal symmetries commute with supersymmetry. Moreover, we have now a large R-symmetry group, $SU(4)_R$. The four Weyl fermions transform in the fundamental of $SU(4)_R$, while the six real scalars in the two times anti-symmetric representation, which is nothing but the fundamental representation of $SO(6)$ (the fact that the scalar fields transform under the fundamental representation of $SO(6)$, which is real, makes the R-symmetry group of the $\mathcal{N}=4$ theory being at most $SU(4)$ and not $U(4)$). Finally, the auxiliary fields are singlets under the R-symmetry group.

Using $\mathcal{N}=1$ superfield formalism the Lagrangian reads

$$\begin{aligned}\mathcal{L}_{\text{SYM}}^{\mathcal{N}=4} &= \frac{1}{32\pi}\,\text{Im}\left(\tau\int d^2\theta \;\text{Tr}\; W^\alpha W_\alpha\right)+\int d^2\theta d^2\bar{\theta}\;\text{Tr}\;\sum_M \overline{\Phi}_M\, e^{2gV}\,\Phi_M + \\ &\quad -\int d^2\theta\sqrt{2}g\;\text{Tr}\;\Phi_1\left[\Phi_2,\Phi_3\right]+h.c.,\end{aligned} \tag{6.15}$$

where, similarly to the $\mathcal{N}=2$ Lagrangian (6.1), the commutator in the third term appears because the three chiral superfields Φ_M transform in the adjoint representation of the gauge group. Notice that in writing the expression (6.15) we made a choice of a specific $\mathcal{N}=1$ supersymmetry generator out of the four Q^I's. This breaks the full $SU(4)_R$ R-symmetry to $SU(3)\times U(1)_R$, meaning that the full non-abelian R-symmetry is not manifest using $\mathcal{N} = 1$ superspace formalism. The three complex scalars belonging to the chiral superfields Φ_M are related to the six real scalars of the $\mathcal{N} = 4$ vector multiplet as

$$\phi_1 = \frac{1}{\sqrt{2}}(X_1 + iX_2), \quad \phi_2 = \frac{1}{\sqrt{2}}(X_3 + iX_4), \quad \phi_3 = \frac{1}{\sqrt{2}}(X_5 + iX_6). \tag{6.16}$$

They transform in the $\underline{3}$ of $SU(3)$ and have R-charge $R = 2/3$ under the $U(1)_R$. In $\mathcal{N} = 1$ language the four gauginos λ^I get split as one gaugino and three Weyl fermions transforming in the adjoint representation of the gauge group

$$\lambda^I = (\lambda, \psi_1, \psi_2, \psi_3). \tag{6.17}$$

The gaugino λ is a singlet of $SU(3)$ and has R-charge $R = 1$ under the $U(1)_R$, while the three ψ_A transform as a triplet of $SU(3)$ and have $U(1)_R$ R-charge $R = -1/3$. Note that from the point of view of the $\mathcal{N} = 1$ supersymmetry which is manifest in the Lagrangian (6.15), the $SU(3)$ symmetry acts as a flavor (non-R) symmetry.

It is an easy but tedious exercise to perform the integration in superspace and get an explicit expression in terms of ordinary fields. Then, one can solve for the auxiliary fields and get an expression where only propagating degrees of freedom are present and where, in terms of the $\mathcal{N} = 4$ basis λ^I, X_i, the invariance under $SU(4)_R$ is manifest. We refrain to perform the calculation here and just report the end result which reads

$$\begin{aligned}\mathcal{L} = \mathrm{Tr}\Big[-\frac{1}{4}F_{\mu\nu}F^{\mu\nu} - i\lambda^I\sigma^\mu D_\mu\overline{\lambda}_I + \frac{1}{2}D_\mu X^i D^\mu X_i\Big] + \frac{\theta_{\mathrm{YM}}}{32\pi^2}\, g^2\mathrm{Tr}F_{\mu\nu}\widetilde{F}^{\mu\nu} + \\ + \frac{1}{2}g f^{abc}\,\Sigma_i^{IJ} X_a^i\,\lambda_{Ib}\,\lambda_{Jc} - \frac{1}{2}g^2\ \mathrm{Tr} \sum_{i,j=1}^{6} [X_i, X_j]^2 ,\end{aligned} \tag{6.18}$$

where sum on dummy indices is understood and the six 4×4 matrices Σ_i are the structure constants of the $SU(4)_R$ R-symmetry group.

From the above expression we can read-off the scalar potential which reads

$$V = \frac{1}{2}g^2\ \mathrm{Tr} \sum_{i,j=1}^{6} [X_i, X_j]^2 . \tag{6.19}$$

So $\mathcal{N} = 4$ SYM enjoys a large moduli space of vacua. At a generic point, very much like pure $\mathcal{N} = 2$ SYM, the gauge group is broken to $U(1)^r$, where r is the rank of G, and the dynamics is that of r copies of $\mathcal{N}=4$ $U(1)$ gauge theory. At the origin, and more generally whenever some VEVs become degenerate the theory gets partially unhiggsed and non-abelian gauge factors survive at low energy.

One might ask whether a $\mathcal{N}=4$ version of non-linear σ-model exists. After all, we are plenty of scalar fields, actually $3n$ complex scalars, if n is the dimension of G. The answer is that there is only one possible σ-model compatible with $\mathcal{N}=4$ supersymmetry, which is nothing but the trivial one, $\mathcal{M}=\mathbb{R}^{6n}$. So, the Lagrangian (6.15) is actually the only possible $\mathcal{N}=4$ Lagrangian one can build. This also implies that, unlike pure $\mathcal{N}=2$ SYM, the moduli space of vacua has a trivial topology. As we will see, this is related to the very much constrained dynamics $\mathcal{N}=4$ SYM enjoys at quantum level.

6.3 On non-renormalization theorems

One of the advantages, in fact *the* advantage of supersymmetry is that it makes quantum corrections much better behaved with respect to ordinary field theories.

Many relevant results about UV properties of supersymmetric field theories were obtained back in the 1980's and can be summarized in terms of powerful non-renormalization theorems. At that time, a very efficient approach was developed to deal with supersymmetric quantum field theories, a version of Feynman rules, known as *supergraph* techniques, which let one work directly with superfields in superspace with no need to expand into component fields. Most non-renormalization theorems were proved using such techniques whose description, however, is beyond the scope of this book. Here I just want to mention what is possibly the main result thus obtained: in a $\mathcal{N}=1$ supersymmetric quantum field theory containing chiral and vector superfields, the most general term that can be generated by loop diagrams has only one Grassman integral over *all* superspace

$$\int d^4x_1\dots d^4x_n\, d^2\theta d^2\bar{\theta}\; G(x_1,\dots,x_n)F_1(x_1,\theta,\bar{\theta})\dots F_n(x_n,\theta,\bar{\theta}), \quad (6.20)$$

where $G(x_1,\dots,x_n)$ is a translationally invariant function and the F_i's are products of superfields and their covariant derivatives. Such term is a D-term and does not contribute to superpotential terms, which are F-terms, implying that the superpotential is tree-level exact, *i.e.* it is not renormalized at any order in perturbation theory! The only possible corrections may arise at the non-perturbative level (and in some cases, namely when only chiral superfields are present, the latter also vanish, as we will see later). On the contrary, because of eq. (6.20), the Kähler potential is renormalized and generically receives corrections at any order in perturbation theory (and non-perturbatively).

Let us try and see what non-renormalization theorems imply for theories with different number of supersymmetries.

Let us first focus on a renormalizable $\mathcal{N} = 1$ action describing a chiral superfield Φ (the generalization to many chiral superfields is straightforward and does not present any relevant difference). There are three supersymmetric contributions to the action. One, the kinetic term, is a D-term, and undergoes renormalizations. Two are F-terms (the mass term and the cubic term) and are hence exact, perturbatively. Concretely

$$\int d^4x d^2\theta d^2\bar{\theta}\ \overline{\Phi}\Phi \to Z_\Phi \int d^4x d^2\theta d^2\bar{\theta}\ \overline{\Phi}\Phi \tag{6.21}$$

$$m \int d^4x d^2\theta\ \Phi^2 + h.c. \to m \int d^4x d^2\theta\ \Phi^2 + h.c. \tag{6.22}$$

$$\lambda \int d^4x d^2\theta\ \Phi^3 + h.c \to \lambda \int d^4x d^2\theta\ \Phi^3 + h.c.\ . \tag{6.23}$$

This means that m and λ do get renormalized but only logarithmically at one loop, instead of quadratically and linearly, respectively. Indeed, from the above equations one concludes that

$$Z_m Z_\Phi = 1 \text{ which implies } m \to Z_\Phi^{-1} m$$
$$Z_\lambda Z_\Phi^{3/2} = 1 \text{ which implies } \lambda \to Z_\Phi^{-3/2} \lambda$$

Something similar happens for the renormalization of the factor e^{2gV}. However, things are more subtle, here. Notice that

$$\int d^2\theta d^2\bar{\theta}\, \overline{\Phi} e^{2gV} \Phi \tag{6.24}$$

is a D-term and therefore it does renormalize. However, what one can see using for example the background field method is that gV does not renormalize. An independent renormalization for g and V leading to a kinetic term of the form

$$\overline{\Phi}\, e^{2Z_g g Z_V^{1/2} V}\, \Phi \tag{6.25}$$

would correspond to counterterms of the form

$$\overline{\Phi} V e^{gV} \Phi, \tag{6.26}$$

which are not gauge invariant and cannot be generated by loop diagrams. This implies that the integral (6.24) should renormalize as the kinetic term (6.21) (not because of supersymmetry, but just due to gauge invariance!), meaning that g and V compensate each other upon renormalization. In other words, gV is not renormalized. Another way to see this is the following. Consider pure SYM. In this theory the only possible counterterm would correspond to something proportional to the action itself

$$\int d^2\theta \ \mathrm{Tr} \ W^\alpha W_\alpha + h.c., \tag{6.27}$$

which would then correspond to a wave-function renormalization of the full Lagrangian (this is certainly there since the integral above is not a F-term, but rather a D-term, as already noticed). This means that one should multiply by the same function both the kinetic terms $dVdV$ as well as the interaction terms $gVVdV$ and g^2V^4 to keep gauge invariance (recall we are considering a non-abelian gauge group). In order for this to be the case one needs that if

$$V \rightarrow Z_V^{1/2} V, \tag{6.28}$$

then

$$g \rightarrow Z_V^{-1/2} g, \tag{6.29}$$

which implies that gV is not renormalized, as anticipated, namely that $Z_g = Z_V^{-1/2}$.

The conclusion is that in renormalizable theories with $\mathcal{N} = 1$ supersymmetry there are only two independent renormalization, Z_Φ and Z_V, which are just logarithmically divergent at one-loop, and correspond to wave-function renormalization of chiral and vector superfields, respectively.

Notice, in passing, that the fact that $Z_V \neq 1$ means that the integral (6.27) is renormalized. This does not contradict non-renormalization theorems, since, as already observed, (6.27) is not a F-term, really, but actually a D-term. That $Z_V \neq 1$ also implies that $Z_g \neq 1$, meaning that in $\mathcal{N} = 1$ SYM theories the gauge coupling runs, and can get corrections at all loops, in general.

What about higher supersymmetry? All what we said, so far, still applies, since any extended supersymmetry model is also $\mathcal{N} = 1$. However, extended supersymmetry imposes further constraints. In what follows, we stick to the notation we have used in sections 6.1 and 6.2 when discussing $\mathcal{N}=2$ and $\mathcal{N}=4$ Lagrangians, respectively.

Let us start from $\mathcal{N}=2$ supersymmetry. For one thing, since V and Φ belong now to the same multiplet, we have that

$$Z_V = Z_\Phi . \tag{6.30}$$

As for the hypermultiplets, from the cubic interaction $gH_1\Phi H_2$ which appears in the superpotential (and which is then tree level exact, being a F-term) we get the following condition

$$Z_g^2 Z_\Phi Z_{H_1} Z_{H_2} = 1. \tag{6.31}$$

The first two contributions cancel since $Z_\Phi = Z_V$ and since gV is not renormalized, $Z_g^2 = Z_V^{-1}$. So we get in the end

$$Z_{H_1} Z_{H_2} = 1. \tag{6.32}$$

Hence the wave-functions of the two chiral superfields making up an hypermultiplet are not independent. All in all, we have then only two independent renormalizations in $\mathcal{N}=2$ supersymmetry, Z_V and, say, Z_{H_1}. This is not the end of the story, though. For massless representations there is the $SU(2)_R$ symmetry rotating the scalar components of H_1 and H_2 into each other. Hence, they should have the same renormalization, which means, using eq. (6.32), that $Z_{H_1} = Z_{H_2} = 1$. The same holds for massive (BPS) representations. In this case the existence of a non-trivial central charge does break the R-symmetry group to $USp(2)$; however, the algebra of such group is the same as that of $SU(2)$ and one can again conclude that $Z_{H_1} = Z_{H_2} = 1$.

It turns out that because of the relation between Z_Φ and Z_g, not only $\mathcal{N}=2$ SYM has a unique renormalization but it is one-loop exact in perturbation theory. In other words, the gauge coupling β-function gets only one-loop contributions, perturbatively. We will derive this important result in a later chapter, when discussing the dynamics of supersymmetric gauge theory. There, we will use a very powerful approach which is based on a crucial property of supersymmetry, known as *holomorphy*. For the time being let us just stress that this one-loop exactness of $\mathcal{N}=2$ SYM gauge coupling does not hold for $\mathcal{N}=1$ SYM, whose gauge coupling receives corrections at all orders in perturbation theory.

Let us finally consider $\mathcal{N}=4$ supersymmetry. Here we have a single superfield, the vector superfield which, in $\mathcal{N}=1$ language, can be seen as one vector superfield V and three chiral superfields Φ_A (all transforming in the adjoint representation of the gauge group). The $SU(3)$ symmetry rotating the three chiral superfields implies that the latter should have all and the same wave-function renormalization

$$Z_{\Phi_1} = Z_{\Phi_2} = Z_{\Phi_3} = Z. \tag{6.33}$$

The $\mathcal{N} = 4$ Lagrangian can be thought as a $\mathcal{N} = 2$ SYM Lagrangian coupled to a hypermultiplet transforming in the adjoint representation of the gauge group, with one of the three Φ_A playing the role of Φ and belonging to the $\mathcal{N} = 2$ vector multiplet. Then, using eq. (6.30), one concludes that Z should equal Z_V, the wave-function of the vector superfield. Plugging this into eq. (6.31), which for the $\mathcal{N} = 4$ Lagrangian (6.15) is

$$Z_g^2 Z_{\Phi_1} Z_{\Phi_2} Z_{\Phi_3} = Z_g^2 Z^3 = 1, \tag{6.34}$$

and recalling that $Z_g = Z_V^{-1/2}$ it finally follows that

$$Z_V = Z_{\Phi_1} = Z_{\Phi_2} = Z_{\Phi_3} = 1, \tag{6.35}$$

meaning that $\mathcal{N} = 4$ SYM is perturbatively finite; in other words, the theory does not exhibit ultraviolet divergences in the correlation functions of canonical fields! Though we are not going to prove it here, it turns out that in fact $\mathcal{N} = 4$ is finite also once non-perturbative corrections are taken into account. More precisely, the latter give finite contributions only, and therefore the theory is believed to be UV finite.

There is yet another important property of $\mathcal{N} = 4$ SYM we would like to mention: the theory is superconformal invariant, and it is so at the full quantum level. Let us see how this goes. A theory whose Lagrangian contains only dimension four operators, like the $\mathcal{N} = 4$ Lagrangian (and many others, in fact) is classically scale invariant. For any relativistic field theory this implies a larger symmetry algebra, the conformal Poincaré algebra which, besides Poincaré generators, includes also dilations and special conformal transformations, the corresponding group being $SO(2,4) \simeq SU(2,2)$. The generators associated to dilations and special conformal transformations, D and K^μ, respectively, act as follows

$$\begin{aligned} D : x^\mu &\longrightarrow \lambda x^\mu \\ K^\mu : x^\mu &\longrightarrow \frac{x^\mu + a^\mu x^2}{1 + 2x^\nu a_\nu + a^2 x^2}, \end{aligned}$$

and have the following commutation relations between themselves and with the generators of the Poincaré algebra

$$[P_\mu, D] = iP_\mu, \quad [P_\mu, K_\nu] = 2i\left(M_{\mu\nu} - \eta_{\mu\nu}D\right)$$
$$[M_{\mu\nu}, K_\rho] = i\left(\eta_{\nu\rho}K_\mu - \eta_{\mu\rho}K_\nu\right), \quad [K_\mu, D] = -iK_\mu, \tag{6.36}$$

with all others commutators vanishing.

Supersymmetry enlarges further the symmetry group. A conformal invariant supersymmetric theory enjoys an even larger algebra, the *superconformal* algebra, which includes, besides dilations and special conformal transformations, also conformal supersymmetry transformations $S^I_\alpha, \overline{S}^I_{\dot\alpha}$ (they appear in the commutator of the supersymmetry charges Q^I_α with the generators of special conformal transformations K^μ), and the generators associated to R-symmetry transformations, T^I_J (where $I, J = 1, \ldots, N$), which are now part of the algebra and do not act just as external automorphisms (they appear in the anti-commutator of the supersymmetry charges with the S^I_α's). The associated supergroup is $SU(2,2|N)$. The non-vanishing (anti)commutators involving the new generators are

$$\left[K_\mu, Q^I_\alpha\right] = 2i\sigma_{\mu\alpha\dot\beta}\overline{S}^{\dot\beta I}, \quad \left[D, Q^I_\alpha\right] = -\frac{i}{2}Q^I_\alpha, \quad \left[D, S^I_\alpha\right] = \frac{i}{2}S^I_\alpha,$$
$$\left[P_\mu, S^I_\alpha\right] = \sigma_{\mu\alpha\dot\beta}\overline{Q}^{\dot\beta I}, \quad \{S^I_\alpha, \overline{S}_{\dot\beta J}\} = 2\sigma^\mu_{\alpha\dot\beta}K_\mu\delta^I{}_J,$$
$$\left[M_{\mu\nu}, S^I_\alpha\right] = -i(\sigma_{\mu\nu})_\alpha^\beta S^{I\beta}, \quad \left[T^I_J, S^K_\alpha\right] = -\delta^K_J S^I_\alpha + \frac{1}{4}\delta^I{}_J S^K_\alpha,$$
$$\{Q^I_\alpha, S_{\beta J}\} = \epsilon_{\alpha\beta}\left(\delta^I{}_J D + T^I_J\right) + \frac{1}{2}\delta^I{}_J\sigma^{\mu\nu}_{\alpha\beta}M_{\mu\nu}. \tag{6.37}$$

The $\mathcal{N} = 4$ SYM action is invariant under this larger symmetry algebra, $SU(2,2|4)$ in this case, but it is certainly not the only theory having this property, at the classical level. Classical superconformal invariance is shared by any supersymmetric Lagrangian made solely by dimension four operators (in other words, with dimensionless, hence classically marginal couplings), well-known examples being the massless WZ model, and in fact any SYM theory, like $\mathcal{N} = 1$ SQCD discussed in the previous chapter.

What makes $\mathcal{N} = 4$ SYM special is that, as we have observed above, the Lagrangian does not renormalize (recall that essential to this proof was the use of the $SU(3)$ subgroup of the R-symmetry group rotating the three scalar superfields Φ^A). In particular, as we have seen before,

$Z_g = 1$. In other words, the $\mathcal{N} = 4$ β-function vanishes identically: the theory remains scale invariant at the quantum level, and the superconformal symmetry $SU(2,2|4)$ is then an exact symmetry of the theory. An equivalent conclusion can be reached by observing that $\mathcal{N} = 4$ SYM is a (very special) $\mathcal{N} = 2$ theory. The $\mathcal{N} = 2$ gauge coupling is one-loop exact and since in $\mathcal{N} = 4$ SYM this is the only coupling appearing in the Lagrangian, it is enough to compute the one-loop β-function for g. One can easily see that such one-loop coefficient vanishes, concluding that the theory is superconformal also at the quantum level. In fact, the equivalence of these proofs lies in the fact that the gauge coupling β-function and the R-symmetry current are in the same supermultiplet, the ($\mathcal{N} = 4$) supercurrent multiplet.

This non-renormalization property is not shared by other theories, in general: typically, the superconformal algebra is broken by quantum corrections and couplings run. For instance, in the massless WZ model the coupling, which is classically marginal, becomes irrelevant quantum mechanically (*i.e.*, it flows to zero and the theory becomes free in the IR). On the contrary, UV-free supersymmetric gauge theories, like pure $\mathcal{N} = 1, 2$ SYM, enjoy dimensional transmutation and a dynamically scale is generated at the quantum level.

What we said above about the finiteness of $\mathcal{N} = 4$ does not mean that any operator has protected dimension. The scaling dimension of canonical fields (gauge fields, gauginos and adjoint scalars) is unaffected by quantum corrections, but this does not happen, in general, to composite gauge invariant operators. Yet, in a superconformal theory there are special operators whose dimension is protected. To see how this comes, let us start considering the conformal algebra (6.36). In unitary theories there is a lower bound for the scaling dimension Δ of a field (e.g. $\Delta \geq 1$ for a scalar field in four dimensions). Since K_μ lowers the scaling dimension of a field, any representation of the conformal algebra should admit an operator with minimal dimension Δ which is annihilated by K_μ (at $x_\mu = 0$). Such states are called conformal primary operators. Since the conformal algebra is a subalgebra of the superconformal algebra, representations of the latter decompose into representations of the former. By definition, a superconformal primary operator is an operator which is annihilated (at $x_\mu = 0$) both by K_μ and $S^I_\alpha, \overline{S}^I_{\dot\alpha}$. From the commutator $[K_\mu, Q^I_\alpha]$ in (6.37) it also follows that any operator which is obtained from a superconformal primary by the action of Q^I_α, and hence sits in the same supermultiplet, is a primary operator of the conformal algebra. Superconformal primary operators which are annihilated by some of the supercharges are called *chiral primaries* and, most importantly, their dimension is fixed by their R-symmetry representation, and as such are protected against quantum corrections. This can be proven from the last (anti)commutation relation in eqs. (6.37), which lets express the scaling dimension Δ of a chiral primary operator (for which $\{Q^I_\alpha, S_{\beta J}\}$ vanishes) in terms of Lorentz and R-symmetry representations which are quantum numbers and as such does not renormalize. By supersymmetry, this implies that in a superconformal theory operators belonging to supersymmetry representations which include a chiral primary operator do not renormalize. For instance, in $\mathcal{N}=4$ SYM, a class of superconformal primaries are all operators made of symmetric traceless products of the scalar fields X_i's, e.g. $\mathrm{Tr}\,(X^{\{i}X^{j\}}) = \mathrm{Tr}\,(X^iX^j) - \frac{1}{6}\delta^{ij}\mathrm{Tr}\,(X^kX^k)$.

As a final comment, let us notice that in $\mathcal{N}=4$ SYM superconformal invariance is/is not realized depending on the point of the moduli space one is sitting. The phase where all scalar field VEVs $\langle X_i \rangle$ vanish is called superconformal phase: at the origin of the moduli space the gauge group remains unbroken and superconformal invariance is preserved. In other words, physical states are not only gauge invariant, but carry unitary representations of $SU(2,2|4)$. On the contrary, at any other point of the moduli space, where gauge symmetry is (partially) broken, superconformal symmetry is also broken since scalar VEVs $\langle X_i \rangle$ set a dimension full scale in the theory.

6.4 Exercises

Exercise 6.1. Derive the potential (6.6) starting from the matter Lagrangian (6.5).

Exercise 6.2. Starting from the expression (6.15), compute the off-shell $\mathcal{N}=4$ Lagrangian and derive, upon integrating out auxiliary fields, the on-shell one, eq. (6.18). For scalars and fermion fields use bases X_i and λ_I, defined in eqs. (6.16) and (6.17), respectively, which carry faithful representations of $SU(4)_R$.

References

[1] A. Bilal, *Introduction to supersymmetry*, Section 8, arXiv:hep-th/0101055.
[2] M. F. Sohnius, Introducing Supersymmetry, Sections 12 and 13, *Phys. Rept.* **128** (1985) 39.

[3] P. Frè, Lectures on special Kahler geometry and electric - magnetic duality rotations, *Nucl. Phys. Proc. Suppl.* **45BC** (1996) 59 [hep-th/9512043].
[4] E. D'Hoker and D. Z. Freedman, *Supersymmetric gauge theories and the AdS/CFT correspondence*, Section 3, arXiv:hep-th/0201253.
[5] M. J. Strassler, *An unorthodox introduction to supersymmetric gauge theory*, Sections 2.3 and 3.1, TASI 2001, arXiv:hep-th/0309149.
[6] R. Argurio, *Introduction on Supersymmetry*, Chapter 6, https://ptm.ulb.be/rargurio/susycourse.pdf
[7] M. T. Grisaru, W. Siegel and M. Rocek, Improved Methods for Supergraphs, *Nucl. Phys. B* **159** (1979), 429.
[8] N. Seiberg, Supersymmetry and Nonperturbative beta Functions, *Phys. Lett. B* **206** (1988), 75.

Chapter 7

Spontaneous Supersymmetry Breaking

If supersymmetry is at all realized in Nature, it must be broken at low enough energy: we do not see any mass degeneracy in the elementary particle spectrum, at least at energies of order 1 TeV or lower. The idea is then that supersymmetry is broken at some scale M_s, such that at energies $E > M_s$ the theory behaves as a supersymmetric theory, while at energies $E < M_s$ it does not. On general ground, there are two ways supersymmetry can be broken, either spontaneously or explicitly.

- *Spontaneous supersymmetry breaking*: the theory is supersymmetric but has a scalar potential admitting (stable, or metastable but sufficiently long-lived) supersymmetry breaking vacua. In such vacua one or more scalar fields acquire a VEV of order M_s, which then sets the scale of supersymmetry breaking.
- *Explicit supersymmetry breaking*: the Lagrangian contains terms which do not preserve supersymmetry by themselves. In order for them not to ruin the nice and welcome UV properties of supersymmetric theories, these terms should have positive mass dimension, in other words they should be irrelevant in the far UV. In this case we speak of *soft* supersymmetry breaking. In such scenario, the scale M_s enters explicitly in the Lagrangian.

As we will show later, soft supersymmetry breaking models can (and typically do) actually arise as low energy effective descriptions of models where supersymmetry is broken spontaneously. Therefore, we will start focusing on spontaneous supersymmetry breaking. Only after we will discuss supersymmetry breaking induced by soft terms.

7.1 Vacua in supersymmetric theories

We have already seen that supersymmetric vacua are in one-to-one correspondence with the zero's of the scalar potential. In other words, the vacuum energy is zero if and only if the vacuum preserves supersymmetry. Hence, non-supersymmetric vacua correspond to minima of the potential which are not zero's. In this case supersymmetry is broken in the perturbative theory based on these positive energy vacua.

Notice how different is spontaneous supersymmetry breaking with respect to spontaneous breaking of ordinary internal symmetries (being them global or local). There, what matters is the location of the minima of the potential in field space, while here it is the absolute value of the potential at such minima. This implies that in, e.g. a supersymmetric gauge theory there can be minima which preserve both gauge symmetry and supersymmetry, others which break both, and others which preserve gauge symmetry and break supersymmetry, or viceversa. A schematic picture of these different situations is reported in Figure 7.1.

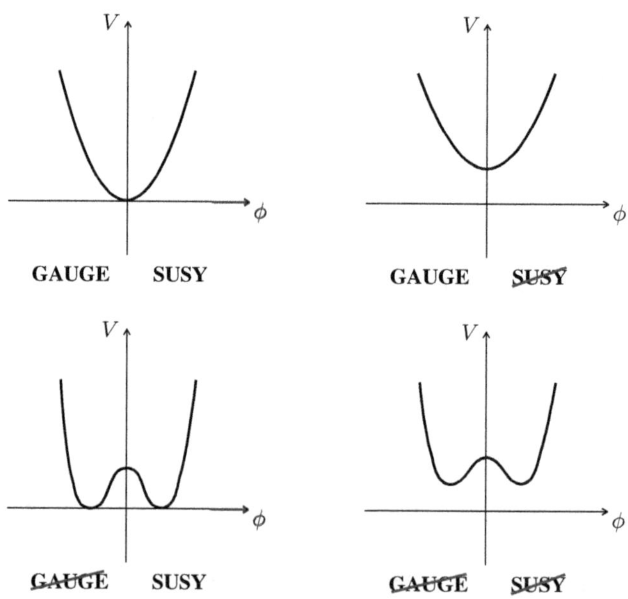

Fig. 7.1 A schematic picture of possible patterns of spontaneous gauge symmetry and supersymmetry breakings. The potential on the upper left has a unique vacuum which preserve both supersymmetry and gauge symmetry. The one on the lower right, instead, admits two vacua breaking both symmetries. The other two represent mixed situations where either supersymmetry or gauge symmetry are broken.

While non-supersymmetric vacua can be either global or local minima of the potential, supersymmetric vacua, if present, are obviously global minima of the potential, since in a supersymmetric theory the scalar potential is a semi-positive definite quantity.

Recall the expression (5.84), that is

$$V(\phi,\overline{\phi}) = \overline{F}F + \frac{1}{2}D^2, \tag{7.1}$$

where

$$\overline{F}_i = \frac{\partial W}{\partial \phi^i}, \quad D^a = -g\left(\overline{\phi}_i (T^a)^i_{\ j}\,\phi^j + \xi^a\right) \tag{7.2}$$

(we focus here on models with canonical Kähler potential; later we will also discuss situations where the Kähler potential is not canonical).

Supersymmetric vacua are described by all possible set of scalar field VEVs satisfying the D-term and F-term equations

$$\overline{F}_i(\phi) = 0, \quad D^a(\overline{\phi},\phi) = 0. \tag{7.3}$$

If there exist more than one solution, it means there are more supersymmetric vacua, generically a moduli space of vacua, if these are not isolated. If there does not exist a set of scalar field VEVs for which eqs. (7.3) are satisfied, then supersymmetry is broken and the minima of the potential are all necessarily positive, $V_{\text{min}} > 0$.

Notice, on the contrary, that on any vacuum, supersymmetric or not, global or local, the following equations always hold

$$\frac{\partial V(\phi,\overline{\phi})}{\partial \phi^i} = 0, \quad \frac{\partial V(\phi,\overline{\phi})}{\partial \overline{\phi}_i} = 0, \tag{7.4}$$

which simply say that vacua sit at extrema of the scalar potential.

An equivalent statement about supersymmetric vacua is that on supersymmetric vacua the supersymmetric variations of fermion fields vanish. This can be seen as follows. Due to Lorentz invariance, on a vacuum any field's VEV or its derivative should vanish, but scalar fields. Recalling how the different field components of a chiral or vector superfield transform under supersymmetry transformations, it follows that on a vacuum state we have

$$\begin{aligned}
&\delta\langle\phi^i\rangle = 0, \quad \delta\langle F^i\rangle = 0, \quad \delta\langle\psi^i_\alpha\rangle \sim \epsilon_\alpha\langle F^i\rangle \\
&\delta\langle F^a_{\mu\nu}\rangle = 0, \quad \delta\langle D^a\rangle = 0, \quad \delta\langle\lambda^a_\alpha\rangle \sim \epsilon_\alpha\langle D^a\rangle.
\end{aligned} \tag{7.5}$$

Therefore, in a generic vacuum the supersymmetric variations of the fermions is not zero: it is actually proportional to the vacuum expectation values of the auxiliary fields. A supersymmetric vacuum state is by definition supersymmetric invariant (!). Hence, from the above equations it follows that on a supersymmetric vacuum also the supersymmetric variations of the fermions should be zero, the latter being equivalent to the D and F-term equations (7.3), as anticipated.

7.2 Goldstone theorem and the goldstino

When a global symmetry is spontaneously broken, Goldstone theorem says that there is a massless mode in the spectrum, the Goldstone field, whose quantum numbers should be related to the broken symmetry. We should expect this theorem to work also for spontaneously broken supersymmetry. In fact, given that supersymmetry is a fermionic symmetry, the Goldstone field should be in this case a Majorana spin 1/2 fermion, the so-called *goldstino*.

Consider the most general supersymmetric Lagrangian with gauge and matter fields, eq. (5.83), and suppose it admits some vacuum where supersymmetry is broken. In this vacuum eqs. (7.4) hold, while (some of) eqs. (7.3) do not. Recalling eqs. (5.82) and (5.84) we have in this vacuum that

$$\frac{\partial V(\phi,\overline{\phi})}{\partial \phi^i} = F^j(\overline{\phi})\frac{\partial^2 W}{\partial\phi^i\partial\phi^j} - gD^a\overline{\phi}_j(T^a)^j_{\ i} = 0. \tag{7.6}$$

On the other hand, since the superpotential is gauge invariant, we have that

$$\delta^a W = \frac{\partial W}{\partial \phi^i}\delta^a\phi^i = \overline{F}_i(T^a)^i_{\ j}\phi^j = 0. \tag{7.7}$$

Combining the former equation with the complex conjugate of the latter evaluated in the vacuum, we easily get a matrix equation

$$M\begin{pmatrix}\langle F^j\rangle \\ \langle D^a\rangle\end{pmatrix} = 0 \quad \text{where } M = \begin{pmatrix}\langle\frac{\partial^2 W}{\partial\phi^i\partial\phi^j}\rangle & -g\langle\overline{\phi}_l\rangle(T^a)^l_{\ i} \\ -g\langle\overline{\phi}_l\rangle(T^b)^l_{\ j} & 0\end{pmatrix}. \tag{7.8}$$

The above equation implies that the matrix M has an eigenvector with zero eigenvalue. Now, this matrix is nothing but the fermion mass matrix of the Lagrangian (5.83)! This can be seen looking at the non-derivative fermion bilinears of (5.83), which on the vacuum get contributions also from the cubic coupling between scalar fields, their superpartners and gauginos, and which can be written as

$$\cdots - \frac{1}{2}\left(\psi^i,\ \sqrt{2}i\lambda^b\right) M \begin{pmatrix} \psi^j \\ \sqrt{2}i\lambda^a \end{pmatrix} + h.c. + \cdots \tag{7.9}$$

Hence, on the supersymmetry breaking vacuum the spectrum necessarily admits a massless fermion, the goldstino. In terms of spin 1/2 particles belonging to the different multiplets, the goldstino ψ^G_α corresponds to the following linear combination

$$\psi^G_\alpha \sim \langle \overline{F}_i \rangle\, \psi^i_\alpha + \langle D_a \rangle\, \lambda^a_\alpha . \tag{7.10}$$

The proof we have provided of the goldstino theorem has been based on the Lagrangian (5.83). In fact, one can provide a similar proof using supersymmetry Ward identities, which does not rely on the existence of an explicit classical Lagrangian. The supersymmetry Ward identity reads

$$\langle \partial^\mu S_{\mu\alpha}(x) \overline{S}_{\nu\dot\beta}(0) \rangle = -\delta^4(x) \langle \delta_\alpha \overline{S}_{\nu\dot\beta} \rangle = -2\, \sigma^\mu_{\alpha\dot\beta} \langle T_{\mu\nu} \rangle\, \delta^4(x), \tag{7.11}$$

where the last equality follows from the current algebra, eq. (4.72) (Schwinger terms cannot have a non-vanishing VEV in a Lorentz invariant vacuum, while this is possible for the energy-momentum tensor, since $T_{\mu\nu} \sim \eta_{\mu\nu}$ which is Lorentz invariant). Integrating eq. (7.11), one gets for the two-point function of the supercurrent that

$$\langle S_{\mu\alpha}(x) \overline{S}_{\nu\dot\beta}(0) \rangle = \cdots + (\sigma_\mu \bar\sigma^\rho \sigma_\nu)_{\alpha\dot\beta} \frac{x_\rho}{x^4} \langle T \rangle, \tag{7.12}$$

where $\langle T \rangle = \eta^{\mu\nu} \langle T_{\mu\nu} \rangle$ and the $\ldots$ are terms which are not relevant to the present discussion. Upon Fourier transforming we finally get

$$\langle S_{\mu\alpha}(k) \overline{S}_{\nu\dot\beta}(-k) \rangle = \cdots + (\sigma_\mu \bar\sigma^\rho \sigma_\nu)_{\alpha\dot\beta}\, k_\rho \frac{\langle T \rangle}{k^2}, \tag{7.13}$$

which shows the presence of a massless pole (the goldstino), proportional to the vacuum energy density, in the supercurrent two-point function. The above equation shows that the goldstino is the lowest energy excitation of the supercurrent, and it is so if and only if the vacuum energy is non-vanishing. This shows, as anticipated, that the goldstino theorem holds universally, *i.e.* also for vacua where supersymmetry is broken in a strongly coupled phase of the theory, where classical arguments may not apply.

7.3 F-term breaking

From our discussion it is clear that given a generic Lagrangian, there are two a priori independent ways we can break supersymmetry: either giving a non-vanishing expectation value to (some) F-terms or to (some) D-terms. We will consider both options in turn.

In this section we will start considering F-term breaking and therefore we assume, for the time being, to deal with a theory with chiral superfields only.

The most general renormalizable Lagrangian of this sort reads

$$\mathcal{L} = \int d^2\theta d^2\bar{\theta}\, \overline{\Phi}_i \Phi^i + \int d^2\theta\, W(\Phi^i) + \int d^2\bar{\theta}\, \overline{W}(\overline{\Phi}_i), \tag{7.14}$$

where

$$W(\Phi^i) = a_i \Phi^i + \frac{1}{2} m_{ij} \Phi^i \Phi^j + \frac{1}{3} g_{ijk} \Phi^i \Phi^j \Phi^k. \tag{7.15}$$

The equations of motions for the auxiliary fields read

$$\overline{F}_i(\phi) = \frac{\partial W}{\partial \phi^i} = a_i + m_{ij}\phi^j + g_{ijk}\phi^j \phi^k, \tag{7.16}$$

and the potential is

$$V(\phi, \overline{\phi}) = \sum_i |a_i + m_{ij}\phi^j + g_{ijk}\phi^j \phi^k|^2. \tag{7.17}$$

Supersymmetry is broken if and only if there does not exist a set of scalar field VEVs such that all F-terms vanish, $\langle F^i \rangle = 0$. This implies that in order for supersymmetry to be broken, it is necessary some a_i to be different from zero. If not, the trivial solution $\langle \phi^i \rangle = 0$ solves all F-equations. So, any model of F-term supersymmetry breaking needs a superpotential admitting linear terms.

Notice that this conclusion applies also for a superpotential with higher non-renormalizable couplings. In fact, it does also in presence of a non-canonical Kähler potential! This can be seen recalling the expression of the scalar potential when a non-canonical Kähler metric is present

$$V(\phi, \overline{\phi}) = \left(K^{-1}\right)^i_{\ j} \frac{\partial W}{\partial \phi^i} \frac{\partial \overline{W}}{\partial \overline{\phi}_j}. \tag{7.18}$$

From this expression it is clear that unless it were singular (something signalling, as already discussed, an inconsistency of the effective theory analysis), a non-trivial Kähler metric could not influence the existence/ non existence of supersymmetric vacua, which is still dictated by the possibility/ impossibility to make the first derivatives of the superpotential vanish. What gets modified by a non-trivial Kähler potential, instead, is the value of the vacuum energy (for non-supersymmetric vacua, only!) and the particle spectrum around a given vacuum (for both supersymmetric and non-supersymmetric vacua).

In what follows we will consider several examples of F-term breaking.

Example 1: The Polonyi model.

Let us consider the theory of a single chiral superfield with canonical Kähler potential and a linear superpotential

$$K = \overline{X}X, \quad W = \lambda X. \tag{7.19}$$

This is the most minimal set-up one can imagine for a F-term supersymmetry breaking model. The potential reads

$$V = \left|\frac{\partial W}{\partial X}\right|^2 = |\lambda|^2, \tag{7.20}$$

Supersymmetry is clearly broken for any $|X|$, and the latter is in fact a flat direction. The supersymmetry breaking scale is set by the modulus of λ itself, $|\lambda| = M_s^2$. In Figure 7.2 we report the (trivial) shape of the scalar potential.

A few comments are in order. First notice that the theory possesses an R-symmetry, the R-charge of X being $R(X) = 2$. At a generic point of the moduli space, then, both supersymmetry and R-symmetry are broken. Second, notice that although supersymmetry is broken, the spectrum is

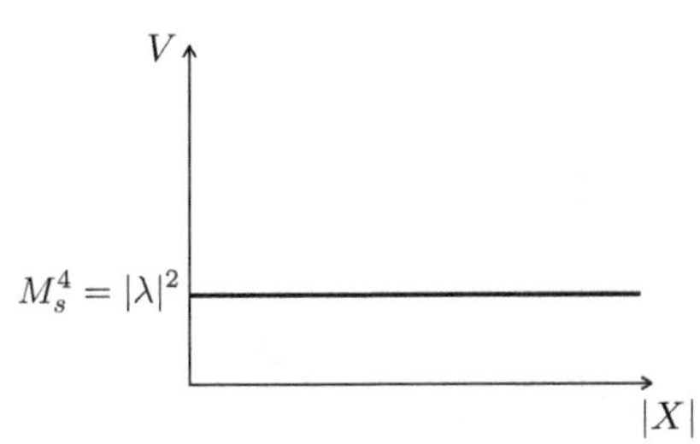

Fig. 7.2 The potential of the Polonyi model.

degenerate in mass: $|X|$, its phase α, and ψ_X are all massless. The fermion field has a good reason to be massless: it is the goldstino predicted by Goldstone theorem. Seemingly, the phase of X is expected to be massless: it is nothing but the goldstone boson associated to the broken R-symmetry. Finally, the modulus of the scalar field $|X|$ is massless since it parametrizes the (non-supersymmetric) moduli space. However, there are no reasons to expect this moduli space to be protected against quantum corrections, since supersymmetry is broken. Hence, generically, one would expect it to be lifted at the quantum level and $|X|$ to get a mass. This is not the case in this simple theory since it is a non-interacting theory and there are no quantum corrections whatsoever. In general, however, things are different: a non-supersymmetric moduli space gets typically lifted at one or higher loops, and the putative moduli get a mass. For this reason, non-supersymmetric moduli spaces are dubbed *pseudo-moduli* spaces, and the moduli parametrizing them, pseudo-moduli. We will see examples of this sort soon.

Let us now consider the following innocent-looking modification of the model above. Let's add a mass term to X,

$$\Delta W = \frac{1}{2} m X^2. \tag{7.21}$$

Things drastically change, since supersymmetry is now restored. Indeed, the F-term equation now reads

$$\overline{F}(X) = mX + \lambda = 0, \tag{7.22}$$

which admits the solution $\langle X \rangle_{\text{SUSY}} = -\frac{\lambda}{m}$. Hence, there is a choice of scalar field VEV which makes the potential $V = |\lambda + mX|^2$ vanish, as illustrated in Figure 7.3.

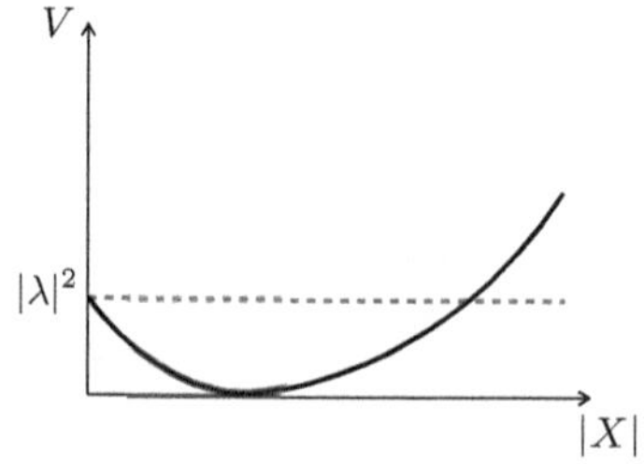

Fig. 7.3 The potential of the massive Polonyi model.

The spectrum is supersymmetric and massive: all fields have mass m. This agrees with physical expectations: ψ_X is no more the goldstino, since this is not expected to be there now; $|X|$ is no more a (pseudo)modulus since the supersymmetric vacuum is isolated; finally, α is not anymore the goldstone boson associated to the broken R-symmetry since the superpotential term ΔW breaks the R-symmetry explicitly. More precisely, $W = \lambda X + \frac{1}{2} m X^2$ does not admit any R-charge assignment for X such that $R(W) = 2$, so the theory does not admit a R-symmetry to start with.

Things might also change (both qualitatively and quantitatively) if one allows the Kähler potential not being canonical. Suppose we keep $W = \lambda X$ but we let the Kähler metric be non-trivial, that is

$$V = \left(K_{X\overline{X}}\right)^{-1} |\lambda|^2 \quad \text{with } K_{X\overline{X}} = \frac{\partial^2 K}{\partial X \partial \overline{X}} \neq 1. \tag{7.23}$$

A non-trivial Kähler metric can deform sensibly the pseudo-moduli space of Figure 7.2. A sample of possible different behaviors, which depend on the asymptotic properties (or singularities) of the Kähler metric, is reported in Figure 7.4.

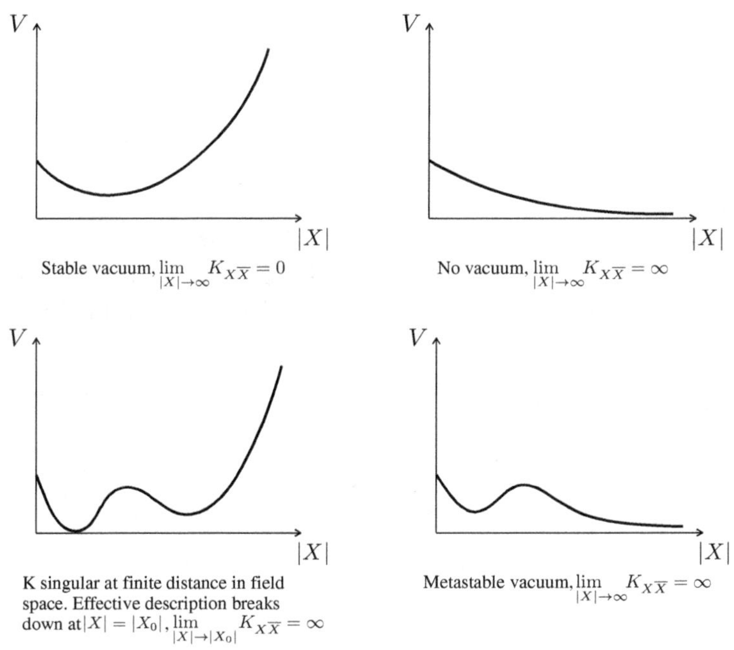

Fig. 7.4 Qualitatively different potentials of non-canonical Polonyi-like models.

Physically, the different behaviors reported in Figure 7.4 should be understood as follows. In presence of a classical pseudo-moduli space like the one in Figure 7.2, the lifting of the pseudo-moduli at quantum level occurs because the massless particles which bring from a vacuum to another get a mass at one loop. Sometime, such effect can be mimicked by a non-canonical Kähler potential. In fact, these seemingly ad-hoc theories can and sometime do arise at low energies as *effective theories* of more complicated UV-renormalizable theories: the mass scale entering the non-canonical Kähler potential is nothing but the UV cut-off of these low energy effective theories.

Let us try to make the above discussion more concrete by considering an explicit example.

Example 2: A Polonyi model with quartic Kähler potential.

Let us consider the following model

$$K = \overline{X}X - \frac{c}{\Lambda^2}\left(\overline{X}X\right)^2, \quad W = \lambda X, \tag{7.24}$$

where $c > 0$. Notice that the R-symmetry is not broken by the non-canonical Kähler potential (7.24), which is R-symmetry invariant. So the $U(1)_R$ symmetry is a symmetry of the theory. The Kähler metric and the scalar potential now read

$$K_{X\overline{X}} = 1 - \frac{4c}{\Lambda^2}\overline{X}X, \quad V = K^{X\overline{X}}|\lambda|^2 = \frac{|\lambda|^2}{1 - \frac{4c}{\Lambda^2}X\overline{X}}. \tag{7.25}$$

The Kähler potential is an instance of models like those in the upper-left diagram of Figure 7.4: the Kähler metric $K_{X\overline{X}}$ vanishes for large enough $|X|$, $|X| \to |\Lambda|/2\sqrt{c}$, which is of the order of the natural cut-off of the theory. The potential, which is depicted in Figure 7.5, admits a (unique)

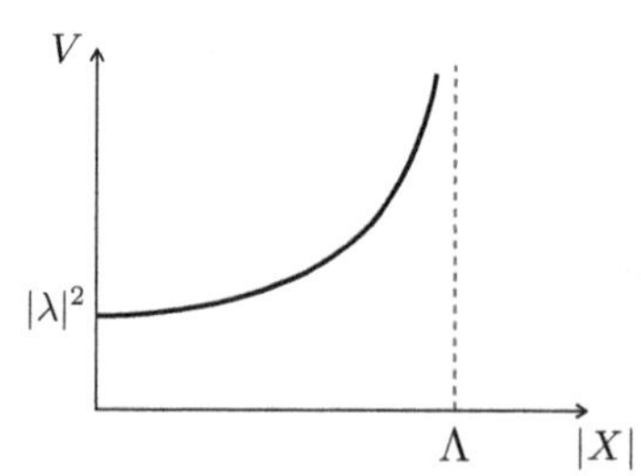

Fig. 7.5 The potential of a Polonyi model with quartic Kähler potential (7.24).

mimimum at $\langle X \rangle = 0$. Therefore, there is an isolated vacuum now and it is a supersymmetry breaking one. One can compute the spectrum around such vacuum and find that ψ_X is consistently massless (it is the goldstino), while now the scalar field is massive, $m_X^2 \sim c|\lambda|^2/\Lambda^2$.

Example 3: Supersymmetry restoration by new degrees of freedom.

Let us now deform the basic Polonyi model by adding a new superfield, Y, while keeping the Kähler potential canonical. The superpotential reads

$$W = \lambda X + \frac{1}{2} h X Y^2. \tag{7.26}$$

Notice that this model has an R-symmetry, with R-charge assignment $R(X) = 2$, $R(Y) = 0$. From the F-equations one can compute the potential which reads

$$V = |\lambda + \frac{1}{2} h Y^2|^2 + |hXY|^2, \tag{7.27}$$

implying that there are two supersymmetric vacua at

$$\langle X \rangle_{\text{SUSY}} = 0, \quad \langle Y \rangle_{\text{SUSY}} = \pm\sqrt{-2\frac{\lambda}{h}}. \tag{7.28}$$

So we see that the additional degrees of freedom have restored supersymmetry. Interestingly, there are other *local* minima of the potential, a pseudo-moduli space in fact, where supersymmetry is broken

$$\langle X \rangle_{\text{SB}} = \text{any}, \quad \langle Y \rangle_{\text{SB}} = 0 \quad \text{where } V = |\lambda|^2. \tag{7.29}$$

The physical interpretation is as follows. For large $\langle X \rangle$, the superfield Y gets a large mass and affects the low energy theory lesser and lesser. The theory reduces effectively to the original Polonyi model, which breaks supersymmetry and whose vacuum energy is indeed $V = |\lambda|^2$. It is a simple but instructive exercise to compute the mass spectrum around the non-supersymmetric minima. The chiral superfield X is obviously massless while Y gets a mass. There is a first (obvious) contribution to both the scalar and the fermion components of Y from $h\langle X \rangle$, and a second contribution which affects only the scalar component of Y coming from F_X, which is non-vanishing. The end result is

$$m_Y^2 = |h\langle X \rangle|^2 \pm |h\lambda|, \quad m_{\psi_Y} = h\langle X \rangle. \tag{7.30}$$

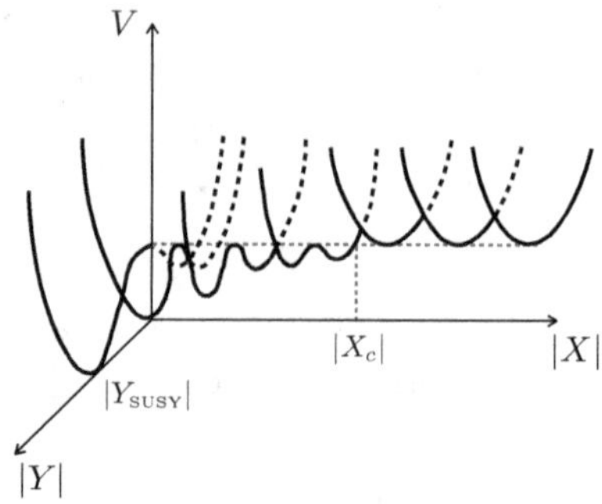

Fig. 7.6 The potential of the model of Example 3: supersymmetry restoration.

From the above expressions, we see that the supersymmetry breaking pseudo-moduli space has a tachyonic mode which develops (and destabilizes the vacuum) for

$$|X|^2 < |\lambda/h| \equiv |X_c|^2. \tag{7.31}$$

In such region the potential decreases along the $\langle Y \rangle$ direction towards the supersymmetry vacua. A qualitative picture of the potential is reported in Figure 7.6.

Example 4: Runaway behavior.

A minimal modification of the above theory gives a completely different dynamics. Let us suppose that the cubic term of the superpotential (7.26) has the square shifted from Y to X. In this case we would have for the superpotential the following ewxpression

$$W = \lambda X + \frac{1}{2} h X^2 Y. \tag{7.32}$$

The F–equations are

$$\overline{F}_X = \lambda + hXY, \quad \overline{F}_Y = \frac{1}{2} h X^2, \tag{7.33}$$

and the scalar potential

$$V = |\frac{1}{2} h X^2|^2 + |hXY + \lambda|^2. \tag{7.34}$$

Differently from previous example, it is not possible to satisfy both F-equations and so there are no supersymmetric ground states now. Notice, in passing, that the R-symmetry is preserved by the superpotential as in the previous example, with charge assignment $R(X) = 2$ and $R(Y) = -2$ this time.

Now the question is: where is the minimum of this supersymmetry breaking potential? An analysis of V shows that the global minimum is reached for $Y \to \infty$. A quick way to see it is to set $X = -\frac{\lambda}{hY}$, which kills the second contribution to the potential, the F_X-equation. By plugging this back into V one gets

$$V = \left| \frac{\lambda^2}{2hY^2} \right|^2 \underset{Y \to \infty}{\longrightarrow} 0. \tag{7.35}$$

In other words, there is no stable vacuum but actually a *runaway* behavior and supersymmetry is restored at infinity in field space.

A more physical way to reach the same conclusion is as follows. From the superpotential (7.32) we see that in the region of large $|Y|$ the effective mass for X becomes larger and larger. Hence the theory can be described by a theory where X is integrated out solving its equation of motion, which for large enough mass reduces to $\partial W / \partial X = 0$. This sets $X = -\lambda/(hY)$ and the Y-dependent only superpotential becomes

$$W_{\text{eff}} = -\frac{\lambda^2}{2hY}, \tag{7.36}$$

which gives the runaway behavior described by the potential (7.35).

Notice that within all models discussed so far, the only renormalizable one which breaks supersymmetry in a stable vacuum is the original Polonyi model (in fact there is an all pseudo-moduli space). This model is however rather uninteresting per sé, since it describes a non-interacting theory. One might wonder whether there exist reasonably simple models which are renormalizable, interacting and break supersymmetry in stable vacua. The simplest such model is the re-known O'Raifeartaigh model, which we now describe.

Example 5: The O'Raifeartaigh model.

Let us consider the theory of three chiral superfields with canonical Kähler potential and a superpotential given by

$$W = \frac{1}{2} h X \Phi_1^2 + m \Phi_1 \Phi_2 - \mu^2 X. \tag{7.37}$$

The superpotential respects the R-symmetry, the R-charge assignment being $R(X) = 2, R(\Phi_1) = 0$ and $R(\Phi_2) = 2$. The F-term equations read

$$\begin{cases} \overline{F}_X = \frac{1}{2}h\phi_1^2 - \mu^2 \\ \overline{F}_1 = hX\phi_1 + m\phi_2 \\ \overline{F}_2 = m\phi_1 \end{cases} \tag{7.38}$$

Clearly the first and the third equations cannot be solved simultaneously. Hence supersymmetry is broken. Let us try to analyze the theory a bit further. There are two dimensionful scales, μ and m. Let us choose in what follows $|\mu| < |m|$ (nothing crucial of the following analysis would change choosing a different regime). In this regime one can show that the minimum of the potential is at

$$\phi_1 = \phi_2 = 0, \quad X = \text{any} \tag{7.39}$$

and the vacuum energy is $V = |\mu^2|^2$. Again, we find a pseudo-moduli space of vacua since X is not fixed by the minimal energy condition. In Figure 7.7 we depict the potential as a function of the scalar fields.

Let us compute the (classical) spectrum around the supersymmetry breaking vacua. The full chiral superfield X is massless, right in the same way as for the Polonyi model (notice that for larger and larger $|X|$ the model gets closer and closer to the Polonyi model since all other fields get heavier and heavier). The massless fermion mode ψ_X is nothing but the goldstino. The only non-vanishing F-term in the vacuum is F_X, so that the goldstino gets contribution only from ψ_X agrees with eq. (7.10). The phase α of the scalar field $X = |X|e^{i\alpha}$ is the Goldstone boson associated to R-symmetry, which is spontaneously broken in the vacuum. Note that while

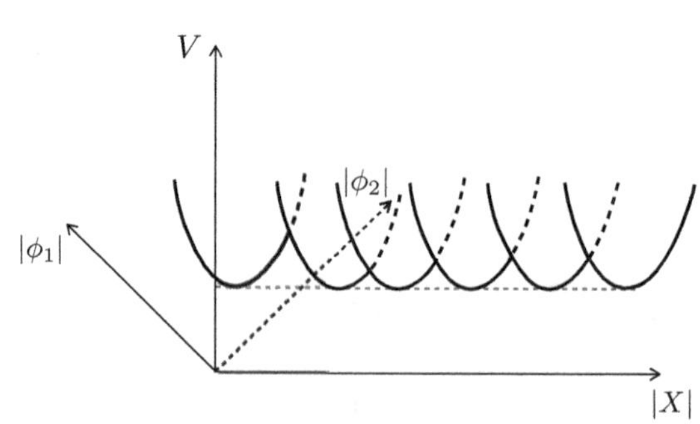

Fig. 7.7 The (classical) potential of the O'Raifeartaigh model.

Φ_2 is charged under the R-symmetry, the phase of ϕ_2 does not contribute to the R-axion, the goldstone boson of the broken R-symmetry, since the VEV of ϕ_2 is vanishing in the supersymmetry breaking vacua (7.39). Finally, $|X|$ is massless since it is a modulus (at least at classical level).

One can easily compute the ($|X|$-dependent) mass spectrum of all other fields and get

$$\begin{aligned} m_0^2(|X|) &= |m|^2 + \frac{1}{2}\eta|h\mu^2| + \frac{1}{2}|hX|^2 \pm \\ &\pm\frac{1}{2}\sqrt{|h\mu^2|^2 + 2\eta|h\mu^2||hX|^2 + 4|m|^2|hX|^2 + |hX|^4} \\ m_{\frac{1}{2}}^2(|X|) &= \frac{1}{4}\left(|hX| \pm \sqrt{|hX|^2 + 4|m|^2}\right)^2 . \end{aligned} \tag{7.40}$$

where $\eta = \pm 1$, giving different masses to the four real scalar modes belonging to Φ_1 and Φ_2. As expected, the spectrum is manifestly non-supersymmetric. Notice that $m_{1/2}^2(|X|) = m_0^2(|X|)|_{\mu^2=0}$. These infinitely many (degenerate in energy) vacua are in fact physically inequivalent, since the mass spectrum depends on $|X|$.

Example 6: A modified O'Raifeartaigh model.

Let us end this overview by considering a modification of the previous model. Let us add a (small) mass perturbation for Φ_2

$$\Delta W = \frac{1}{2}\epsilon\, m\Phi_2^2 \quad \text{with } \epsilon \ll 1. \tag{7.41}$$

Notice that this term breaks the R-symmetry enjoyed by the original O'Raifeartaigh model. The only F-equation which gets modified is the one for Φ_2 which now reads

$$\overline{F}_2 = m\phi_1 + \epsilon m\phi_2. \tag{7.42}$$

The presence of the second term removes the conflict we had before between this equation and the F-equation for X. Hence we can solve all F-term equations simultaneously and supersymmetry is not broken anymore. The (two) supersymmetric vacua are at

$$X = \frac{m}{h\epsilon}, \quad \phi_1 = \pm\sqrt{\frac{2\mu^2}{h}}, \quad \phi_2 = \mp\frac{1}{\epsilon}\sqrt{\frac{2\mu^2}{h}}. \tag{7.43}$$

For $\epsilon \ll 1$ these vacua are far away, in field space, from where the supersymmetry breaking vacua of the O'Raifeartaigh model sit (the VEV of ϕ_2 becomes larger and larger and hence very far from $\phi_2 = 0$, the value of

ϕ_2 in O'Raifeartaigh model supersymmetry breaking vacua). In fact, near the origin of field space the potential of the present model is practically identical to the one of the original O'Raifeartaigh model and one can show by direct computation that a classically marginal pseudo-moduli space of supersymmetry breaking minima is present, there.

Computing the mass spectrum near the origin one gets now

$$m_0^2(|X|) = \frac{1}{2}\Big\{|hX|^2 + |m|^2\left(2+|\epsilon|^2\right) + \eta|h\mu^2| \pm \sqrt{[|hX|^2+ +|m|^2\left(2+|\epsilon|^2\right) + \eta|h\mu^2|]^2 - 4|m|^2\left[|hX\epsilon - m|^2 + \eta|h\mu^2|\left(1+|\epsilon|^2\right)\right]}\Big\}$$

$$m_{\frac{1}{2}}^2(|X|) = m_0^2(|X|)|_{\mu^2=0}. \tag{7.44}$$

A close look to the above spectrum shows that in order for mass squared eigenvalues being all positive, the following inequality should be satisfied

$$\left|1 - \frac{\epsilon h X}{m}\right|^2 > \left(1+|\epsilon|^2\right)\left|\frac{h\mu^2}{m^2}\right|. \tag{7.45}$$

For small ϵ and μ/m, the marginally stable region described by the above inequality includes a large neighborhood around the origin, and the tachyonic mode develops only for $|X|$ (parametrically) larger than a critical value $|X_c|$

$$|X| < |X_c|. \tag{7.46}$$

Notice that this is quite the opposite of what we got in Example 4, where the marginally stable region was above a critical value; as we will see, this difference has crucial consequences at the quantum level.

Note, finally, that for $\epsilon \to 0$ one gets that $X_c, X_{\text{SUSY}} \to \infty$ and the supersymmetric vacua are pushed all the way to infinity. This is consistent with the fact that for $\epsilon = 0$ one recovers the original O'Raifeartaigh model where supersymmetric vacua are not present. A rough picture of the potential is given in Figure 7.8.

For future reference, let me notice the following interesting fact. In all models we have been considering so far, the existence of (stable) supersymmetry breaking vacua was always accompanied by the existence of an R-symmetry in the theory (think of the original Polonyi model in Example 1, the model in Example 2, the O'Raifeartaigh model of Example 5 and, to some extent, the model of Example 4). Its presence, however, does not seem to be a sufficient condition for supersymmetry breaking: think of the model of Example 3 which does possess an R-symmetry but

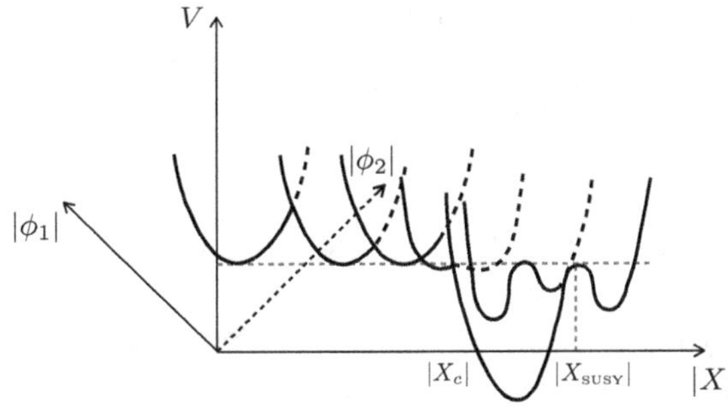

Fig. 7.8 The (classical) potential of the modified O'Raifeartaigh model.

does not break supersymmetry. On the contrary, whenever superpotential terms explicitly breaking the R-symmetry were introduced (the massive Polonyi model of Example 1 or Example 6), supersymmetric vacua were found. Finally, every time we found locally stable supersymmetry breaking vacua (again Example 6), in the vicinity of such vacua an approximate R-symmetry, which the theory did not possess as an exact symmetry, was recovered (essentially, in Examples 6 the superpotential perturbation responsible for the explicit breaking of the R-symmetry becomes negligible near the marginally stable supersymmetry breaking vacua). All this suggests some sort of relation between R-symmetry and supersymmetry breaking. We will discuss this issue later in this chapter, and put such apparent connection on a firm ground.

7.4 Pseudo-moduli space: quantum corrections

In most supersymmetry breaking models we discussed, we found a pseudo-moduli space of non supersymmetric vacua. Associated to this, we found a massless scalar mode $|X|$. While the masslessness of the goldstino and of the Goldstone boson associated to R-symmetry breaking are protected by symmetries, there are no symmetries protecting the pseudo-modulus from getting a mass. There isn't any symmetry relating the (degenerate in energy) non supersymmetric vacua. Therefore, by computing quantum corrections, one might expect this field to get a mass, somehow.

Let us stress the difference with respect to a moduli space of supersymmetric vacua. Think about the harmonic oscillator. When we quantize the bosonic harmonic oscillator, the energy of the ground state gets a $\frac{1}{2}\hbar\omega$ contribution. On the contrary, if the ground state is fermionic, the contribution is the same in modulus but with opposite sign (fermions tend to push the energy down). In a supersymmetric situation, the mass degeneracy between bosonic and fermionic degrees of freedom provides equal but opposite contribution to the vacuum energy and the total energy hence remains zero. In a non-supersymmetric vacuum the degeneracy is not there anymore (think about the spectrum we computed in the O'Raifeartaigh model) so one expects things to change. In what follows we will try to make this intuition concrete by computing the one-loop Coleman-Weinberg effective potential for both the O'Raifeartaigh and the modified O'Raifeartaigh models. In practice, what we have to do is to compute corrections in the coupling h at one loop in the background where the pseudo-modulus $|X|$ has a non-vanishing VEV.

For a supersymmetric theory the Coleman-Weinberg potential reads

$$V_{\text{eff}} = \frac{1}{64\pi^2}\,\text{STr}\,\mathcal{M}^4 \log\frac{\mathcal{M}^2}{\Lambda^2} = \frac{1}{64\pi^2}\left(\text{Tr}\,m_B^4 \log\frac{m_B^2}{\Lambda^2} - \text{Tr}\,m_F^4 \log\frac{m_F^2}{\Lambda^2}\right), \tag{7.47}$$

where $\mathcal{M} = \mathcal{M}(|X|)$ is the full tree level mass matrix, STr stands for supertrace, *i.e.* a trace where bosonic and fermionic states contribute with opposite sign, m_B and m_F correspond to boson and fermion masses respectively, and Λ is a UV cut-off.

There are a few terms missing in the expression (7.47) of the effective potential, if compared to a generic non-supersymmetric theory. Let us consider them in turn. First, we miss the cosmological constant term

$$\sim \Lambda^4. \tag{7.48}$$

This term is missing since it only depends on the spectrum, and not on the masses of the different modes. In a supersymmetric theory the spectrum admits an equal number of bosonic and fermionic degrees of freedom, no matter whether one is in a supersymmetric or non supersymmetric vacuum. Since bosons and fermions contribute opposite to this term, this degeneracy ensures this term to be vanishing. A second term which is missing is the one proportional to

$$\sim \Lambda^2\,\text{STr}\,\mathcal{M}^2. \tag{7.49}$$

This is not expected to vanish in our supersymmetry breaking vacuum since particles have different masses. In other words, the mass spectrum is not supersymmetric along the pseudo-moduli space, recall for instance eqs. (7.40) describing the non-supersymmetric mass spectrum of the

O'Raifeartaigh model. However, an explicit computation shows that also this term is vanishing. This is not specific to this model. As we will show in the next chapter, every time supersymmetry is broken spontaneously at tree level, provided the Kähler potential is canonical and in the absence of FI terms, cancellations occur so to give $\text{STr}\,\mathcal{M}^2 = 0$.

The only divergent term present in the expression (7.47) is proportional to

$$\sim \log \Lambda^2 \ \text{STr}\ \mathcal{M}^4. \tag{7.50}$$

This term does not vanish in general but it does not depend on $|X|$. As such, as we will see momentarily, it can be reabsorbed in the renormalization of the tree level vacuum energy $|\mu^2|^2$. The upshot is that the only non-trivial $|X|$-dependent term in (7.47) is the finite term

$$\sim \ \text{STr}\ \mathcal{M}^4 \log \mathcal{M}^2. \tag{7.51}$$

Let us start considering the O'Raifeartaigh model. We should simply plug the tree level masses (7.40) into formula (7.47). A lengthy but straightforward computation shows that V_{eff} is a monotonic increasing function of $|X|$ and can hence be expanded in a power series in $|X|^2$. For small $|X|$ we get

$$V_{\text{eff}}(|X|) = V_0 + m_X^2 |X|^2 + \mathcal{O}(|X|^4) \tag{7.52}$$

where

$$V_0 = |\mu^2|^2 \left[1 + \frac{|h|^2}{32\pi^2}\left(\log \frac{|m|^2}{\Lambda^2} + v(y) + \frac{3}{2}\right) + \mathcal{O}(h^4)\right], \tag{7.53}$$

with

$$y = \left|\frac{h\mu^2}{m^2}\right| < 1 \quad \text{and} \quad v(y) = -\frac{y^2}{12} + \mathcal{O}(y^4), \tag{7.54}$$

and

$$m_X^2 = \frac{1}{32\pi^2}\left|\frac{h^4\mu^4}{m^2}\right| z(y) \quad \text{where } z(y) = \frac{2}{3} + \mathcal{O}(y^2). \tag{7.55}$$

The minimum of the potential is at $|X| = 0$, and besides the tree level contribution $|\mu^2|^2$ it gets a contribution proportional $\sim |h^2|$ which is just a constant, $|X|$-independent shift. As anticipated, the UV cut-off dependence can be reabsorbed in a renormalization of the vacuum energy. Indeed, we can define a running coupling

$$\mu^2(E) \equiv \mu^2_{bare}\left[1 + \frac{|h|^2}{64\pi^2}\left(\log\frac{E^2}{\Lambda^2} + \frac{3}{2}\right) + \mathcal{O}(h^4)\right] \tag{7.56}$$

getting

$$V_0 = |\mu^2(E = m)|^2\left(1 + \frac{|h|^2}{32\pi^2}v(y) + \mathcal{O}(h^4)\right) \tag{7.57}$$

and the Λ-dependence has disappeared from the potential.

The upshot of this analysis is that loop corrections have lifted the classical pseudo-moduli space, leaving just one isolated non supersymmetric vacuum. In this vacuum the scalar field X gets a (one-loop) mass while ψ_X, which is the goldstino, remains massless (notice, in passing, that in the unique supersymmetry breaking vacuum R-symmetry is preserved, as in Example 2). The shape of the potential in the X-direction becomes at all similar to that of Figure 7.5.

Let us now see what quantum corrections say about the marginally stable supersymmetry breaking vacua of the modified O'Raifeartaigh model, the one including the superpotential perturbation (7.41). We will just briefly sketch the main results. The interested reader could try to work out all data in detail. One should again evaluate the Coleman-Weinberg potential using the tree level spectrum computed near the origin of field space, where the putative marginally stable vacua live, eqs. (7.44). Plugging the latter into formula (7.47), what one finds is that, again, the vacuum degeneracy is lifted and a (locally) stable non-supersymmetric vacuum survives at $|X| = |X_{\text{Meta}}|$ where in our regime, $\epsilon \ll 1$, X_{Meta} is near the origin and very far, in field space, from the two supersymmetric vacua sitting at $|X| = |X_{\text{SUSY}}|$. More precisely we get

$$V_{\text{eff}}(|X|) = V_0 + m_X^2|X - X_{\text{Meta}}|^2 + \mathcal{O}(\epsilon^2, |X - X_{\text{Meta}}|^4), \tag{7.58}$$

where $X_{\text{min}} \sim \frac{\epsilon m}{h}f(y) + \mathcal{O}(\epsilon^3)$. The spectrum in the supersymmetry breaking vacuum enjoys a massless fermion ψ_X, the goldstino, while the X-field gets a (ϵ-independent) mass, as in the original O'Raifeartaigh model. The effective potential, once projected into X-direction, looks roughly like that in Figure 7.9.

One might ask whether such locally stable supersymmetry breaking minimum is of any physical relevance. An estimate of its lifetime τ can

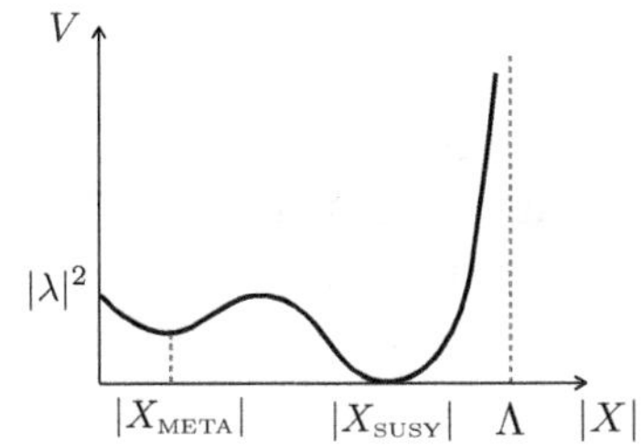

Fig. 7.9 The effective potential of the modified O'Raifeartaigh model project onto the $|X|$ direction.

be given looking at the decay rate

$$\Gamma \sim e^{-S_B} \tag{7.59}$$

(recall that $\tau \sim 1/\Gamma$) where S_B is the so-called bounce action, the difference between the Euclidean action of the tunneling configuration and that of remaining in the metastable vacuum. Its exact form depends on the details of the potential, but a simple estimate can be given in the so-called thin wall approximation, which is justified when $|X_{\rm SUSY} - X_{\rm Max}|^4 \gg V_{\rm Meta}$, and which is the case here. In this approximation the bounce action reads

$$S_B \sim \frac{\langle \Delta X \rangle^4}{\Delta V} \quad \text{where } \langle \Delta X \rangle = \langle X \rangle_{\rm SUSY} - \langle X \rangle_{\rm Meta}, \; \Delta V = V_{\rm Max} - V_{\rm Meta}. \tag{7.60}$$

An explicit computation shows that $S_B \sim \epsilon^{-\alpha}$ with $\alpha > 0$. This implies that for $\epsilon \ll 1$ the bounce action can be made arbitrary large so, in this limit, the locally stable vacuum can be made parametrically long-lived. The upshot of this analysis is that at the quantum level the classically marginal pseudo-moduli space of the modified O'Raifeartaigh model is lifted but a local, parametrically long-lived supersymmetry breaking vacuum survives. It is an instructive exercise, which is left to the reader, to repeat this quantum analysis for the classically marginal pseudo-moduli space of Example 3. In this case, the pseudo-moduli space gets completely lifted, and no locally stable supersymmetric minimum survives quantum corrections.

Let us close this section stressing that nothing we said (and computed) about quantum corrections, pseudo-moduli lifting, etc... affects the supersymmetry breaking mechanism itself. All models we have been discussing so far, if breaking supersymmetry, were doing it at tree level. We have not encountered examples where supersymmetry was unbroken at tree level and one-loop quantum corrections induced supersymmetry breaking. Everything coming from the one-loop potential can and does modify the classical supersymmetry breaking vacua (which are not protected by supersymmetry), while it leaves completely unaffected the supersymmetric ones, if any. This agrees with non-renormalization theorems and our claims about the robustness of supersymmetric moduli spaces against (perturbative) quantum corrections.

We will have more to say about F-term breaking in due time. Let us pause a bit now, and consider the other possibility we have alluded to, *i.e.* spontaneous supersymmetry breaking induced at tree level by D-terms.

7.5 D-term breaking

In a generic theory, where chiral and vector superfields are present, in absence of FI terms it is F-term dynamics which governs supersymmetry breaking. This is because it so happens that whenever one can set all F-terms to zero, using (global) gauge invariance acting on the scalar fields one can set to zero all D-terms, too. So, if one wants to consider genuine D-term breaking, one should consider FI terms, hence abelian gauge factors. In what follows, we will review the most simple such scenario, where two massive chiral superfields with opposite charge are coupled to a single $U(1)$ factor, and a FI term is present in the Lagrangian. The Lagrangian reads

$$\begin{aligned}\mathcal{L} = \frac{1}{32\pi}\,\mathrm{Im}\left(\tau\int d^2\theta\, W^\alpha W_\alpha\right) + \int d^2\theta\, d^2\bar\theta\, \left(\xi V + \overline{\Phi}_+ e^{2eV}\,\Phi_+ + \right.\\ \left. + \overline{\Phi}_- e^{-2eV}\,\Phi_-\right) + m\int d^2\theta\, \Phi_+\Phi_- + h.c.,\end{aligned} \tag{7.61}$$

where under a gauge transformation the two chiral superfields transform as $\Phi_\pm \to e^{\pm ie\Lambda}\Phi_\pm$. There is also an anomaly free R-symmetry under which the chiral superfields $\Phi_\pm$ have $R(\Phi_\pm) = 1$ (this implies that fermion fields have vanishing R-charge and so they do not contribute to the ABJ anomaly).

The equations of motion for the auxiliary fields read

$$\begin{cases}\overline{F}_\pm = m\phi_\mp \\ D = -\frac{1}{2}\left[\xi + 2e\left(|\phi_+|^2 - |\phi_-|^2\right)\right]\end{cases} \tag{7.62}$$

It is clearly impossible to satisfy all auxiliary fields equations, due to the presence of the FI parameter ξ. Hence supersymmetry is broken, as anticipated. The scalar potential reads

$$\begin{aligned} V &= \frac{1}{8}\left[\xi + 2e\left(|\phi_+|^2 - |\phi_-|^2\right)\right]^2 + m^2\left(|\phi_+|^2 + |\phi_-|^2\right) \\ &= \frac{1}{8}\xi^2 + \left(m^2 - \frac{1}{2}e\,\xi\right)|\phi_-|^2 + \left(m^2 + \frac{1}{2}e\,\xi\right)|\phi_+|^2 + \\ &\quad + \frac{1}{2}e^2\left(|\phi_+|^2 - |\phi_-|^2\right)^2 . \end{aligned} \tag{7.63}$$

The vacuum structure and the low energy dynamics depends on the sign of $m^2 - \frac{1}{2}e\,\xi$. There is a qualitative difference depending on such sign.

- $m^2 > \frac{1}{2}e\,\xi$. All terms in the potential are positive and the minimum of V is at $\langle\phi_\pm\rangle = 0$, where $V = \frac{1}{8}\xi^2$. Supersymmetry is broken but gauge symmetry is preserved. The only auxiliary field which gets a non-vanishing VEV is D, so in this case one speaks of *pure* D-term breaking. We are in a situation like the one depicted in the upper right diagram of Figure 7.1.

  One can compute the mass spectrum and find agreement with expectations. The two fermions belonging to the two chiral multiplets have (supersymmetric) mass m and hence form a massive Dirac fermion. The two scalar fields ϕ_+ and ϕ_- have masses $\sqrt{m^2 + 1/2e\xi}$ and $\sqrt{m^2 - 1/2e\xi}$, respectively. Finally, both the photon A_μ and the photino λ remain massless. The former, because gauge symmetry is preserved, the latter because supersymmetry is broken and a massless fermionic mode, the goldstino, is expected (in this case the goldstino gets contribution from the photino only, since the only non-vanishing auxiliary field VEV is that of the D-field, $\psi^G_\alpha \sim \langle D\rangle\lambda_\alpha$).
- $m^2 < \frac{1}{2}e\,\xi$. Now the sign of the mass term for ϕ_- is negative. The minimum of the potential is at $\langle\phi_+\rangle = 0$, $\langle\phi_-\rangle = \sqrt{\frac{\xi}{2e} - \frac{m^2}{e^2}} \equiv h$. Hence, both supersymmetry and gauge symmetry are spontaneaously broken. Both the D-field and F_+ get a VEV: in this case we have a so-called *mixed* D and F-term breaking. The value of the potential at its minumum is $V = \frac{1}{8}\xi^2 - \frac{1}{2}e^2h^4$. We are in a situation of the type depicted in the lower right diagram of Figure 7.1.

  In order to compute the mass spectrum one should expand the potential around $\langle\phi_+\rangle = 0$ and $\langle\phi_-\rangle = h$. A lengthy but simple computation gives the following answer. The complex scalar field ϕ_+ has mass $m_{\phi_+} = \sqrt{2}m$. The real part of ϕ_- gets a mass $m_{\phi_-^R} = \sqrt{2}eh$, while the imaginary part disappears from the spectrum (it is eaten by the photon, which becomes massive, with mass $m_{A_\mu} = \sqrt{2}eh$). The three fermions mix between themselves (there is a mixing induced from Yukawa couplings). One eigenfunction is massless, and is nothing but the goldstino $\psi^G_\alpha \sim \langle D\rangle\lambda_\alpha + \langle \overline{F}_+\rangle\psi_{+\alpha}$. The other two, $\widetilde{\psi}_\pm$, get equal mass $m_{\widetilde{\psi}_\pm} = \sqrt{2e^2h^2 + m^2} = \sqrt{e\xi - m^2}$.

Since one of the scalar fields gets a non-vanishing VEV, in this regime, one would naively expect to find an R-axion. The spectrum we derived, however, does not have such a massless scalar. The reason is that in fact the vacuum $\langle\phi_+\rangle = 0, \langle\phi_-\rangle = h$ does preserve an R-symmetry, a linear combination of the (broken) UV R-symmetry and of the global part of the (broken) $U(1)$ gauge symmetry under which the chiral superfields have charges $R(\Phi_+) = 2, R(\Phi_-) = 0$. So, consistently, there should not be an R-axion in the spectrum.

Figure 7.10 gives a summary of the mass spectrum of the FI model as a function of $\frac{1}{2}e\xi$ which, at fixed m, is the order parameter of the supersymmetry breaking transition.

One of the most attractive features of supersymmetric fields theories is the stability of masses under quantum corrections. In models where the FI mechanism plays a role, the physical mass spectrum depends on ξ which is not protected, a priori, since it appears in a D-term. This is different from models where F-terms are responsible for the supersymmetry breaking dynamics, since these are superpotential terms and protected by non-renormalization theorems. Therefore, it is important to investigate the

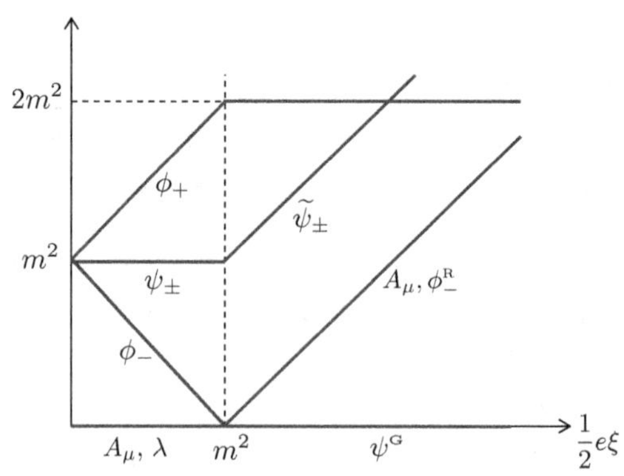

Fig. 7.10 The mass spectrum of the Fayet-Iliopoulos model as a function of the FI parameter ξ.

circumstances under which the FI term does not get renormalized. It can be shown that the contribution renormalizing the FI term is proportional to the trace of the $U(1)$ generator taken over all chiral superfields present in the model. This trace is proportional to the gravitational anomaly. Therefore, we can conclude that the FI term does not renormalize for theories free of gravitational anomalies.

7.6 Indirect criteria for supersymmetry breaking

We have already alluded to some possible relation between supersymmetry breaking and R-symmetry. In what follows, we will try to make this intuition precise and, more generally, present a few general criteria one can use to understand whether a theory might or might not break supersymmetry, without having a precise knowledge of the details of the theory itself. These criteria might be useful as guiding principles when trying to construct models of supersymmetry breaking in a bottom-up approach and, at the same time, they allow to have a handle on theories which are more involved than the simple ones we analyzed in previous sections. Finally, having some general criteria, possibly being valid also beyond the realm of perturbative physics might also be useful when one has to deal with theories in strongly coupled phases, where a perturbative, semi-classical approach is not possible, and where the direct study of the zero's of the potential is not easy or even not possible.

7.6.1 *Supersymmetry breaking and global symmetries*

Let us consider a supersymmetric theory which has a spontaneously broken global symmetry and which does not admit (non compact) classical flat directions. This theory, generically, breaks supersymmetry. This can be easily proven as follows. Since there is a broken global symmetry, the theory admits a goldstone boson (a massless particle with no potential). If supersymmetry were unbroken then one should expect a scalar companion of this goldstone boson, which, being in the same multiplet of the latter, would not admit a potential either. But then, the theory would admit a flat direction, contrary to one of the hypotheses. This is a *sufficient* condition for supersymmetry breaking.

In the above reasoning we have assumed that the second massless scalar corresponds to a non-compact flat direction. This is typically the case since the Goldstone boson is the phase of the order parameter, and its scalar companion corresponds to a dilation of the order parameter, and therefore represents indeed a non-compact flat direction.

Consider now a theory of F chiral superfields Φ^i with superpotential W. Supersymmetry is unbroken if

$$\overline{F}_i = \frac{\partial W}{\partial \phi^i} = 0, \quad \forall i = 1, 2, \ldots, F. \tag{7.64}$$

These are F holomorphic conditions on F complex variables. Therefore, if the superpotenital W is *generic*, one expects (typically distinct) solutions to exist. Hence, supersymmetry is unbroken. By the superpotential being generic we mean the following. The superpotential is, in general, a function of the Φ^i's of degree, say, n. It is generic if all possible polynomials of degree n or lower compatible with the symmetries of the theory are present.

Suppose now that W preserves some global non-R symmetry. Hence, W is a function of singlet combinations of the Φ^i's. It is easy to see that in terms of these reduced number of variables, eqs. (7.64) impose an equal number of independent conditions. Suppose, for definiteness, that the global symmetry is a $U(1)$ symmetry and call q_i the corresponding charge of the i-th chiral superfield Φ^i. Hence, we can rewrite the superpotential as e.g.

$$W = W(X_i) \quad \text{where } X_i = \Phi_i \Phi_1^{-q_i/q_1}, \quad i = 2, 3, \ldots, F. \tag{7.65}$$

If we now consider eqs. (7.64) we have

$$\begin{cases} j \neq 1 & \frac{\partial W}{\partial \phi^j} = \phi_1^{-q_j/q_1} \frac{\partial W(X_i)}{\partial X_j} = 0 \\ j = 1 & \frac{\partial W}{\partial \phi^1} = \frac{\partial W(X_i)}{\partial X_k} \frac{\partial X_k}{\partial \phi_1} = 0, \quad k = 2, \ldots, F. \end{cases} \tag{7.66}$$

We see that the equation for $\overline{F}_1$ is automatically satisfied if the others $F-1$ are satisfied. Hence, having a system of $F-1$ holomorphic equations in $F-1$ variables, generically the system allows for solutions. The same reasoning holds for a generic global symmetry. A global symmetry (under which the superpotential is a singlet) diminishes the number of independent variables, but it diminishes also the number of independent F-equations by the same amount. Hence, again, if W is generic, eqs. (7.64) can be solved and supersymmetry is unbroken.

Suppose now that the global symmetry under consideration is a R-symmetry. The crucial difference here is that the superpotential is charged under this symmetry, $R(W) = 2$. Let us call r_i the R-charge of the i-th superfield Φ^i. We can now rewrite the superpotential as

$$W = \Phi_1^{2/r_1} f(X_i) \quad \text{where } X_i = \Phi_i \Phi_1^{-r_i/r_1}, \quad i = 2, 3, \ldots, F. \tag{7.67}$$

If we now compute eqs. (7.64) we get

$$\begin{cases} j \neq 1 & \frac{\partial W}{\partial \phi^j} = \phi_1^{\frac{2-r_j}{r_1}} \frac{\partial f(X_i)}{\partial X_j} = 0 \\ j = 1 & \frac{\partial W}{\partial \phi^1} = \frac{2}{r_1} \phi_1^{\frac{2}{r_1}-1} f(X_i) + \phi_1^{\frac{2}{r_1}} \frac{\partial f(X_i)}{\partial X_k} \frac{\partial X_k}{\partial \phi_1} = 0. \end{cases} \tag{7.68}$$

Once the first $F-1$ equations are satisfied, the remaining one reduces to $f(X_i) = 0$, which is not at all trivial. So now we have F independent equations in $F-1$ variables and hence, generically, solutions do not exist. So we conclude that supersymmetry is broken, generically. The upshot is that the existence of an R-symmetry is a *necessary* condition for supersymmetry breaking, if the potential is generic (and, if it is then spontaneously broken, it is a sufficient condition if there are no classical flat directions).

This is known as the Nelson-Seiberg criterium. The O'Raifeartaigh model meets this criterium. It possesses an R-symmetry (which is then spontaneously broken along the pseudo-moduli space), the superpotential is generic, and it breaks supersymmetry. The modified O'Raifeartaigh model instead admits supersymmetry preserving vacua. Indeed, the R-symmetry is absent since the mass perturbation ΔW breaks it explicitly. So, one would expect the model not to break supersymmetry. And in fact it doesn't. We have noticed, though, that somewhere else in the space of scalar field VEVs this model admits non-supersymmetric vacua which, if the mass perturbation is small enough, we have proven to be long-lived. In this region the R-breaking perturbation is negligible and an approximate R-symmetry (the O'Raifeartaigh model's original one) is recovered. This property is in fact not specific to the modified O'Raifeartaigh model, but is a generic feature of supersymmetry breaking metastable vacua.

Summarizing, a rough guideline in the quest for supersymmetry breaking theories can be as follows:

No R-symmetry	$\longrightarrow$	SUSY unbroken
R-symmetry	$\longrightarrow$	SUSY (maybe) broken
Approximate R-symmetry	$\longrightarrow$	SUSY (maybe) broken locally, restored elsewhere

Since necessary conditions are quite powerful tools, let me stress again one important point. The existence of an R-symmetry is a necessary condition for supersymmetry breaking under the assumption that the superpotential is generic. If this is not the case, supersymmetry can be broken even if the R-symmetry is absent. Another possibility, which typically occurs when gauge degrees of freedom are present in the Lagrangian, is that R-symmetry is absent, but then it arises as an accidental symmetry in the low energy effective theory. Also in this case supersymmetry can be broken even if R-symmetry was absent in the UV Lagrangian. We will see examples of this sort later.

7.6.2 *Topological constraints: the Witten Index*

Another powerful criterium exists which helps when dealing with theories with complicated vacuum structure and for which it is then difficult to determine directly whether supersymmetry is broken, *i.e.* to find the zero's of the potential. This criterium, which provides a necessary condition for supersymmetry breaking, has to do with the so-called Witten index, which, for a supersymmetric theory, is a topological invariant quantity.

The Witten index, let us dub it I_W, is an integer number which measures the difference between the number of bosonic and fermionic states, for any given energy level. In a supersymmetric theory, for any positive energy level there is an equal number of bosonic and fermionic states. This is obvious if supersymmetry is unbroken, but it also holds if supersymmetry is broken: every state is degenerate with the state obtained from it by adding a zero-momentum goldstino (which is certainly there, if supersymmetry is broken). In other words, a state $|\Omega\rangle$ is paired with $|\Omega + \{p_\mu = 0 \text{ goldstino}\}\rangle$. On the contrary, zero-energy states can be unpaired since, due to the supersymmetry algebra, such states are annihilated by the supercharges. Therefore, in a supersymmetric theory the Witten index can get contribution from the zero energy states only, regardless the vacuum one is considering preserves supersymmetry or it does not.

Strictly speaking the above argument holds only if we put the theory in a finite volume. In an infinite volume, when supersymmetry is broken one has to deal with IR singularity issues. In particular, the (broken) supercharge diverges and acting with it on a physical state gives a non-normalizable state. This can be seen as follows. From the current algebra (4.72) one sees that

$$E\,\eta^{\mu\nu} = \langle T^{\mu\nu} \rangle = \frac{1}{4}\,\bar{\sigma}^{\mu\,\dot{\alpha}\alpha} \langle \{Q_\alpha, \overline{S}^{\nu}_{\dot{\alpha}}\} \rangle . \tag{7.69}$$

This shows that if the vacuum energy is non vanishing, the vacuum is transformed into one goldstino states if acted with Q_α (recall that the supercurrent creates a goldstino when acting on the vacuum). In an infinite volume a non-vanishing vacuum energy density corresponds to an infinite total energy, implying that the supercharge diverges and that the zero-momentum goldstino state is not defined (the corresponding state does not exist in the Hilbert space). Putting the theory in a finite volume is a way to regularize (to preserve supersymmetry and translational invariance one should impose periodic boundary conditions on all fields). Therefore, in what follows we will start considering the theory in a finite volume V and only later take the infinite volume limit. As we will see, what is relevant for the argument we want to convey is not affected by these issues and holds true also when $V \to \infty$. Our goal is to discuss whether supersymmetry is or is not broken. Working at finite volume could then be worrisome since ordinary global symmetries are typically unbroken in a finite volume, regardless their fate in the infinite volume limit (in a finite volume states can mix and this usually ensures that the ground state is invariant under all internal symmetries). Supersymmetry behaves differently, though, since its breaking just depends on whether the energy of the ground state is positive or zero, and in a finite volume the ground state energy can certainly be positive.

A theory in a finite volume has a discrete energy spectrum, all states in the Hilbert space are discrete and normalizable and can be counted unambiguously. Our goal is to compute the Witten index in such finite volume theory. For this, we just need to count the number of zero energy states (more precisely, the difference between bosonic and fermionic zero energy states), and we can then restrict to the zero three-momentum subspace of the Hilbert space. In a supersymmetric theory the energy of any state is semi-positive definite hence, from the relativistic equation $m^2 = E^2 - |\vec{p}|^2$, it follows that the energy is larger or equal than the three-momentum, so zero energy states have $\vec{p} = 0$. Setting $|\vec{p}| = 0$ we are excluding from the counting massless states with $E > 0$. These states, though, do not contribute to the Witten index. Massive states, on the other hand, never contribute to it since they necessarily have $E > 0$, regardless the value of $|\vec{p}|$. So only massless states with $|\vec{p}| = 0$ contribute to the Witten index. The upshot is that restricting the Hilbert space to the subspace $|\vec{p}| = 0$ does not hart and so this is what we will do in what follows. Nicely, in such subspace the supersymmetry algebra simplifies. Going to a basis of appropriately rescaled hermitian linear combination of the supersymmetry generators, the supersymmetry algebra (2.61) reduces to

$$\{\mathcal{Q}_i, \mathcal{Q}_j\} = \delta_{ij} P_0 \ , i, j = 1, \ldots, 4, \tag{7.70}$$

where

$$\mathcal{Q}_1 = \frac{1}{2}\left(Q_1 + \overline{Q}_{\dot{1}}\right), \ \mathcal{Q}_2 = \frac{i}{2}\left(Q_1 - \overline{Q}_{\dot{1}}\right)$$
$$\mathcal{Q}_3 = \frac{1}{2}\left(Q_2 + \overline{Q}_{\dot{2}}\right), \ \mathcal{Q}_4 = \frac{i}{2}\left(Q_2 - \overline{Q}_{\dot{2}}\right). \tag{7.71}$$

This implies that $\mathcal{Q}_1^2 = \mathcal{Q}_2^2 = \mathcal{Q}_3^2 = \mathcal{Q}_4^2 = H$, where $H = P_0$ is the Hamiltonian of the system.

Suppose to have a bosonic state $|b\rangle$ for which $\mathcal{Q}^2|b\rangle = E|b\rangle$, where $\mathcal{Q}$ is one of the $\mathcal{Q}_i$'s. Then the fermionic state obtained from $|b\rangle$ as

$$|f\rangle \equiv \frac{1}{\sqrt{E}} \mathcal{Q}|b\rangle, \tag{7.72}$$

has also energy E. This does not apply to zero-energy states since they are annihilated by $\mathcal{Q}$, and hence are not paired. So, as anticipated, the Witten index can receive contributions only from zero energy states. In Figure 7.11 we report the general form of the spectrum of a supersymmetric theory.

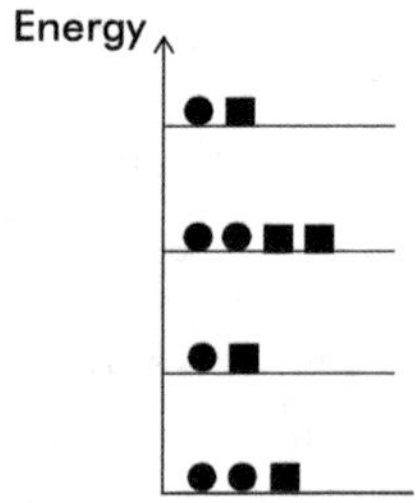

Fig. 7.11 The spectrum of a supersymmetric theory in a finite volume. Circles indicate bosons, squares indicate fermions. The zero energy level is the only one where there can exist a different number of circles and squares.

In order to appreciate its topological nature, let us define the index a bit more rigorously. A supersymmetric theory is a unitary representation of the Poincaré superalgebra on some Hilbert space $\mathcal{H}$. Let us assume that

$$\mathcal{H} = \bigoplus_{E\geq 0} \mathcal{H}_E. \tag{7.73}$$

The Witten index is defined as

$$I_W(\beta) = \mathrm{STr}_{\mathcal{H}} e^{\beta H} \equiv \mathrm{Tr}_{\mathcal{H}}(-1)^F e^{\beta H}, \quad \beta \in \mathbb{R}^+. \tag{7.74}$$

It follows that

$$\begin{aligned} I_W(\beta) &= \sum_{E\geq 0} e^{\beta E}\, \mathrm{Tr}_{\mathcal{H}_E}(-1)^F = \sum_{E\geq 0} e^{\beta E}\left[n_B(E) - n_F(E)\right] = \\ &= n_B(0) - n_F(0) = \mathrm{Tr}_{\mathcal{H}_0}(-1)^F = I_W(0). \end{aligned} \tag{7.75}$$

We have been rewriting what we have already shown to hold. The point is that this way it is clear that the index does not depend on β: its value does not vary if we vary β. More generally, one can prove that the Witten index does not depend on any parameter, like in particular coupling constants, and can then be computed in appropriate corners of the parameter space (say at weak coupling) and the result one gets is exact. In other words, the Witten index is a topological invariant.

Suppose one starts from a situation like the one depicted in Figure 7.11. Varying the parameters of the theory, like masses, couplings, etc..., it may very well be that some states move around in energy. The point is that they must do it in pairs, in a supersymmetric theory. Hence, it can happen that a pair of non-zero energy states moves down to zero energy; or, viceversa, that some zero energy states may acquire non-zero energy. But again, this can only happen if an equal number of bosonic and fermionic zero energy states moves towards a non-zero energy level. The upshot is that the Witten index does not change. This is summarized pictorially in Figure 7.12.

What is this useful for? The crucial point is that the Witten index measures the difference between zero-energy states only. Suppose it is different from zero, $I_W \neq 0$. This means that there exists some zero-energy state, hence supersymmetry is unbroken. But, because of the topological nature of I_W, this conclusion holds at any order in perturbation theory and even non-perturbatively! A theory with non-vanishing Witten

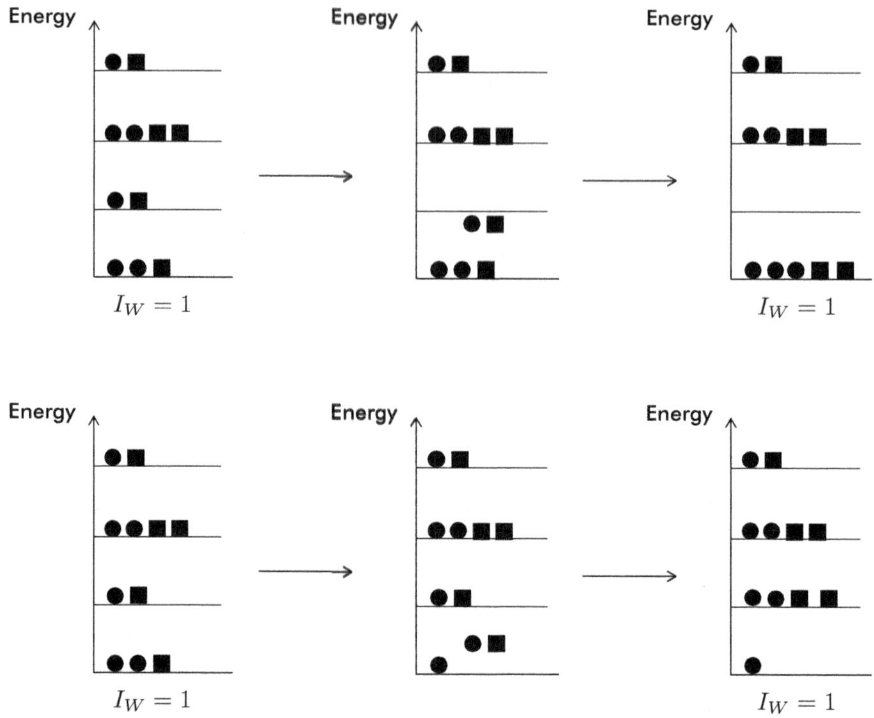

Fig. 7.12 Supersymmetric theory dynamics. Upon modifications of parameters of the theory, the number of zero energy states can change, but the Witten index remains the same.

index cannot break supersymmetry. Suppose instead that $I_W = 0$. Now one cannot conclude anything, just that the number of bosonic and fermionic zero-energy state is the same; but one cannot tell whether this number is zero (broken supersymmetry) or different from zero (unbroken supersymmetry).

So we conclude that having a non-vanishing Witten index is a sufficient condition for the existence of supersymmetric vacua, and that having it vanishing is a *necessary* condition for supersymmetry breaking. And a robust one, since I_W is an exact quantity.

Few comments are in order at this point.

First, the fact that we have been working in a finite volume does not question our main conclusions. If supersymmetry is unbroken in an arbitrary finite volume it means the ground state energy $E(V)$ is zero for any V. Since the large-V limit of zero is still zero, supersymmetry is unbroken also in the infinite volume limit. If one can explicitly compute the Witten index at finite volume and find that it is not vanishing, one can safely conclude that supersymmetry is not broken even in the actual theory, *i.e.* at infinite volume. On the contrary, the converse is not necessarily true. It might be that supersymmetry is broken at finite volume and restored in the infinite volume limit. Suppose that $I_W = 0$ and that one knows that supersymmetry is broken, that is the minimal energy states have positive energy. The energy density goes as $E(V)/V$ and it may very well be that for $V \to \infty$ the increase of E is not enough to compensate for the larger and larger volume. So the energy density can very well become zero in the infinite volume limit and supersymmetry restored. But this does not hurt much, since all what the vanishing of the Witten index provides is a necessary condition for supersymmetry breaking, not a sufficient one.

A second comment regards the relation between classical and quantum results. Suppose one can explicitly check at tree level that a given theory has non-vanishing Witten index, $I_W \neq 0$. For what we said above, this implies that supersymmetry is unbroken classically and that it cannot be broken whatsoever, neither perturbatively nor non-perturbatively. On the contrary, if $I_W = 0$ at tree level and we know that supersymmetry is unbroken classically, it can very well be that (non-perturbative) quantum effects may break it.

The theorem we have discussed may find very useful applications. For one thing, it turns out that pure SYM theories have non-vanishing Witten index (for SYM with gauge group G, the index equals the dual Coxeter number of G, which for, e.g. SU and Sp groups is just $r + 1$, where r is the rank of G). So pure SYM theories cannot break supersymmetry. As a corollary, SYM theories with massive matter (like massive SQCD) cannot break supersymmetry either. This is because for low enough energy all massive fields can be integrated out and the theory flows to pure SYM, which has non-vanishing index. More generally, non-chiral theories, for which a mass term can be given to all matter fields, are not expected to break supersymmetry.

What about chiral theories instead? Chiral theories behave differently. In this case some chiral superfield cannot get a mass anyway, and so these theories cannot be obtained from deformation of vector-like theories, as massive SQCD. Hence, one cannot conclude that these theories cannot break supersymmetry. As we will see when discussing models of dynamical supersymmetry breaking, most known examples of theories breaking supersymmetry are, in fact, chiral theories.

There is a subtlety in all what we said so far, which is sort of hidden in some of our claims. We said that the Witten index is robust against any continuous change of parameters. But it turns out that a perturbation that changes the asymptotic behavior of the potential may induce a change in I_W. This is related to the topological nature of the index, which makes it depending on boundary effects. Consider the following simple potential for a scalar field ϕ,

$$V(\phi) = \left(m\phi - g\phi^2\right)^2 . \tag{7.76}$$

For $g = 0$ low energy states correspond to $\phi \sim 0$. For $g \neq 0$ low energy states may correspond to $\phi = 0$ but also to $\phi \sim m/g$ (no matter how small g is). So we see here that $I_W(g = 0) \neq I_W(g \neq 0)$. What is going on? The point is that switching on and off g changes the asymptotic behavior of V for large ϕ (that is, at the boundary of field space): in the large ϕ region, for $g = 0$ $V \sim \phi^2$ while for $g \neq 0$ $V \sim \phi^4$. The punchline is that the Witten index is invariant under any change in the parameters of a theory in which, in the large field regime, the potential changes by terms no bigger than the terms already present. If this is not the case, the Witten index can change discontinuously. In other words, I_W is independent of numerical values of parameters as long as these are non-zero. When sending a set of parameters to zero, or switching on some couplings which were absent, one should check that the asymptotic behavior of the potential is unchanged, in order to avoid new states coming in from (or going out to) infinity. Coming back to our SQCD example, we see that for massive SQCD the potential in the large field regime is quadratic, while for massless SQCD is flat: the two theories do not have a priori the same Witten index. Therefore, while massive SQCD is expected to be in the same equivalence class of pure SYM (as far as supersymmetry breaking is concerned), this is not guaranteed for massless SQCD. In other words, no conclusions can be drawn for the massless regime by the analysis in the massive regime. We will see explicit examples of this phenomenon in later chapters.

7.6.3 *Genericity and metastability*

Before concluding this section, there is yet another important conclusion we can draw from all what we have learned. Both the Nelson-Seiberg criterium and Witten index argument seem to favor, at least statistically, supersymmetry breaking into metastable vacua.

For one thing, thinking about R-symmetry one might have the impression to fall into a vicious circle. Having an R-symmetry (which is a necessary condition for supersymmetry breaking, if the superpotential is generic) forbids a mass term for the gaugino which, being a fermion in a real representation and having R-charge $R(\lambda) = 1$, would have a R-symmetry breaking mass term. But we do not see any massless gauginos around, so gauginos should be massive. If we have an R-symmetry which is spontaneously broken, we could generate a mass for gauginos but we should also have an R-axion, which is not observed. This suggests that R-symmetry should be broken explicitly. But then, generically, we cannot break supersymmetry! The conclusion is that asking for stable vacua compatible with phenomenological observations implies one should look for non-generic theories, which are obviously much less than generic ones. If one accepts, instead, that supersymmetry might be broken in metastable vacua, then R-symmetry would not be an exact symmetry but only an approximate one. In this case gaugino mass and supersymmetry breaking would be compatible, at least in the metastable vacuum, and also the R-axion would be massive. It is worth noticing that in concrete models there is, not unexpectedly, some tension between the magnitude of gaugino mass and the lifetime of such supersymmetry breaking vacua. A long lifetime would imply massive but light gauginos and R-axion and, on the positive side, a parametrically long-lived lightest supersymmetric particle, *i.e.* a valuable dark matter candidate.

Regardless the R-axion problem mentioned above (which in fact can also get a mass by gravitational effects), there exists another argument favoring metastability. This is related with the computations we performed in section 7.4 when we studied one-loop corrections of the O'Raifeartaigh model. We have seen that quantum corrections lift the classical pseudomoduli space, leaving one unique supersymmetry breaking vacuum. However, such vacuum is (the only) one where R-symmetry is in fact not broken, so gauginos cannot get a mass! This is not a specific feature of the O'Raifeartaigh model but applies to any model where the R-charges of superfields are either 0 or 2. It has been proven that a necessary condition for having the true vacuum to break the R-symmetry is to have fields in the Lagrangian having R-charge different from 0 and 2. Models of this kind exist and have been constructed. However, in all such models supersymmetry preserving vacua also exist. Hence, supersymmetry breaking vacua where also R-symmetry is in the end broken, are actually metastable.

Also Witten index argument favors metastability, statistically. If accept we leave in a metastable vacuum, it means we allow for the existence of supersymmetry preserving vacua elsewhere in field space. Hence, all theories with non vanishing Witten index would not be anymore excluded from the landscape of possible supersymmetry breaking and phenomenologically sensible theories. For example, non-chiral theories would be back in business. A notable such example will be presented in Chapter 11.

The punchline is that, generically, it may be more likely we leave in a metastable vacuum rather than in a fully stable one. Just... we need to ensure that its lifetime is long enough to be safe!

7.7 Exercises

Exercise 7.1. Consider a theory of n chiral superfields Φ^i with superpotential (7.15). Prove that, for an interacting theory (that is, some g_{ijk} should be non-vanishing), in order to have spontaneous supersymmetry breaking one needs at least three chiral superfields. Derive the generic form of the corresponding three-superfield superpotential.

Exercise 7.2. Compute the one-loop effective potential on the classically marginal non-supersymmetric vacua of the model in Example 3. What is the fate of these vacua after quantum corrections are taken into account?

Exercise 7.3. Compute the mass spectrum of the FI model both in the pure D-term as well as in the mixed D and F-terms breaking phases, and check explicitly that the spectrum satisfies the so-called supertrace mass formula, that is $\mathrm{STr}\,\mathcal{M}^2 = 0$.

Exercise 7.4. Consider a theory of three chiral superfields with canonical Kähler potential and superpotential

$$W = \frac{1}{2}h_1 X \Phi_1^2 + \frac{1}{2}h_2 \Phi_2 \Phi_1^2 + fX$$

Show the existence of a classical moduli space of supersymmetry breaking vacua. Compute the one-loop corrections to the tree-level result and show that the moduli space is not lifted at one-loop. Can you find a simple reason to explain such a behavior?

Exercise 7.5. Consider all models of F-term breaking of section 7.3 and discuss whether and how the Nelson-Seiberg criterium applies.

References

[1] A. Bilal, *Introduction to supersymmetry*, Section 6, arXiv:hep-th/0101055.

[2] M. F. Sohnius, Introducing Supersymmetry, Sections 6 and 11, *Phys. Rept.* **128** (1985) 39.

[3] K. A. Intriligator, N. Seiberg, D. Shih, Dynamical SUSY breaking in metastable vacua, Appendix A, *JHEP* **0604** (2006) 021 [hep-th/0602239].

[4] K. A. Intriligator, N. Seiberg, Lectures on Supersymmetry Breaking, Section 2, *Class. Quant. Grav.* **24** (2007) S741-S772 [hep-ph/0702069].

[5] K. A. Intriligator, N. Seiberg and D. Shih, Supersymmetry Breaking, R-Symmetry Breaking and Metastable Vacua, Section 3, *JHEP* **0707** (2007) 017 [arXiv:hep-th/0703281].

[6] L. O'Raifeartaigh, Spontaneous Symmetry Breaking for Chiral Scalar Superfields, *Nucl. Phys. B* **96** (1975), 331.

[7] S. R. Coleman, The Fate Of The False Vacuum. 1. Semiclassical Theory, *Phys. Rev. D* **15** (1977) 2929 [Erratum-ibid. D **16** (1977) 1248].

[8] M. J. Duncan and L. G. Jensen, Exact tunneling solutions in scalar field theory, *Phys. Lett.* **B291** (1992) 109.

[9] P. Fayet and J. Iliopoulos, Spontaneously Broken Supergauge Symmetries and Goldstone Spinors, *Phys. Lett. B* **51** (1974), 461.

[10] S. Weinberg, *The quantum theory of fields. Vol. 3: Supersymmetry*, Chapters 27.6, 29.1 and 29.2, Cambridge University Press (2000).

[11] E. Poppitz and S. P. Trivedi, Dynamical supersymmetry breaking, Sections 3 and 4.4, *Ann. Rev. Nucl. Part. Sci.* **48** (1998) [arXiv:hep-th/9803107].

[12] Y. Shadmi and Y. Shirman, Dynamical supersymmetry breaking, Sections III.A and III.B, *Rev. Mod. Phys.* **72** (2000) 25 [arXiv:hep-th/9907225].

[13] A. E. Nelson and N. Seiberg, R symmetry breaking versus supersymmetry breaking, *Nucl. Phys. B* **416** (1994), 46-62 [arXiv:hep-ph/9309299 [hep-ph]].

[14] E. Witten, Constraints On Supersymmetry Breaking, *Nucl. Phys. B* **202** (1982) 253.

[15] D. Shih, Spontaneous R-symmetry breaking in O'Raifeartaigh models, *JHEP* **0802** (2008) 091 [hep-th/0703196].

Chapter 8

Supersymmetric Extensions of the Standard Model

In this chapter we will elaborate a little bit on how the machinery we have been constructing so far, can be used to describe physics beyond the Standard Model, under the hypothesis that this is described by a supersymmetric theory.

The basic idea is that the Standard Model should be viewed as an effective theory, valid only up to some scale, and that Nature, at energy above such scale, is described by some suitable $\mathcal{N} = 1$ supersymmetric extension of the Standard Model itself. The most economic option we can think of would be a $\mathcal{N} = 1$ Lagrangian which just includes known particles (gauge bosons, Higgs fields, leptons and quarks) and their superpartners. In fact, strictly speaking this does not apply to the Higgs sector, which should be doubled, not to spoil the anomaly-free properties of the Standard Model, since the Higgs fermionic partner, the higgsino, would introduce gauge anomalies. Therefore, two Higgs multiplets are needed, having opposite $U(1)_Y$ charge. Another way to realize the need of two Higgs doublets in supersymmetric extensions of the Standard Model is to notice that the Higgs field H gives mass to down quarks (and charged leptons) and $\overline{H}$ to up quarks. Due to holomorphy of the superpotential, $\overline{H}$ cannot enter W and so one needs a second, independent doublet to give up quarks a mass. These two chiral superfields are dubbed H_u and H_d and the minimal extension of the Standard Model, MSSM.

Within such minimal supersymmetric extension, one might then ask whether is it possible to break supersymmetry spontaneously and be consistent with phenomenological constraints and expectations. Addressing this question will be our main concern, here.

So far we have discussed possible supersymmetry breaking scenarios at tree level: the Lagrangian is supersymmetric but the classical potential is such that the vacuum state breaks supersymmetry (or at least there exist metastable but sufficiently long-lived supersymmetry breaking vacua, besides supersymmetric ones). Can these scenarios, *i.e.* either F or D-term supersymmetry breaking at tree level, occur in such a minimal extension of the Standard Model and give a sensible phenomenology? In what follows we will claim the answer is no: things cannot be as simple as that.

8.1 Towards dynamical supersymmetry breaking

From a purely theoretical view point there is at least one point of concern as far as tree level supersymmetry breaking. As discussed in the first chapter, there are several reasons to prefer sparticle masses around the TeV scale or so. This scale is not much different from the electro-weak scale and as such much smaller than any natural UV cut-off one can think of, like the Planck mass. If supersymmetry is broken at tree level, the mass setting the scale of supersymmetry breaking would be some mass parameter entering the bare Lagrangian. For instance, in the O'Raifeartaigh model that we discussed in previous chapter this scale is μ, the coefficient of the linear term in the superpotential. This way, we would have a scenario where an unnaturally small mass scale has been introduced in a theory in order to solve the unnatural hierarchy between the electro-weak scale and, say, the Planck scale, and avoid a fine-tuning problem for the Higgs mass. What we gain introducing supersymmetry would then just be that this small parameter, put by hand into the Lagrangian, would be protected against quantum corrections. Is this a satisfactory solution of the hierarchy problem?

It would be much more natural for this small mass parameter to be explained in some dynamical way. This is possible, in fact. In order to understand how it comes, we should first recall two facts.

Due to non-renormalization theorems, in a supersymmetric theory the superpotential is tree-level exact in perturbation theory, meaning that its full structure looks like

$$W_{\text{eff}} = W_{\text{tree}} + W_{\text{non-pert}}, \tag{8.1}$$

where the subscript np stands for non-perturbative. As we already observed, this implies that if supersymmetry is unbroken at tree level, then it cannot be broken at any order in perturbation theory, but only non-perturbatively.

The second piece of knowledge we need comes from a well-known property that many gauge theories share, *i.e.* dimensional transmutation. Due to the running of the gauge coupling, which becomes bigger and bigger towards the IR, any UV-free gauge theory possesses an intrinsic scale, Λ, which governs the strong-coupling IR dynamics of the theory, is RG-invariant and is exponentially suppressed with respect to the scale M_X at which the theory is weakly coupled. Its one-loop expression reads

$$\Lambda \sim M_X e^{-\frac{8\pi^2}{b_1 g^2(M_X)}} \ll M_X, \tag{8.2}$$

where b_1 is the one-loop coefficient of the β function.

Suppose now we have some complicated supersymmetric gauge theory which does not break supersymmetry at tree level (so all F-terms coming from $W_{\rm tree}$ are zero), but whose strong coupling dynamics generates a contribution to the superpotential $W_{\rm np}$ which does provide a non-vanishing F-term. This F-term will be order the dynamical scale Λ, since it should vanish in the classical limit $\Lambda \to 0$, and so will the scale of supersymmetry breaking. This would imply

$$M_s \sim \Lambda \ll M_X, \tag{8.3}$$

hence giving a natural hierarchy between M_s and the UV scale M_X (which can be the GUT scale or any other scale of the UV-free theory under consideration). This idea is known as Dynamical Supersymmetry Breaking (DSB) and can be regarded as the most natural way we can think of supersymmetry breaking in a fully satisfactory way.

We will discuss several DSB models in a subsequent chapter. For now, let me just anticipate that what we learned about tree-level supersymmetry breaking in previous chapter will be of great help also as far as DSB is concerned. As we will see, in DSB models the effective superpotential has typically an O'Raifeartaigh-like structure: at low enough energy gauge degrees of freedom typically disappear from the low energy spectrum (because of confinement, higgsing and alike) and the effective theory ends-up being a theory of chiral superfields, only. The analysis will then follow the one of the previous chapter, but with the great advantage that the mass parameter setting the scale of supersymmetry breaking and sparticle masses has been dynamically generated (and with the complication that in general the Kähler potential will be non-canonical, of course).

8.2 The supertrace mass formula

There is yet another reason making tree-level supersymmetry breaking not welcome in the MSSM. This is more phenomenological in nature, and related to the so-called supertrace mass formula.

Let us consider the most general $\mathcal{N} = 1$ renormalizable Lagrangian (5.81) and suppose that supersymmetry is spontaneously broken at tree level. Suppose we want to compute the trace over all bosonic and fermionic fields of the mass matrix squared in an arbitrary supersymmetry breaking vacuum. To this aim, let us suppose that, generically, all F and D auxiliary fields have some non-vanishing vacuum expectation value.

- *Vectors*
  If F and D-fields are non vanishing, it means that some scalar fields ϕ^i have acquired a non-vanishing VEV. If such fields are charged under the gauge group, a mass for some vector fields will be induced since $\overline{D_\mu\phi^i}D_\mu\phi^i \supset g^2\overline{\phi}T^aT^b\phi\, v_{a,\mu}v_b^\mu$. Hence, we have for the mass matrix squared of vector bosons

$$\left[(\mathcal{M}_1)^2\right]^{ab} = 2g^2\langle\overline{\phi}T^a\rangle\langle T^b\phi\rangle = 2\langle D_i^a\rangle\langle D^{bi}\rangle, \tag{8.4}$$

  where the lower index on the mass matrix refers to the spin, which is one for vectors and we have defined $D_i^a = \partial D^a/\partial\phi^i$, $D^{ai} = \partial D^a/\partial\overline{\phi}_i$.
- *Fermions*
  The fermion mass matrix can be easily read from the Lagrangian (5.81) to be

$$\mathcal{M}_{1/2} = \begin{pmatrix} \langle\overline{F}_{ij}\rangle & \sqrt{2}i\langle D_i^b\rangle \\ \sqrt{2}i\langle D_j^a\rangle & 0 \end{pmatrix}, \tag{8.5}$$

  where, as usual, $\overline{F}_{ij} = \partial^2 W/\partial\phi^j\partial\phi^i$. The matrix squared reads

$$\mathcal{M}_{1/2}\mathcal{M}_{1/2}^\dagger = \begin{pmatrix} \langle\overline{F}_{il}\rangle\langle F^{lj}\rangle + 2\langle D_i^b\rangle\langle D_b^j\rangle & -\sqrt{2}i\langle\overline{F}_{il}\rangle\langle D_b^l\rangle \\ \sqrt{2}i\langle D_l^a\rangle\langle F^{lj}\rangle & 2\langle D_l^a\rangle\langle D_b^l\rangle \end{pmatrix} \tag{8.6}$$

  with $F^{ij} = \partial^2\overline{W}/\partial\overline{\phi}_j\partial\overline{\phi}_i$.

- *Scalars*
  The scalar mass matrix squared is instead

$$(\mathcal{M}_0)^2 = \begin{pmatrix} \left\langle \frac{\partial^2 V}{\partial\phi^i \partial\overline{\phi}_k} \right\rangle & \left\langle \frac{\partial^2 V}{\partial\phi^i \partial\phi^l} \right\rangle \\ \left\langle \frac{\partial^2 V}{\partial\overline{\phi}_j \partial\overline{\phi}_k} \right\rangle & \left\langle \frac{\partial^2 V}{\partial\overline{\phi}_j \partial\phi^l} \right\rangle \end{pmatrix}. \tag{8.7}$$

  Recalling that $V = \overline{F}_i F^i + \frac{1}{2} D^a D_a$, one can write it as

$$(\mathcal{M}_0)^2 = \begin{pmatrix} \langle\overline{F}_{ip}\rangle\langle F^{kp}\rangle + \langle D^{ak}\rangle\langle D^a_i\rangle + \langle D^a\rangle D^{ak}_i & \langle F^p\rangle\langle\overline{F}_{ilp}\rangle + \langle D^a_i\rangle\langle D^a_l\rangle \\ \langle\overline{F}_p\rangle\langle F^{jkp}\rangle + \langle D^{aj}\rangle\langle D^{ak}\rangle & \langle\overline{F}_{lp}\rangle\langle F^{jp}\rangle + \langle D^{aj}\rangle\langle D^a_l\rangle + \langle D^a\rangle D^{aj}_l \end{pmatrix}, \tag{8.8}$$

  with $D^{ai}_j = -gT^{ai}_j$, $\overline{F}_{ijk} = \partial^3 W/\partial\phi^i\partial\phi^j\partial\phi^k$, $F^{ijk} = \partial^3\overline{W}/\partial\overline{\phi}_i\partial\overline{\phi}_j\partial\overline{\phi}_k$.

Taking the trace over gauge and flavor indexes of the three matrices (8.4), (8.6) and (8.8) we get

$$\begin{aligned} \text{Tr } (\mathcal{M}_1)^2 &= 2\langle D^a_i\rangle\langle D^{ai}\rangle \\ \text{Tr } \mathcal{M}_{1/2}\mathcal{M}^\dagger_{1/2} &= \langle\overline{F}_{il}\rangle\langle F^{li}\rangle + 4\langle D^a_i\rangle\langle D^i_a\rangle \\ \text{Tr } (\mathcal{M}_0)^2 &= 2\langle\overline{F}_{il}\rangle\langle F^{il}\rangle + 2\langle D^{ai}\rangle\langle D^a_i\rangle - 2g\langle D^a\rangle\text{Tr}\, T^a \end{aligned}$$

and finally for the supertrace

$$\text{STr}\, \mathcal{M}^2 = 3\text{Tr}\, (\mathcal{M}_1)^2 + \text{Tr}\, (\mathcal{M}_0)^2 - 2\text{Tr}\, (\mathcal{M}_{1/2})^2 = -2g\langle D^a\rangle\, \text{Tr}\, T^a. \tag{8.9}$$

This formula puts severe phenomenological constraints. First notice that, because of the trace on gauge generators, the rhs is non vanishing only in presence of $U(1)$ factors. If this is the case, then one needs non trivial FI terms to let it being non-vanishing, since we know that if $\xi = 0$ then also $\langle D^a\rangle = 0$. Now, suppose supersymmetry were broken spontaneously at tree level in the MSSM. We have only two $U(1)$ factors we can play with, the hypercharge generator $U(1)_Y$ and, eventually, $U(1)_{\text{em}}$. The latter cannot be of any use since if the corresponding FI parameter ξ were non-vanishing, some squarks or sleptons would get a VEV and hence would break $U(1)_{\text{em}}$. As for the hypercharge, this again cannot work, since the trace of $U(1)_Y$ taken over all chiral superfields vanishes in the Standard Model (this is just telling us that the Standard Model is free of gravitational anomalies). The upshot is that within the MSSM, formula (8.9) reduces to

$$\mathrm{STr}\ \mathcal{M}^2 = 0. \tag{8.10}$$

It is easy to see that this formula is hardly compatible with observations. Since supersymmetry commutes with internal quantum numbers eq. (8.10) should hold separately for each value of unbroken conserved quantity (*i.e.*, color and electric charge). This would imply that for any given Standard Model set of fields with equal charge we should observe at least a real component of a sparticle with a mass smaller than all particles with the same charge. Take for instance a charged $SU(3)$ sector. Gluons are massless, since $SU(3)$ is unbroken. From (8.4) it then follows that $\langle D_i^a \rangle = \langle D^{bi} \rangle = 0$, which, by (8.5), implies that the corresponding gluinos are also massless. Then, in such charged sector, only quarks and squarks can contribute non-trivially to eq. (8.10). Since they contribute with opposite sign, the squarks cannot all be heavier than the heaviest quark, and some must be substantially lighter. Take, for instance, the color-triplet sector with electric charge $e = -1/3$, to which down, strange and bottom quarks belong. We get $m_d^2 + m_s^2 + m_b^2 \simeq (5\mathrm{GeV})^2$. In order to satisfy eq. (8.10) we need scalar partners to satisfy $\sum_i m_{\phi_i}^2 \simeq 2(5\mathrm{GeV})^2$, which implies that a charged scalar with mass smaller than 7 GeV should exist. This is clearly excluded experimentally.

The upshot of this discussion is that we should give up with the idea that the whole story is as simple as just tree level supersymmetry breaking in the MSSM.

8.3 Beyond the MSSM

The supertrace mass formula derived above comes from a tree-level analysis and we know that masses get modified by loop effects. So, one thing that one could hope is that at quantum level things change enough as to reconcile with phenomenological constraints. However, within the MSSM such modifications are small since the Standard Model is a weakly interacting theory at the electro-weak scale, so this would not help much.

A way to avoid the supertrace mass formula severe constraints, while still keeping the MSSM, would be to allow for supersymmetry breaking beyond tree-level, that is dynamical supersymmetry breaking. We do have a dynamical scale we can play with in the Standard Model, the $SU(3)$ strong coupling scale Λ_{QCD}. This would provide violations of eq. (8.10) of order $\Lambda_{\rm QCD}$. These violations would then be of order 300 MeV, which is by far too low for accommodating any sensible phenomenology. Hence, this option could not work either.

The punchline is that in order to describe beyond the Standard Model physics we need something more than just the MSSM. We might need new particles and fields and/or new strong interactions. The options one can play with are many, and understanding the correct path of supersymmetry breaking beyond the Standard Model has been, and still is, a matter of concern and great challenge for theoretical physicists. There are, however, at least two basic properties a competitive model should have. Supersymmetry should be broken dynamically, to generate the low scale we need in a natural way. Second, in order to avoid the unpleasant constraints coming from the supertrace mass formula, we should better rely on non-renormalizable couplings, or loop effects, to transmit this breaking to the MSSM. Indeed, as we will see shortly, in both cases this would give corrections to MSSM kinetic terms which could provide, in turn, large violations of eq. (8.10), hence allowing for phenomenologically meaningful sparticle spectra. Moreover, besides invalidating formula (8.10), such an option would also have the free-bonus of providing an extra suppression between the supersymmetry breaking scale of the underlying UV theory and the scale of MSSM sparticle masses. Hence, the primordial supersymmetry breaking scale would not need to be comparable with electro-weak scale. It could be sensibly higher.

8.4 Spurions, soft terms and the messenger paradigm

Let us deviate, for a while, from what we have been saying so far, and come back to what we said at the very beginning of previous chapter about possible mechanisms for supersymmetry breaking. We have a second option we have not yet considered: explicit supersymmetry breaking by soft terms. Let us suppose we add explicit supersymmetry breaking terms to the MSSM Lagrangian. In order to save the nice UV properties of supersymmetry, these terms should be UV irrelevant. For instance, if we were to add non-supersymmetric dimensionless couplings, like Yukawa couplings and scalar quartic couplings, we would certainly destroy the pattern of UV cancellations which makes supersymmetry solving, e.g. the hierarchy problem. We can instead add mass terms, and more generally, positive dimension couplings, like cubic scalar couplings. These would simply tell us below which scale UV cancellations will stop working. Such soft supersymmetry breaking Lagrangian will schematically be of the form

$$\mathcal{L}_{\text{soft}} = m_\lambda \lambda\lambda + b\,\phi\phi + a\,\phi^3 + h.c. - m^2\overline{\phi}\phi, \tag{8.11}$$

where λ represents gauginos and ϕ any possible scalar of the MSSM. The first term provides masses for gauginos (wino, zino, photino, gluino). The second term, known as B-term, may arise in the Higgs sector only and couples the up and down scalar Higgs H_u and H_d. The third, known as A-term, corresponds to cubic gauge and flavor singlet combinations of MSSM scalars, e.g. Higgs and left and right squark components. A-terms are in one-to-one correspondence with Yukawa couplings (which belong to the supersymmetric part of the MSSM Lagrangian): each quark and lepton is just substituted by its scalar partner. Finally, the fourth term provides masses for scalar particles (squarks, sleptons and Higgs particles).

All terms appearing in eq. (8.11) are UV irrelevant and renormalizable, and it was indeed shown that the full Lagrangian

$$\mathcal{L} = \mathcal{L}_{\text{MSSM}} + \mathcal{L}_{\text{soft}} \tag{8.12}$$

is free of quadratic divergences to all orders in perturbation theory. Notice that such a Lagrangian would automatically solve the supertrace mass formula problem. A Lagrangian like the one above would violate eq. (8.10) precisely by terms of order the sparticle masses, see the expression (8.11), which is, by construction, compatible with observations.

There are a number of very important issues one should discuss regarding the Lagrangian (8.12), including problems some of the soft terms could pose, like the so-called supersymmetry flavor, CP and fine-tuning problems, to name a few. These put severe constraints on the precise form of $\mathcal{L}_{\text{soft}}$, but we will not discuss any such issues here. What we want to do instead, is to reconnect to our previous discussion and show how such a rather ad hoc Lagrangian as (8.12), where supersymmetry is broken explicitly, can actually be generated by spontaneous supersymmetry breaking in a larger theory, which includes fields and interactions beyond the MSSM ones.

First, let us recall the idea of *spurion* fields. In a supersymmetric theory any constant, non-zero value for the lowest component of a superfield (a VEV) does not break supersymmetry. Hence, in a supersymmetric Lagrangian each coupling constant can be promoted to a background superfield, a spurion, with non-vanishing such lowest component VEV. Let us take, for instance, the WZ model

$$\mathcal{L} = \int d^2\theta d^2\bar{\theta}\, Z\overline{\Phi}\Phi + \int d^2\theta \left(\frac{1}{2}M\Phi^2 + \frac{1}{6}y\Phi^3\right) + h.c., \tag{8.13}$$

and think of Z as a real background superfield while M and y as chiral background superfields. If only their lowest components have a non-vanishing VEV, this is nothing but the WZ model itself.

Interestingly, one can include supersymmetry breaking terms in the above Lagrangian by allowing these superfields having higher (scalar) component VEVs. We can set in general

$$\begin{aligned}\langle Z\rangle &= 1 + \theta^2 B + h.c. + c\,\theta^2\bar{\theta}^2\\ \langle M\rangle &= \mu - \theta^2 F_M\\ \langle y\rangle &= y - \theta^2 F_y.\end{aligned}$$

Plugging these expressions into the Lagrangian (8.13), after integrating out the auxiliary field F_Φ we find for the potential

$$\begin{aligned}V = V_{\text{SUSY}} - \left(c - |B|^2\right)\overline{\phi}\phi + \Big[&\left(F_M + B\mu\right)\phi^2 +\\ &+ \left(\frac{1}{3}F_y + \frac{1}{2}By\right)\phi^3 + h.c.\Big],\end{aligned} \tag{8.14}$$

where $V_{\text{SUSY}} = |\mu\phi + \frac{1}{2}y\phi^2|^2$. We see that the non-supersymmetric contribution to the potential exactly reproduces the second, third and fourth soft terms of the Lagrangian (8.11), upon the trivial identifications

$$\begin{aligned}m^2 &= c - |B|^2\\ b &= F_M + B\mu\\ a &= \frac{1}{3}F_y + \frac{1}{2}By.\end{aligned}$$

Following the same logic for the SYM action

$$\mathcal{L} = \int d^2\theta\, \tau\, W_a^\alpha W_\alpha^a, \tag{8.15}$$

one can seemingly reproduce gaugino masses by promoting the complexified gauge coupling τ to a chiral superfield and provide a non-vanishing VEV for its F-term

$$\langle\tau\rangle = \tau + \theta^2 m_\lambda. \tag{8.16}$$

Applying this logic to the Lagrangian (8.12), one can actually write all soft terms by means of spurion couplings, and rewrite (8.12) using a pure supersymmetric formalism. This turns out to be a very convenient thing to do when it comes to compute the divergence structure of the theory and prove, e.g. the absence of quadratic divergences.

Although phenomenologically viable and logically consistent, this picture is still not completely satisfactory. The Lagrangian (8.12) depends on more than 100 (a priori) free parameters, masses, phases, mixing angles, etc.... This means that there are few unambiguous predictions one can really make. One might want to find some organizing principle, where these many parameters could be naturally explained in terms of some simpler underlying theory.

Here is where we can close the gap between soft term breaking and spontaneous supersymmetry breaking. It is enough to promote spurions to fully fledged superfields with their own Lagrangian and kinetic terms. By some suitable and for the time being unspecified mechanism, they acquire non-vanishing F and D-terms spontaneously, and then generate soft terms by their interactions with the MSSM fields via couplings of the kind (8.13)/(8.15). This is the basic idea of the so-called *messenger paradigm*: one imagines a fully renormalizable theory where supersymmetry is broken spontaneously in some hidden sector and then communicated to the MSSM fields radiatively. After integrating out heavy fields, this will generate (non-renormalizable) effective couplings precisely as those in the Lagrangian (8.13) and (8.15), with non-vanishing F and D-components for some fields. These F and D-terms will then give rise to soft terms through a procedure like the one above. This way, all specific properties that MSSM supersymmetry breaking soft terms should have, will be ultimately

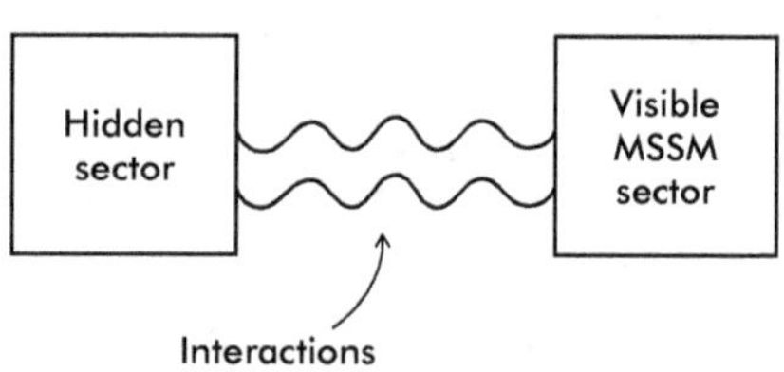

Fig. 8.1 The messenger paradigm: supersymmetry is broken in a hidden sector and then communicated to the visible MSSM sector (or any viable supersymmetric extension of the Standard Model) via interactions felt by the MSSM particles.

generated (and explained) by a larger theory in which supersymmetry breaking occurs spontaneously.

8.5 Mediating the breaking

What are the possible ways in which a scenario as the one outlined above can actually be realized?

An obvious candidate as messenger of supersymmetry breaking is gravity, since any sort of particle couples universally to it. Notice that gravity is inherently non-renormalizable, at least as it manifests itself at energies lower than the Planck scale. Therefore, couplings like those appearing in eqs. (8.13) and (8.15) are precisely what one expects, in this scenario.

Another possibility is that supersymmetry breaking is mediated by gauge interactions. We can imagine that supersymmetry is broken in the hidden sector and that some fields, known as *messenger fields*, feeling (or directly participating in, this is a model-dependent property) supersymmetry breaking are also charged under Standard Model gauge interactions. Gauginos will directly couple to such messenger fields and get a mass at one-loop. Scalar sparticles, instead, would get mass at two loops, interacting with messenger fields via intermediate MSSM vector superfields, to which gauginos belong to. In this scenario, soft terms will be generated after integrating out heavy fields, ending-up again with effective couplings of the kind (8.13) and (8.15). Obviously, the main source of mediation can be gauge interactions only in a regime where the always present gravity mediation is suppressed.

In what follows, we are not going to discuss these two mediation mechanisms in detail, nor any of their diverse phenomenological benchmarks, neither the many variants of the basic models which have appeared in the literature, with their pros and cons. What we want to do is just to give a rough idea on how these two mediation mechanisms work and show, in particular, how they can naturally generate, at low energy, spurion-like couplings with MSSM fields and, eventually, give rise to soft terms.

8.5.1 *Gravity mediation*

From a low energy point of view, one can parameterize the effect of unknown physics at the Planck scale $M_{\rm Pl}$ by higher order operators, suppressed by $M_{\rm Pl}$. Suppose that some hidden sector field X gets a non-vanishing F-term, that is

$$\langle X \rangle = 0, \quad \langle F_X \rangle \neq 0. \tag{8.17}$$

The most general form of the Lagrangian describing the gravitational interaction between X and the visible sector fields will be something like

$$\mathcal{L}_{\rm int} = \int d^2\theta d^2\bar{\theta} \left(\frac{c}{M_{\rm Pl}^2} \overline{X} X \overline{Q}_i Q^i + \frac{b'}{M_{\rm Pl}^2} \overline{X} X H_u H_d + \frac{b}{M_{\rm Pl}} \overline{X} H_u H_d + \cdots + h.c. \right) + \\ + \int d^2\theta \left(\frac{s}{M_{\rm Pl}} X W_a^\alpha W_\alpha^a + \frac{a}{M_{\rm Pl}} X Q^i H_u \widetilde{Q}_i + \dots \right) + h.c., \tag{8.18}$$

where ... refer to higher order operators. In the above expression Q_i's represent all matter superfields as well as the two Higgs doublets, while H_u and H_d obviously refer to the up and down Higgs only. For the sake of simplicity, we have taken all order one dimensionless coefficients in each term to be the same, that is i-independent.

Plugging the values (8.17) into the above Lagrangian one gets all possible MSSM soft terms! The first term on the rhs of eq. (8.18) gives rise to non-supersymmetric masses for all sfermions (squarks, sleptons and scalar Higgs particles), while the second and third terms provide mass terms for the scalar Higgs only (more below). The first term of the second line provides gaugino masses. Finally, the last term generates all A-terms. We see that we get a rather simple pattern of soft terms. Up to order one coefficients, they share one and the same mass scale

$$m_{\rm soft} \sim \frac{\langle F_X \rangle}{M_{\rm Pl}}. \tag{8.19}$$

Imposing $m_{\rm soft}$ to be order the TeV scale we see that in a gravity mediated scenario the primordial supersymmetry breaking scale, the so-called intermediate scale, is order

$$M_s = \sqrt{\langle F_X \rangle} \sim \sqrt{m_{\rm soft}\, M_{\rm Pl}} \sim 10^{11}\ {\rm GeV}, \tag{8.20}$$

somewhat in between the electro-weak scale and the Planck scale (as anticipated, sensibly higher than $m_{\rm soft}$, the scale of MSSM sparticle masses).

In the general framework outlined in section 8.4, a gravity mediation scenario can be visualized as the following steps, from a theory of quantum gravity (let us assume, for definiteness, this theory to be string theory whose field theory limit is reached by sending the string scale α' to 0, but this choice does not make a difference in the general logic) down to a (M)SSM low energy effective Lagrangian

$$\mathcal{L}_{\mathrm{QG}} \xrightarrow[\alpha'\to 0]{} \mathcal{L}^{\mathrm{non-ren}}_{\mathrm{SUSY}} \xrightarrow[\text{heavy fields}]{\text{integrate out}} \mathcal{L}^{\mathrm{non-ren}}_{\mathrm{spurion}} \xrightarrow[\text{breaking) VEVs}]{\text{spurion (SUSY}} \mathcal{L}_{\mathrm{MSSM}}+\mathcal{L}_{\mathrm{soft}}, \tag{8.21}$$

where the mass scale in $\mathcal{L}_{\mathrm{soft}}$ is given by (8.19).

Let us spend a few more words on Higgs mass terms. From the Lagrangian (8.18) we see three contributions to scalar Higgs mass. The first gives rise to mass terms for the up and down Higgs, respectively (they are proportional to $\overline{H}_u H_u$ and $\overline{H}_d H_d$). The second term is a B-term, which gives rise to a quadratic term mixing H_u and H_d. Finally, as for the third term, notice that it can be re-written as

$$\int d^2\theta d^2\bar{\theta}\, \frac{b}{M_{\mathrm{Pl}}}\overline{X} H_u H_d = b\frac{\langle \overline{F}_X\rangle}{M_{\mathrm{Pl}}}\int d^2\theta H_u H_d. \tag{8.22}$$

This contribution is a so-called μ-term contribution and upon integration in chiral superspace it gives a quadratic contribution similar in structure to the first term.

Notice that all these three couplings are needed in order to trigger electro-weak symmetry breaking (and should all be of the same order of magnitude). The first such terms gives masses to scalar Higgs particle, and it can actually give a negative mass square to some of them, something we certainly need to trigger spontaneous symmetry breaking. The second one is also necessary. One can show the B-term to be proportional to $\sin 2\beta$ where $\tan\beta$ is the ratio between the VEVs of the up and down Higgs, $\tan\beta = \langle H_u\rangle/\langle H_d\rangle$. Clearly, if $B = 0$, either the up or the down Higgs do not get a VEV, and therefore one cannot provide masses to all Standard Model particles. Finally, the μ-term is the only possible contribution which can provide higgsino a mass, and therefore should certainly be there.

The way we have rewritten the μ-term makes it clear that it can also be (and generically is) generated from a perfectly supersymmetric superpotential coupling in the MSSM Lagrangian

$$W = \mu_{\mathrm{SUSY}} H_u H_d. \tag{8.23}$$

There is no a priori reason why the above term, which comes from a supersymmetric contribution and is then not related to the dynamics driving the breaking of supersymmetry, should come to be the same scale of the soft terms, as it should. In principle, it could be any scale between m_{soft} and M_{Pl}. This is the famous μ (or $\mu/B\mu$) problem: how to avoid large μ-terms and at the same time have them the same order of magnitude of B-terms. Gravity mediation provides an elegant and simple way to solve this problem. First, one can impose some Peccei-Quinn-like discrete symmetry on the MSSM Lagrangian which forbids a tree level μ-term in the superpotential, $\mu_{\text{SUSY}} = 0$. This is achieved giving charge 1 under this symmetry to H_u and H_d and charge -1/2 to all other chiral superfields (quarks and leptons). This way $\mu_{\text{SUSY}} = 0$ and both the μ- and the B-terms are generated radiatively. The non-trivial thing is to make them be the same order of magnitude. However, as we have seen above, this is exactly what happens in a gravity mediation scenario: up to coefficients of order unity, all soft terms, including B- and μ-terms, are of the same order, eq. (8.19)!

A typical problem of gravity mediation scenarios is instead the so-called supersymmetry flavor problem. In order not to spoil the excellent agreement between flavor changing neutral current (FCNC) effects predicted by the Standard Model and known experimental bounds, any sort of new physics should not induce extra (relevant) FCNC contributions. In order for this to be the case the interactions mediating supersymmetry breaking better be flavor-blind. Gravity as seen at low energy is certainly flavor blind. However, general relativity is just an effective theory and there is no guaranty that a UV completed quantum theory of gravity is flavor blind (actually quite the opposite, given that global symmetries are believed not to be possible in quantum gravity). Therefore, in general, in gravity mediation scenarios one has to confront with the flavor problem. We will not discuss this further here and refer to the references at then end of the chapter for details. Let us just remark that there exist different proposals on how to overcome this problem, the most compelling and natural one being the so-called anomaly mediation scenario.

8.5.2 *Gauge mediation*

Any gauge mediation model is characterized by the assumption that there exist messenger fields. The latter, by definition, are those hidden sector fields which are charged under the Standard Model gauge group. The basic idea of gauge mediation is as follows.

Messengers couple (in a model-dependent way) to hidden sector supersymmetry breaking dynamics and this affects their mass matrix which, besides a supersymmetric contribution (which is supposed to be large enough not to make messengers appear at energies of order the electro-weak scale), receives a non-supersymmetric contribution. By coupling radiatively with MSSM fields, supersymmetry breaking is communicated to MSSM fields and provides all desired soft terms. For instance, gauginos get a mass at one-loop while squarks, sleptons feel supersymmetry breaking at two loops through ordinary $SU(3) \times SU(2)_L \times U(1)_Y$ gauge boson and gauginos interactions. One of the beauties of gauge mediation as opposed to gravity mediation, is that gauge mediation supersymmetry breaking can be understood entirely in terms of loop effects in a renormalizable framework. Hence, it has a high level of reliability and calculability.

There are different schemes for gauge mediation, e.g. minimal, direct and semi-direct gauge mediation, which differ, ultimately, by the way the messenger mass matrix is affected by the hidden sector supersymmetry breaking dynamics. This provides different patterns for the MSSM soft terms texture. As an exemplification, in what follows we will briefly discuss minimal gauge mediation (MGM) which is a simple, still rich enough scenario to let one get a feeling on how things work. In MGM all complicated hidden sector dynamics is parameterized in terms of a single chiral superfield X which couples to the messenger sector through a tree-level superpotential coupling. The messenger sector is made of two set of chiral superfields Φ and $\widetilde{\Phi}$ transforming in complex conjugate representation of the SM gauge group, so not to generate gauge anomalies. The interaction term is as simple as

$$W = X\widetilde{\Phi}\Phi. \tag{8.24}$$

A rough scheme of MGM is depicted in Figure 8.2.

The spurion-like field X inherits non-vanishing F and lower component term VEVs from the hidden sector,

$$\langle X \rangle = M - \theta^2 \langle F_X \rangle. \tag{8.25}$$

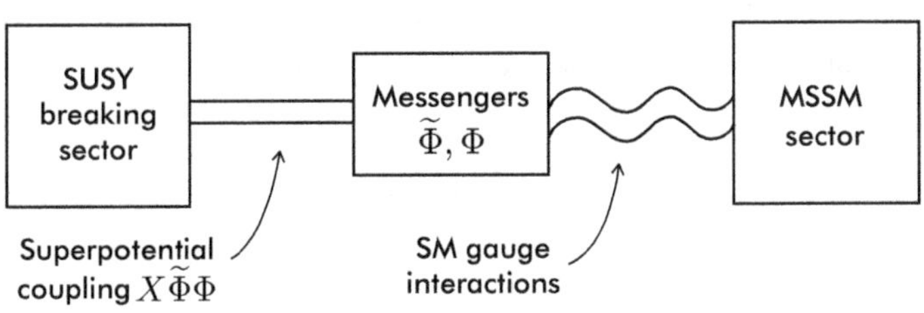

Fig. 8.2 Minimal gauge mediation. Messengers feel supersymmetry breaking via a cubic coupling with a spurion-like chiral superfield X which has a non-vanishing F-term VEV inherited from the hidden sector non-supersymmetric dynamics.

Once plugged into the messenger Lagrangian, this gives a split messenger mass spectrum

$$m^2_{\phi,\tilde{\phi}} = M^2 \pm \langle F_X \rangle, \quad m_{\psi,\tilde{\psi}} = M. \tag{8.26}$$

While fermions receive only the supersymmetric contribution, scalars receive both supersymmetric and non-supersymmetric contributions. Recalling that messenger fields are charged under the SM gauge group we see there is a stability bound which forces us to take $M^2 > \langle F_X \rangle$ (if not, some messenger scalars would get a non-vanishing VEV and would break part of the SM gauge group). If M is large enough we can then integrate the messengers out and the effective low energy theory at scale lower than M breaks supersymmetry.

The net low energy effect boils down to radiative corrections to gauginos' propagator, which get a mass at one loop, while gauge bosons remain massless since they are protected by gauge invariance. Via intermediate Standard Model gauge coupling interactions, also MSSM scalar fields get a non-supersymmetric mass contribution, though at two-loop order. Feynman diagrams contributing to gaugino and scalar masses are reported in Figures 8.3 and 8.4, respectively.

The gaugino mass computation is rather easy, since only one type of diagram contributes. In the limit of small $\langle F_X \rangle / M^2$ the end result can be organized in a series expansion and reads, to leading order in $\langle F_X \rangle / M^2$

$$m_\lambda \sim \frac{g^2}{16\pi^2} \frac{\langle F_X \rangle}{M} \left[1 + \mathcal{O}\left(\left| \frac{\langle F_X \rangle}{M^2} \right|^2 \right) \right]. \tag{8.27}$$

Summing-up all two-loop contributions renormalizing scalar masses is instead quite laborious even if conceptually straightforward. However, the end result is surprisingly simple and, again in the limit of small $\langle F_X \rangle / M^2$,

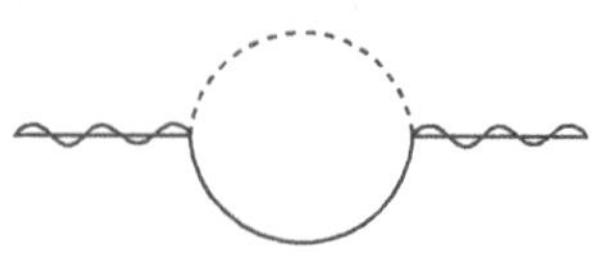

Fig. 8.3 The one-loop diagram providing gaugino mass. Grey lines are MSSM fields (just gauginos in this case), green lines are messenger fields. Dashed lines correspond to scalar fields and continuous lines to fermion fields.

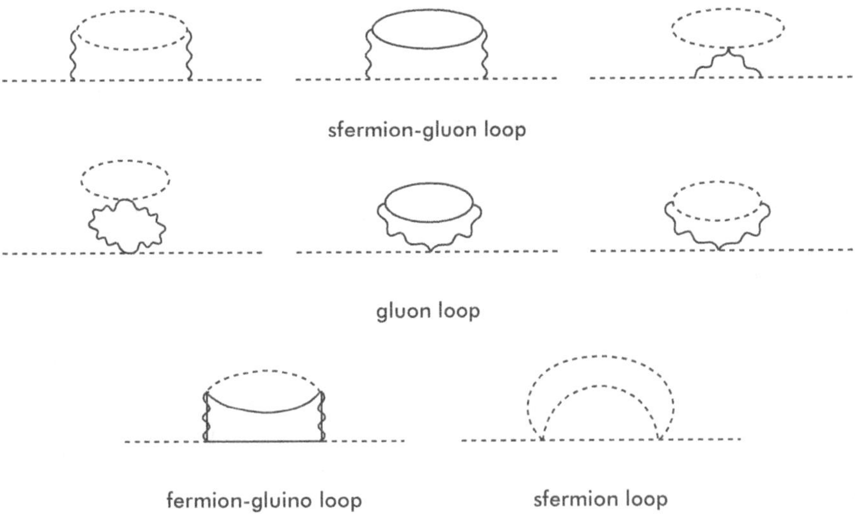

Fig. 8.4 Two-loops diagrams providing sfermion masses. There are four different class of diagrams, the first three originating from a specific MSSM scalar field one-loop diagram by inserting messenger loop corrections, as indicated. The last is a two-loop diagram which comes from D-terms and mixes MSSM and messenger scalars. Conventions are as in Figure 8.3.

reads

$$m_{\rm sf}^2 \sim \left(\frac{g^2}{16\pi^2}\right)^2 \left|\frac{\langle F_X\rangle}{M}\right|^2 \left[1+\mathcal{O}\left(\left|\frac{\langle F_X\rangle}{M^2}\right|^2\right)\right]. \tag{8.28}$$

In principle, there is also a one-loop contribution to sfermion masses originating from the quartic scalar coupling involving two sfermions and two scalar messengers, the same vertex which gives raise to the two-loop contribution in the last diagram of Figure 8.4. Due to contraction on gauge indexes, this can be non-vanishing for abelian factors only, like e.g. $U(1)_Y$. This contribution would be proportional to the hypercharge of the corresponding sfermions and therefore may induce tachyonic mass contributions which would be problematic, phenomenologically. Therefore, a symmetry in the messenger sector is usually imposed in order to avoid such dangerous one-loop contributions.

A-terms are also generated radiatively via the insertion of the renormalized gaugino propagator of Figure 8.3 inside a fermion-higgsino-gaugino loop to which a Higgs field and two sfermions can be attached as external legs, as shown in Figure 8.5. Overall, this is again a two-loop effect.

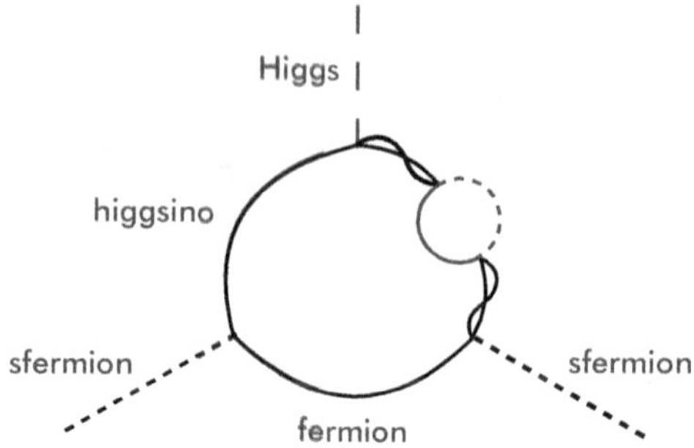

Fig. 8.5 The two-loop diagram generating A-terms in gauge mediation.

Note, finally, that B and μ-terms cannot be generated by any of the diagrams in Figure 8.4 and require a separate discussion, as we will see shortly.

In agreement with the general philosophy advocated in section 8.4, one can get these same results working within the effective low energy theory valid at scales smaller than M, which is obtained by integrating messenger fields out. At $E < M$ the effect of the messengers is taken care of in the wave function renormalization of gauge and matter kinetic terms of the MSSM fields. Soft terms arise from derivatives in the X-field of the renormalized gauge and matter kinetic functions, $Z_V(X,\mu)$ and $Z_Q(X,\overline{X},\mu)$, which can be evaluated at a scale μ by solving the RG equations. For example, soft masses read

$$m_\lambda \sim \left.\frac{\partial \ln Z_V(X,\mu)}{\partial \ln X}\right|_{X=M} \frac{\langle F_X\rangle}{M} \tag{8.29}$$

$$m_{\rm sf}^2 \sim \left.\frac{\partial^2 \ln Z_Q(X,\overline{X},\mu)}{\partial \ln X \partial \ln \overline{X}}\right|_{X=M} \left|\frac{\langle F_X\rangle}{M}\right|^2, \tag{8.30}$$

and similar formulae hold for the A-terms. This is a powerful method which is not specific to gauge mediation but works whenever supersymmetry breaking is communicated by renormalizable perturbative interactions. We refer to the bibliography at the end of the chapter for more details.

We see from eqs. (8.27)–(8.28) that in MGM all soft terms come naturally of the same order of magnitude

$$m_{\rm soft} \sim \frac{g^2}{16\pi^2}\frac{\langle F_X\rangle}{M}. \tag{8.31}$$

$$\frac{\langle F_X\rangle}{M} \sim 10^5 {\rm GeV}, \tag{8.32}$$

which implies that in MGM the primordial supersymmetry breaking scale M_s is bounded from below as

$$M_s = \sqrt{\langle F_X \rangle} \sim 10\sqrt{m_{\text{soft}} M} \geq 10^5 \text{ GeV}, \tag{8.33}$$

where in the last inequality we used that $M^2 \geq \langle F_X \rangle$ and eq. (8.32) (the lower bound for M_s is reached for $M^2 \sim \langle F_X \rangle$).

The analogous of the structure (8.21) is now

$$\mathcal{L}^{\text{ren}}_{\text{SUSY}} \xrightarrow[\text{heavy fields}]{\text{integrate out}} \mathcal{L}^{\text{non-ren}}_{\text{spurion}} \xrightarrow[\text{breaking) VEVs}]{\text{spurion (SUSY}} \mathcal{L}_{\text{MSSM}} + \mathcal{L}_{\text{soft}}, \tag{8.34}$$

where the mass scale in $\mathcal{L}_{\text{soft}}$ is given in this case by (8.31).

As we have already observed, gravity mediation is an always present contribution to supersymmetry breaking mediation mechanisms (*i.e.* the field X would also interact gravitationally with the visible sector via a Lagrangian like (8.18), in general). Hence, it is only when its contribution is suppressed with respect to that of gauge mediation that the latter can play a role. In order for gravity effects to be negligible, say to contribute no more than 1/1000 to soft mass squared, one gets an upper bound for the scale M

$$\frac{g^2}{16\pi^2}\frac{\langle F_X \rangle}{M} \geq 10^{3/2}\frac{\langle F_X \rangle}{M_{\text{Pl}}} \longrightarrow M \leq \frac{g^2}{16\pi^2} 10^{-3/2} M_{\text{Pl}} \sim 10^{15} \text{ GeV}. \tag{8.35}$$

Using the relation $M_s \sim 10\sqrt{m_{\text{soft}} M}$ this gives an upper bound for M_s of order 10^{10} GeV. Together with the lower bound (8.33) this implies that the supersymmetry breaking scale M_s can range from 10^5 to up to 10^{10} GeV, in gauge mediation scenarios.

Let us close this brief overview on gauge mediation saying a few words about flavor and μ problems. We are in a sort of reversed situation with respect to gravity mediation. Gauge interactions are intrinsically flavor-blind. Hence, gauge mediation does not provide any further FCNC contribution to the Standard Model and the flavor problem is then automatically solved in this framework. On the contrary, the μ problem is much harder. One can again avoid a supersymmetric μ-term by means of some discrete symmetry to be imposed on the Higgs sector supersymmetric Lagrangian. What is problematic, though, is to generate radiatively μ- and B-terms of the same order of magnitude. The two-loop diagrams in Figure 8.4 do not provide B- and μ-terms and one should then argue for a direct coupling between the Higgs and the messenger sectors. The simplest possible model one can think of, does not work. Allowing a cubic coupling between H_u, H_d and the field X

$$W_{\mathrm{H}} = \lambda_{\mathrm{H}} X H_u H_d, \tag{8.36}$$

one could in principle generate both a μ and a B-term from supersymmetry breaking dynamics but they do not come of the same order of magnitude. In order for the μ term being of the order of other soft masses, as it should be, we need

$$\mu = \lambda_{\mathrm{H}} M \sim 1 \text{ Tev}. \tag{8.37}$$

This implies that λ_{H} is order 10^{-2} or smaller. This enhances the B-term. Indeed, recalling that $\langle F_X \rangle \leq M^2$, the non-supersymmetric to supersymmetric mass ratio contribution coming from the superpotential coupling (8.36) is

$$\frac{B}{\mu^2} \sim \frac{\lambda_{\mathrm{H}} \langle F_X \rangle}{\lambda_{\mathrm{H}}^2 M^2} \sim \frac{\langle F_X \rangle}{\mu M} \sim 10^2, \tag{8.38}$$

where in the last step we used the fact that $\langle F_X \rangle / M \sim 10^5$GeV. This gives an unacceptably large B-term. This problem is not specific to MGM nor to the actual way we have generated μ- and B-terms here. It is a problem which generically plagues any gauge mediation scenario. Even though several proposals has been put forward to solve the μ problem in gauge mediation, it is fair to say that a fully satisfactory and natural framework to solve this problem is not yet available.

It is finally worth stressing that the simple mass pattern (8.27)–(8.28) is not a generic feature of gauge mediation but specific to MGM only. Indeed, in another popular scheme, direct gauge mediation, the soft spectrum tends to be split, that is gauginos are typically suppressed with respect to scalar particles.

A generic, model-independent prediction of gauge mediation scenarios, instead, is that the gravitino is the lightest supersymmetric particle. Gravitinos interact only gravitationally and get a mass due to higgsing of order $\langle F_X \rangle / M_{\mathrm{Pl}}$. Therefore, while in a gravity mediation scenario gravitinos have a mass of the same order of magnitude of all other soft terms, in gauge mediation they are suppressed, since

$$m_{3/2} \sim \frac{\langle F_X \rangle}{M_{\mathrm{Pl}}} = \frac{\langle F_X \rangle}{M} \frac{M}{M_{\mathrm{Pl}}} \ll \frac{\langle F_X \rangle}{M} = m_{\mathrm{soft}}. \tag{8.39}$$

Let me conclude this brief overview stressing again what is the main point of this all business. What all these mediation models are about is to provide a theory of the soft terms, a predictive pattern for these extra terms that one can (and has to) add to the MSSM Lagrangian or any desired supersymmetric extension of the Standard Model. We have been trying to give an idea on how things might work, and reviewed few aspects of the most basic mediation mechanisms. A throughout analysis of the phenomenology of these schemes and their variants is not our goal here and we refer to the bibliography at the end of the chapter for a more detailed analysis. In the remainder of this book we will instead focus on the hidden sector dynamics, trying to deepen our understanding of supersymmetric dynamics at strong coupling. Besides its intrinsic interest (and the far reaching consequences in our understanding of strong coupling regimes of gauge theories in general), this will also allow us to study concrete models of dynamical supersymmetry breaking.

8.6 Exercises

Exercise 8.1. Derive the gaugino mass formula (8.27) from the Feynman diagram of Figure 8.3.

Exercise 8.2. Compute the contribution of two diagrams arbitrarily chosen out of those depicted in Figure 8.4 to the sfermion mass formula (8.28).

References

[1] A. Bilal, *Introduction to supersymmetry*, Section 6.4, arXiv:hep-th/0101055.

[2] S. P. Martin, *A Supersymmetry Primer*, Sections 4, 5.3, 6.4, 6.6, 6.7 and 6.8, World Scientific Singapore (2010) [arXiv:hep-ph/9709356v5].

[3] M. A. Luty, *2004 TASI lectures on supersymmetry breaking*, Sections 5, 7, 8, 9, 10 and 11, arXiv:hep-th/0509029.

[4] Y. Shirman, *TASI 2008 Lectures: Introduction to Supersymmetry and Supersymmetry Breaking*, Sections 6 and 7, arXiv:0907.0039 [hep-ph].

[5] L. Girardello and M. T. Grisaru, Soft Breaking of Supersymmetry, *Nucl. Phys. B* **194** (1982), 65.

[6] G. F. Giudice and R. Rattazzi, Theories with gauge-mediated supersymmetry breaking, Sections 2, 3 and 6, *Phys. Rept.* **322** (1999) 419 [arXiv:hep-ph/9801271].

[7] J. Terning, *Modern supersymmetry: Dynamics and duality*, Chapters 6.1, 6.2 and 16, Oxford University Press (2006).

Chapter 9

Non-perturbative Effects and Holomorphy

It is time now to look at the non-perturbative regime of supersymmetric field theories and show how much more we can understand of it as compared to theories without supersymmetry. This will be the topic we will be concerned with in the remaining chapters of this book. In this first chapter our main goal will be to introduce *holomorphy*, or better put holomorphy, which is an intrinsic property of supersymmetric theories, at work. Before doing that, however, there are a few standard non-perturbative field theory results we need to review.

9.1 Instantons in a nutshell

Gauge theories might contain a so-called θ-term, which is

$$\mathcal{S}_\theta = \frac{\theta_{\rm YM}}{32\pi^2}\int d^4x\ {\rm Tr}\, F_{\mu\nu}\widetilde{F}^{\mu\nu} \quad \text{where } \widetilde{F}^{\mu\nu} = \frac{1}{2}\epsilon^{\mu\nu\rho\sigma}F_{\rho\sigma}. \tag{9.1}$$

This term is locally a total space-time derivative, since

$$\frac{1}{2}\int d^4x\ {\rm Tr}\, F_{\mu\nu}\widetilde{F}^{\mu\nu} = \int d^4x\, \epsilon^{\mu\nu\rho\sigma}\partial_\mu {\rm Tr}\left[A_\nu\partial_\rho A_\sigma + \frac{2}{3}A_\nu A_\rho A_\sigma\right]. \tag{9.2}$$

This implies that the θ-term does not have any effect on the classical equations of motion. However, when quantizing a theory one has to average over all fluctuations, not just those satisfying the classical equations of motions and therefore the θ-term can in fact be relevant in some cases.

In quantum field theory all information about physical observable (the spectrum and the S-matrix) can be obtained from correlation functions of operators, which are defined by the Feynman path integral. The most convenient formulation is where these quantities are analytically continued in Euclidean space. Feynman rules can be derived from the path integral but the latter is believed to contain more information, including effects which are non-perturbative in the coupling constant. For a generic gauge theory, the generating functional in Euclidean space reads, schematically

$$Z[J] = \int \mathcal{D}\Phi \, \exp\left(-\frac{1}{g^2} S[\Phi] + \int d^4x J \, \Phi\right), \tag{9.3}$$

where Φ represents a set of fields with source J, $S[\Phi]$ is the Euclidean action and g the dimensionless gauge coupling. The basic idea of semi-classical approximation (which corresponds to the limit of weak coupling, $g^2 \to 0$) is that the path integral is dominated by configurations of lowest Euclidean action and one should proceed expanding around these confugurations. The simplest are perturbative vacua, namely minima of the classical potential, and the expansion is just the loop expansion. However, there can exist other minima with finite action and one should expand in fluctuations around them, too. Note that for a configuration of finite action, if it exists, the leading semi-classical contribution goes as e^{-S/g^2} so it is highly suppressed at weak coupling (and fluctuations lead to corrections which are further suppressed by further powers of g^2).

A class of configurations of this kind, to which the term (9.1) is sensitive to, are *instantons*. Instantons are classical solutions of the Euclidean action that — as any configuration with finite YM action — approach pure gauge as $|x| \to \infty$. Recall that under a gauge transformation by a group element $U(x) = \exp[\lambda^a(x)T_a]$ the gauge connection $A_\mu(x) = gA_\mu^a(x)T_a$ and the field strength $F_{\mu\nu}(x) = \partial_\mu A_\nu(x) - \partial_\nu A_\mu(x) + [A_\mu(x), A_\nu(x)]$ transform as follows

$$A_\mu \to U A_\mu U^{-1} + U\partial_\mu U^{-1} \tag{9.4}$$

$$F_{\mu\nu} \to U F_{\mu\nu} U^{-1}, \tag{9.5}$$

where x-dependence is everywhere understood. Note that if $F_{\mu\nu}$ vanishes then A_μ is a gauge-transform of 0, meaning that for some U we have that $A_\mu = U\partial_\mu U^{-1}$. In order to make the YM action finite $F_{\mu\nu}$ should go to 0 at infinity faster than $1/r^2$. For this to happen A_μ does not need to go as $\mathcal{O}(1/r^2)$ but is enough it is a pure gauge, that is

$$A_\mu \to U\partial_\mu U^{-1} + \mathcal{O}(1/r^2). \tag{9.6}$$

For these configurations the integral (9.2) turns out to be an integer, the so-called instanton number of the configuration, that is

$$\mathcal{S}_\theta = \frac{\theta_{\rm YM}}{32\pi^2} \int d^4x \, {\rm Tr}\, F_{\mu\nu} \widetilde{F}^{\mu\nu} = n\, \theta_{\rm YM} \quad \text{where } n \in \mathbb{Z}. \tag{9.7}$$

One way to see that this is an integer is to notice that since (9.2) is a total derivative, it can be expressed as an integral on S^3 at infinity, meaning that the configuration involves a map $U(x) : S^3_\infty \to G$. This is nothing but the third homotopy group of G, which for simple groups is $\pi_3(G) = \mathbb{Z}$, as advocated above. The integer n in (9.7) tells us the number of times that $U(x)$ winds around the three-sphere at infinity and, for this reason, the instanton number is also known as winding number.

The instanton number is a topological quantity, in the sense that it does not change upon continuous deformations of the gauge field configuration. Moreover, since the action enters the path integral as $\int \mathcal{D}\phi e^{iS_\theta}$, the θ-angle indeed behaves as an angle, in the sense that the shift

$$\theta_{\rm YM} \to \theta_{\rm YM} + 2\pi, \tag{9.8}$$

is a symmetry of the theory.

An instanton field configuration, *i.e.* a configuration for which $n \neq 0$ in (9.7), interpolates between two vacua of the gauge theory. These vacua are gauge equivalent to the usual vacuum with zero gauge potential but the corresponding gauge transformation cannot be deformed to the identity (it a so-called *large gauge transformations*). Because if that were the case it would have been possible to let the field strength vanish in all space-time, contradicting eq. (9.7). The fact that $F_{\mu\nu}$ cannot vanish identically for configurations with $n \neq 0$ implies that there is an energy associated with such interpolating gauge field configuration, an energy barrier and an associated quantum mechanical amplitude proportional to $e^{-\mathcal{S}_E}$ where $\mathcal{S}_E$ is the Euclidean action of the field configuration. Finite action solutions in Euclidean space have the interpretation of mediating quantum tunneling effects. Instantons are nothing but just such interpolating field configurations.

It should be said that the wording we have used might be somewhat misleading. We have been speaking about different but gauge equivalent (!) vacua. Since gauge transformations describe a redundancy of the theory, there are no *different* vacua that the instantons can interpolate between. What one should more properly say is that whenever a gauge group has a non-trivial third homotopy group, there exist non-trivial gauge field configurations, the instantons, which interpolate between vacua which are gauge equivalent to the trivial one but they are so by a large gauge transformation and, because of this, carry a finite energy. A nice quantum mechanical analogue is that of a particle moving on a vertically oriented circle and subject to a constant gravitational force, *i.e.* the quantum pendulum.

Instantons have an intrinsic non-perturbative nature. Recall that the RG-equation for the gauge coupling g reads

$$\mu\frac{\partial g}{\partial \mu} = -\frac{b_1}{16\pi^2}g^3 + \mathcal{O}(g^5), \tag{9.9}$$

where b_1 is a numerical coefficient which depends on the theory. The solution of this equation at one loop is

$$\frac{1}{g^2(\mu)} = -\frac{b_1}{8\pi^2}\log\frac{\Lambda}{\mu}. \tag{9.10}$$

where the scale Λ is defined as the scale where the one-loop coupling diverges. It sets the scale where higher-loop and non-perturbative effects should be taken into account. For any scale μ_0 we have that

$$\Lambda \equiv \mu_0 e^{-\frac{8\pi^2}{b_1 g^2(\mu_0)}}. \tag{9.11}$$

It is important to stress that Λ does not depend on the energy scale: it is a RG-invariant quantity. Indeed

$$\begin{aligned}\frac{\partial\Lambda}{\partial\mu_0} &= e^{-\frac{8\pi^2}{b_1 g^2(\mu_0)}} + \mu_0\left[-\frac{8\pi^2}{b_1 g^3(\mu_0)}\frac{2}{\mu_0}\left(\frac{b_1}{16\pi^2}g^3(\mu_0) + \mathcal{O}(g^5)\right)\right]e^{-\frac{8\pi^2}{b_1 g^2(\mu_0)}}\\ &= e^{-\frac{8\pi^2}{b_1 g^2(\mu_0)}} + \mu_0\left(-\frac{1}{\mu_0}\right)e^{-\frac{8\pi^2}{b_1 g^2(\mu_0)}} = 0.\end{aligned} \tag{9.12}$$

up to higher-order corrections. This can be reiterated order by order in perturbation theory, getting the same result, namely that $\partial\Lambda/\partial\mu_0 = 0$.

An important point regarding instantons is that there exists a lower bound on their Euclidean action. Indeed, we have that

$$0 \leq \int d^4x \,\mathrm{Tr}\left(F_{\mu\nu} \pm \widetilde{F}_{\mu\nu}\right)^2 = \int d^4x \left[2\,\mathrm{Tr}\,F_{\mu\nu}F^{\mu\nu} \pm 2\,\mathrm{Tr}F_{\mu\nu}\widetilde{F}^{\mu\nu}\right] \tag{9.13}$$

which implies

$$\int d^4x \,\mathrm{Tr}\,F_{\mu\nu}F^{\mu\nu} \geq \left|\int d^4x \,\mathrm{Tr}\,F_{\mu\nu}\widetilde{F}^{\mu\nu}\right| = 32\pi^2 n, \tag{9.14}$$

where the last equality holds for an instanton configuration with instanton number n. This implies that there is a lower bound to an instanton action:

instanton contributions to amplitudes are suppressed at least by (just multiply above equation by $1/4g^2$)

$$e^{-\mathcal{S}_{\text{inst}}} = \left(e^{-\frac{8\pi^2}{g^2(\mu)}}\right)^n = \left(\frac{\Lambda}{\mu}\right)^{n\,b_1} \tag{9.15}$$

where in the last step we have used eq. (9.10). This shows that instantons are inherently non-perturbative effects, since they vanish for $\Lambda \to 0$, and are very weak, if not negligible, in the perturbative regime.

9.2 Anomalies in a nutshell

Anomalies are classical symmetries of the action which are broken by quantum effects. In other words, we have

$$\partial_\mu j^\mu = 0 \xrightarrow[\text{corrections}]{\text{quantum}} \partial_\mu j^\mu \neq 0, \tag{9.16}$$

where j^μ is the current associated to the anomalous symmetry.

In what follows we will focus on chiral anomalies, that is anomalies associated to chiral currents. These arise in field theories in which fermions with chiral symmetries (symmetries under which opposite chirality fermions transform in complex conjugate representations) are coupled to gauge fields. Let us consider, e.g. a Weyl fermion ψ coupled to a gauge field A_μ with action

$$\mathcal{S} = \int d^4x\,\overline{\psi}\,\overline{\sigma}^\mu(\partial_\mu + iA_\mu)\psi. \tag{9.17}$$

This action is invariant under a $U(1)$ global symmetry which rotates $\psi \to e^{i\alpha}\psi$ and $\overline{\psi} \to e^{-i\alpha}\overline{\psi}$. The corresponding conserved current can be computed by Nöether method. Considering a x-dependent transformation, that is $\alpha = \alpha(x)$, we get for the action

$$\mathcal{S} \longrightarrow \mathcal{S} - \int d^4x\,\partial_\mu\alpha(x)\cdot(\overline{\psi}\overline{\sigma}_\mu\psi) = \mathcal{S} + \int d^4x\,\alpha(x)\cdot\partial_\mu(\overline{\psi}\overline{\sigma}_\mu\psi), \tag{9.18}$$

where in the second step we have integrated by parts. Since the functional variation of the action with respect to $\alpha(x)$ should vanish, it follows from the last expression above that $\partial^\mu(\overline{\psi}\overline{\sigma}_\mu\psi) = 0$.

The point is that the current $j_\mu = \overline{\psi}\overline{\sigma}_\mu\psi$ is conserved classically but it is not quantum mechanically. This can be seen by computing loop diagrams involving three external currents. Let us consider a set of free Weyl fermions transforming in some representation R of some global symmetry group G

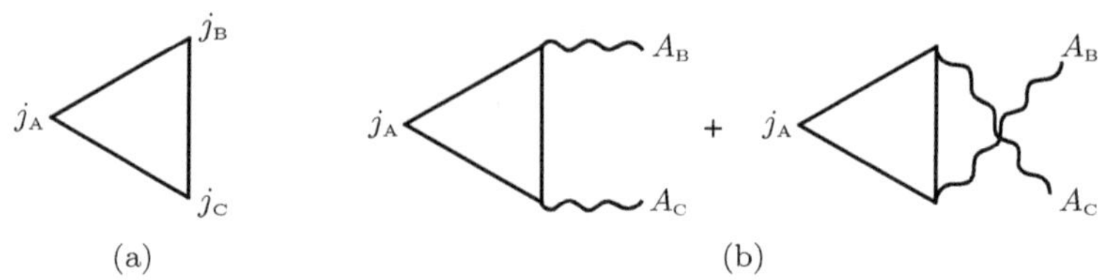

Fig. 9.1 One-loop diagrams contributing to correlators of one global current with two global or or local currents. Diagram (a) does not provide any anomaly. Diagrams (b), instead, contribute to the anomaly of the global current j_A.

and call j_A^μ the associated currents. Let us compute the three-point function of the currents j_A at one loop, diagram a of Figure 9.1. This will give something like

$$\langle j_A^\mu(x_1) j_B^\nu(x_2) j_C^\rho(x_3)\rangle = \operatorname{Tr}(t_A t_B t_C)\, f_{\mu\nu\rho}(x_i), \tag{9.19}$$

where the trace comes from contraction of the group generators around the loop. As we are going to see in the following, this correlator has important properties but it does not provide by itself any anomaly: the corresponding classical conservation law is not violated quantum mechanically.

Suppose now to gauge some (or all) global currents, by coupling the original Lagrangian to gauge fields as

$$\mathcal{L} = \mathcal{L}_{\text{free}} + \sum_B A_\mu^B\, j_B^\mu, \tag{9.20}$$

and let us compute the correlator $\langle j_A A_B A_C\rangle$. The one-loop diagrams contributing to such correlator are now diagrams b of Figure 9.1. By differentiating the result one gets

$$\partial_\mu\, j_A^\mu \sim \operatorname{Tr}(t_A \{t_B, t_C\})\, F^{B\mu\nu}\widetilde{F}^C_{\mu\nu}, \tag{9.21}$$

which says that the current is not anymore conserved (notice, on the contrary, that as anticipated the rhs would vanish if none of the symmetries were gauged). Remarkably, it turns out that this result is not corrected at higher loops, so the anomaly can be entirely evaluated by a one-loop computation!

This same result can be derived following the approach pioneered by Fujikawa, in which the anomaly arises from the non-invariance of the measure of the fermionic path integral under chiral rotations. This is also the approach which makes more manifest the geometrical and topological nature of anomalies (and the connection with the Atiyah-Singer index theorem).

An important fact about anomalies is that local currents cannot be anomalous, since they would imply violation of unitarity of the theory (we know how to couple spin-1 fields in a way respecting unitarity to conserved currents, only). Hence a quantum field theory, in order to make sense, should not have any gauge anomaly. On the contrary, global chiral currents can be anomalous. So the currents j^μ_A in eq. (9.21) should be global currents, since if this were not the case we would have had a violation of unitarity in the quantum theory.

One other important thing one can learn looking at eq. (9.21) is that the anomaly coefficient vanishes for real and pseudoreal representations. Indeed, for real or pseudoreal representations we have that $t_A = -(t_A)^T$ and it then easily follows that

$$\mathrm{Tr}_r\left(t_A\left\{t_B, t_C\right\}\right) = -\,\mathrm{Tr}_r\left(t_A\left\{t_B, t_C\right\}\right). \tag{9.22}$$

Therefore, only massless chiral fermions can contribute to the anomaly coefficient (massive fermions always transform in the $R+\overline{R}$ representation, in order for the mass term to be invariant under the symmetry). This result, once applied to local currents, which cannot be anomalous, provides severe restrictions on the massless fermion content of a quantum field theory.

Finally, from the above computation it is also clear that only abelian symmetries can be anomalous. Suppose to have a theory with gauge group G with generators t_A, a global symmetry group $\widetilde{G}$ with generators $\tilde{t}_A$, and a set of Weyl fermions ψ^i_α, transforming in the representations $(r_i, \tilde{r}_i)$ of the gauge and global symmetry groups, respectively. In this case, the anomaly computation gives

$$\partial_\mu\, j^\mu_A \sim \sum_i \mathrm{Tr}_{\tilde{r}_i}\tilde{t}_A\, \mathrm{Tr}_{r_i}\left(t_B t_C\right) F^{B\mu\nu}\widetilde{F}^C_{\mu\nu}. \tag{9.23}$$

Since $\mathrm{Tr}_{\tilde{r}_i}\tilde{t}_A = 0$ for any simple algebra, only abelian factors $U(1) \subset \widetilde{G}$ can be anomalous. On the other hand, $\mathrm{Tr}_{r_i}\left(t_B t_C\right) = C(r_i)\delta_{BC}$, where $C(r_i)$ is the quadratic invariant (Casimir) of the representation r_i. Working everything out (and paying attention to numerical coefficients!), one finally gets for an abelian group

$$\partial_\mu\, j^\mu = -\frac{A}{16\pi^2} F^{\mu\nu}_B \widetilde{F}^B_{\mu\nu}, \tag{9.24}$$

where $A = \sum_i q_i C(r_i)$ is the anomaly coefficient, q_i being the $U(1)$ global charges of fermionic fields ψ_i.

This result makes it manifest the connection between anomalies and instantons. Integrating eq. (9.24) in space-time we get

$$\int d^4x\, \partial_\mu\, j^\mu = \int dt\, d^3x \left(\partial_0\, j^0 - \partial_i\, j^i\right) = \int dt \partial_0 Q = \Delta Q$$

$$\int d^4x \frac{A}{16\pi^2} F_B^{\mu\nu} \widetilde{F}_{\mu\nu}^B = 2An,$$

from the left and right hand sides, respectively. In the first relation we have used the definition of charge Q and the fact that $\int d^3x \partial_i j^i = 0$, while in the second relation we have used eq. (9.7). Putting everything together we finally get

$$|\Delta Q| = 2A\, n, \tag{9.25}$$

where n is the instanton number and ΔQ the amount of charge violation due to the anomaly. So we see that anomalous symmetries are violated by a specific amount, given by eq. (9.25), in an instanton background. This also shows that anomalies are IR effects, since the violation is very mild at weak coupling.

From eq. (9.18) one sees that the effect of the anomalous $U(1)$ symmetry corresponds to a shift in the θ-angle as

$$\psi_i \rightarrow e^{iq_i\alpha}\psi_i \quad \Longrightarrow \quad \theta_{\rm YM} \rightarrow \theta_{\rm YM} - 2\alpha A. \tag{9.26}$$

So the anomalous breaking can be seen as an explicit breaking: a term in the action, the θ-term in fact, is not invariant under the anomalous symmetry.

Notice that if we perform a $U(1)$ transformation but promote $\theta_{\rm YM}$ to a spurion field and assign to it transformation properties as to compensate for the shift, then the anomalous $U(1)$ is promoted to an actual symmetry of a larger theory (where the complexified gauge coupling is promoted to a dynamical field). This symmetry, however, is spontaneously broken by the coupling constant VEV ($\theta_{\rm YM}$ in this case). As we will later see, this way of looking at anomalous symmetries can be efficiently used to put constraints on the construction of low-energy effective Lagrangians.

9.3 't Hooft anomaly matching condition

The correlator (9.19) does not provide any anomaly if only global currents are involved. However, it does contain very important information, as originally pointed out by 't Hooft. This is because, as we review below, correlators as (9.19) compute scale independent information about a quantum field theory and as such provide a powerful tool to understand some of its non-perturbative properties.

Let us consider a Lagrangian $\mathcal{L}$ defined at some scale μ, with some non-anomalous global symmetry group G generated by currents j^μ_A. Compute the triangle diagram for three global currents (which is not an anomaly) and call $A_{\rm UV}$ the result. Now weakly gauge the global symmetry group G by adding new gauge fields A^A_μ and define a new Lagrangian

$$\mathcal{L}' = \mathcal{L} - \frac{1}{4g^2}{\rm Tr}F_{\mu\nu}F^{\mu\nu} + j^\mu_A A^A_\mu . \tag{9.27}$$

This theory is inconsistent since it has a gauge anomaly, $A_{\rm UV}$, because we have gauged G. We can make the theory consistent by adding some spectator free massless fermion fields ψ_s (spectator in the sense that they couple only through the G-gauge coupling) transforming in representations of G so to exactly cancel the anomaly, *i.e.* $A_s = -A_{\rm UV}$. The resulting theory

$$\mathcal{L}'' = \mathcal{L} - \frac{1}{4g^2}{\rm Tr}F_{\mu\nu}F^{\mu\nu} + \overline{\psi}_s \,\partial\!\!\!/\, \psi_s + \left(j^\mu_{s,A} + j^\mu_A\right) A^A_\mu , \tag{9.28}$$

where $j_{s,A}$ are the currents associated to the spectator fermions ψ_s, is non-anomalous, and it is so for any value of the gauge coupling. Consider this anomaly-free theory at some scale $\mu' < \mu$. Since the spectator fermion fields and gauge fields can be made arbitrarily weakly coupled by taking $g \to 0$, the IR dynamics of the enlarged theory (9.28) is just the IR dynamics of the original theory plus the arbitrarily weakly coupled spectator theory. Therefore, A_s should be the same and since the theory is anomaly free, we should have that $A_{\rm IR} + A_s = 0$, which implies

$$A_{\rm IR} = A_{\rm UV}. \tag{9.29}$$

Taking $g \to 0$ spectators fields completely decouple and (9.29) should still hold. The punchline is that in a quantum field theory anomaly coefficients associated to global currents are scale independent quantities, and their UV and IR values should match. This is known as 't Hooft anomaly matching condition.

A simple equation such as (9.29) puts severe constraints on the IR dynamics of a quantum field theory, in particular as far as its massless spectrum: it implies that a theory with global conserved currents but with 't Hooft anomaly (that is, a non-vanishing triangular anomaly associated to these global currents), does not have a mass gap.

There exist two possible scenarios of this sort. If the global symmetry is preserved, a non-vanishing 't Hooft anomaly implies the existence of massless (typically composite) fermions in the effective IR theory, so to match the anomaly. But it might happen that there does not exist any choice of quantum numbers for composite states to match this anomaly. This suggests that the global symmetry is spontaneously broken. But then the theory is again gapless since by Goldstone theorem massless scalars, the goldstone bosons, are expected to exist. In this case one can match the anomaly (9.29) by coupling the theory to background gauge fields and see that a term in the effective action can (and should) be added which reproduces the UV anomaly. This is known as gauged Wess-Zumino-Witten term. It is worth noting that it is precisely this latter argument which originally suggested that the $SU(3)_L \times SU(3)_R$ global symmetry of QCD should be spontaneously broken, the pions being the corresponding (pseudo) Goldstone bosons. In that case, rather than a background gauge field, an abelian symmetry is weakly gauged, which is nothing but the electromagnetic $U(1)$ gauge symmetry, and a term proportional to $\pi^0 F^{\mu\nu}\widetilde{F}_{\mu\nu}$ can and should be added to the pion Lagrangian which matches the anomaly.

Let us notice again that all these results can be claimed without doing any sort of non-perturbative computation; just a one-loop one!

9.4 Holomorphy

We now want to discuss a property of supersymmetric theories, known as holomorphy, which plays a crucial role when it comes to understand the quantum properties of supersymmetric theories and to what extent they differ from non-supersymmetric ones.

Let us first briefly recall the concept of Wilsonian effective action.

When dealing with effective theories we deal with effective actions. The transition from a fundamental (bare) Lagrangian down to an effective one, involves integrating out high-momentum degrees of freedom. The effective action (aka Wilsonian action) is defined from the bare action $\mathcal{S}_{\mu_0}$ defined at some UV scale μ_0, as

$$e^{i\mathcal{S}_\mu} = \int_{\phi(p)\,,\,p>\mu} \mathcal{D}\phi\, e^{i\mathcal{S}_{\mu_0}} \tag{9.30}$$

where $\mathcal{S}_\mu$ is the effective action, of which we review below few basic properties.

The Wilsonian action correctly describes a theory's degrees of freedom at energies below a given scale μ (the cut-off). It is local on length scales larger than $1/\mu$, and describes in a unitary way physical processes involving energy-momentum transfers less than μ. As far as processes are concerned:

- at energies $E \sim \mu$, the effective couplings and masses are given by the tree-level couplings in the effective action (effects of all higher energy degrees of freedom have already been integrated out),
- at energies $E \ll \mu$ there will be quantum corrections due to fluctuations of modes of the fields in $\mathcal{S}_\mu$ with energies between E and μ.

The upshot is that the Wilsonian action $\mathcal{S}_\mu$ is the action which describes the physics at the scale μ by its *classical* couplings.

Supersymmetry puts severe restrictions on the structure of the Wilsonian action, more specifically to the superpotential (*i.e.*, the F-term part). One way to see this, is as follows. Any parameter in a supersymmetric Lagrangian can be thought of as a VEV of a superfield. This implies, in particular, that each coupling (masses, Yukawa couplings, etc...) appearing in the classical superpotential can be thought of as the lowest component VEV of a (very heavy) chiral superfield (in other words, the theory one is considering can be viewed as an effective theory of a bigger theory where these fields have been integrated out and they act as spurions at low energy).

This implies that the superpotential is not only holomorphic in the fields but also in the couplings and so is the effective superpotential in the Wilsonian action. The couplings of the effective action will be functions of the couplings of the UV theory, and these should be holomorphic functions of such UV couplings.

This important result can also be proven by means of supersymmetric Ward identities. More specifically, they imply that all coupling constants appearing in the tree level superpotential must only appear holomorphically in quantum corrections to the superpotential (which is basically equivalent to what's above).

This property is important since, as we will see, promoting coupling constants to chiral superfields one can often extend symmetries of the superpotential and put severe constraints on the form (and sometime the very existence) of quantum corrections. Holomorphy makes the restrictions on possible quantum corrections allowed by supersymmetry apparent. It provides a supersymmetric version of selection rules.

In the forthcoming sections we will see how much a prominent role does holomorphy plays in constraining the dynamics of supersymmetric filed theories. Before moving on, however, let us discuss a toy model to make it more concrete the above idea about supersymmetric selection rules. Suppose we have a given supersymmetric theory where in the superpotential an operator $\mathcal{O}$ appears, namely $W_{\text{tree}} \supset \lambda\,\mathcal{O}$, where λ is some coupling. If we promote λ to a spurion we gain an extra $U(1)$ symmetry in the enlarged theory which includes the spurion superfield, the $U(1)$ charge of λ being $Q(\lambda) = 1$ and that of $\mathcal{O}$ being $Q(\mathcal{O}) = -1$. This symmetry is however spontaneously broken by the superpotential term in which the spurion enters with its VEV.

Suppose we are interested in the appearance of a given operator of charge -10 among quantum corrections, $\mathcal{O}_{-10}$. In general, we expect it to appear at tenth or higher order as

$$\Delta W \sim \lambda^{10}\,\mathcal{O}_{-10} + \lambda^{11}\overline{\lambda}\,\mathcal{O}_{-10} + \cdots + \lambda^{10} e^{-1/|\lambda|^2}\,\mathcal{O}_{-10}, \tag{9.31}$$

where we have assumed that the classical limit, $\lambda \to 0$ is well defined and so we do not allow any negative powers of λ to appear. Holomorphy implies that *only* the first term can be generated. All other terms cannot be there since are non-holomorphic in the coupling (both λ and $\overline{\lambda}$ appear). A corollary of the above discussion is that any operator with positive $U(1)$ charge is also disallowed. Indeed, we cannot have negative powers of λ because we are supposing the theory is well defined in the classical limit, while any power of $\overline{\lambda}$ is forbidden by holomorphy. Notice that the latter property is due to supersymmetry, and it is not shared by an ordinary field theory.

What this toy model shows is that holomorphy in the coupling constants, usual selection rules for symmetries under which coupling constants may transform and the requirement of smoothness of physics in various weak-coupling limits, can provide severe constraints on the structure of the effective superpotential of a supersymmetric quantum field theory. In the remainder of this lecture we will discuss several such examples.

Let us close this section by recalling that the Wilsonian effective action is not what we usually call the effective action Γ. The latter is obtained by integrating out *all* degrees of freedom down to $\mu = 0$ and it is the generating functional of 1PI graphs and calculates the Green functions of the original UV theory. It is not holomorphic in the coupling constants and suffers from holomorphic anomalies. It is not the correct thing to look at in asymptotically free gauge theories since it is not well defined. The two effective actions are the same only if there are no interacting massless particles, which are those making the 1PI effective action Γ suffer from IR divergences.

9.5 Holomorphy and non-renormalization theorems

Remarkably, using holomorphy one can prove many known non-renormalization theorems (and go beyond them, as we will see).

Example 1: the Wess-Zumino (WZ) model. The tree level superpotential of the WZ model has the following structure

$$W_{\text{tree}} = \frac{1}{2} m\Phi^2 + \frac{1}{3}\lambda\Phi^3. \tag{9.32}$$

The question one might ask is: what is the form of the effective superpotential W_{tree}, once quantum corrections (both perturbative and non-perturbative) are taken into account? Let us try to answer this question using holomorphy. First, promote m and λ to spurion superfields. This makes the theory enlarging its symmetries by a flavor $U(1)$ symmetry and a R-symmetry, according to the table below

$$\begin{array}{ccc} & U(1)_R & U(1) \\ \Phi & 1 & 1 \\ m & 0 & -2 \\ \lambda & -1 & -3 \end{array} \tag{9.33}$$

The superpotential has (correctly) R-charge 2 and flavor $U(1)$ charge 0. Notice that both symmetries are spontaneously broken whenever the spurion superfields, m and λ, have a non-vanishing lower component VEV.

Because of what we discussed in the previous section, the effective (that is, exact) superpotential should be a holomorphic function of Φ, m and λ, and have R-charge equal to 2 and flavor charge equal to 0. Its most general form can be written as a function of $\lambda\Phi/m$ as follows

$$W_{\text{eff}} = m\Phi^2 f\left(\frac{\lambda\Phi}{m}\right) = \sum_{n=-\infty}^{\infty} a_n \lambda^n \, m^{1-n} \, \Phi^{n+2}, \tag{9.34}$$

where $f_{\text{free}} = \frac{1}{2} + \frac{1}{3}\lambda\Phi/m$, and a_n are arbitrary coefficients. Note that due to holomorphy of the superpotential neither $\overline{\lambda}$ nor $\overline{m}$ appear in (9.34).

The form of f can be fixed as follows. First, in the classical limit, $\lambda \to 0$, we should recover the tree level result. This implies that there cannot appear negative powers of λ; hence $n \geq 0$ and, in order to agree with (9.32) at tree-level, $a_0 = \frac{1}{2}$ and $a_1 = \frac{1}{3}$. Taking also the massless limit at the same time, $m \to 0$, restricts n further, *i.e.* $n \leq 1$. The upshot is that the effective superpotential should be nothing but the tree level one: holomorphy (plus some obvious physical requirements, more below) tells us that the superpotential of the WZ model is not renormalized at any order in perturbation theory *and* non-perturbatively!

The requirement about finiteness in the massless limit requires a few more comments. Taking the massless limit at finite λ does not lead to a weakly-coupled theory, so one could not use smoothness arguments so naively. However, taking both $m, \lambda \to 0$ such that $m/\lambda \to 0$ we do achieve the result above, since the theory is free in this case. One may still wonder whether this conclusion is correct since in this limit there is a massless particle and so the effective theory should have some IR divergences. This is not the case since we do not run the RG-flow down to $\mu = 0$: there are no IR divergences in the Wilsonian effective action, as opposed to the 1PI effective action.

Another, equivalent way to see the absence of negative powers of m in the effective superpotential is to observe that all terms with $n \geq 0$ are generated by *tree-level* diagrams only, in the UV theory (it is a matter of number of vertices and propagators), see Figure 9.2. All diagrams of the kind of the one depicted in Figure 9.2 are not 1PI for $n > 1$; they cannot be produced from loops, and they should not be included in the effective action for finite m. So the integer n in eq. (9.34) is indeed restricted to be either 0 or 1. It is easy to get convinced that in a non-supersymmetric context, where e.g. $\overline{\lambda}$ can enter the expansion (9.34), operators with negative powers of λ can come from 1PI diagrams and hence appear in the effective action.

What we have just proven, namely that the superpotential of the WZ model is not renormalized at any order in perturbation theory and non-perturbatively, is not specific to the WZ model. It actually applies to all models where only chiral superfields are present: in these cases, that is in

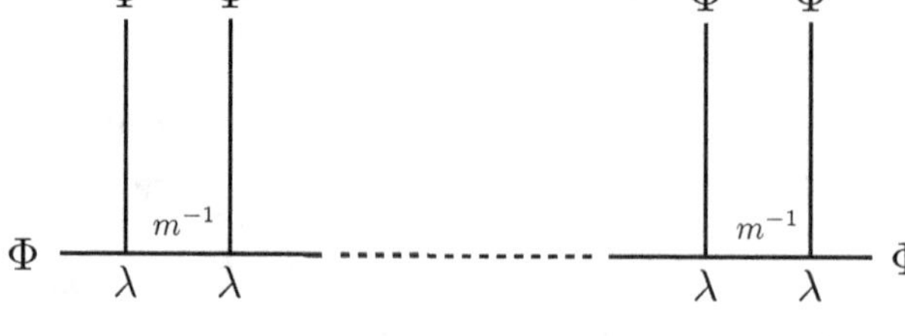

Fig. 9.2 Tree level (super)graph producing terms of the series (9.34) for $n > 1$.

the absence of gauge interactions, the tree-level superpotential is an exact quantity (regardless its specific form, which might even contain classically irrelevant operators!).

Example 2: As a second example, we want to illustrate what holomorphy can tell us about the running of gauge couplings in supersymmetric gauge theories. Let us focus, for definiteness, on SQCD. Recall that this is a supersymmetric gauge theory with gauge group $SU(N)$, F flavors described by F pairs of chiral superfields $(Q, \widetilde{Q})$ transforming in the fundamental respectively anti-fundamental representation of the gauge group and no tree-level superpotential. At the classical level, the global symmetries are as detailed below

	$SU(F)_L$	$SU(F)_R$	$U(1)_B$	$U(1)_A$	$U(1)_{R_0}$
Q^i_a	F	$\bullet$	1	1	a
$\widetilde{Q}^b_j$	$\bullet$	$\overline{F}$	-1	1	a
λ	$\bullet$	$\bullet$	0	0	1

where conventions on indexes are the same as in §5.3.1 when we discussed the (classical) moduli space of massless SQCD. The R-charges of Q and $\widetilde{Q}$ above are chosen to be the same since under charge conjugation (which commutes with supersymmetry), $Q \leftrightarrow \widetilde{Q}$. For later convenience, we have also written down the charges of the gaugino field. The axial current and the R_0 current are anomalous, the anomaly coefficients being

$$A_A = \frac{1}{2}(+1)F + \frac{1}{2}(+1)F = F$$

$$A_{R_0} = \frac{1}{2}\left[(a-1)F + (a-1)F\right] + N = N + (a-1)F$$

These two anomalous symmetries admit an anomaly-free combination (which is obviously an R-symmetry) with current

$$j^R_\mu = j^{R_0}_\mu + \frac{(1-a)F - N}{F} j^A_\mu, \qquad (9.35)$$

under which the matter fields have the following charges $R(Q^i_a) = R(\widetilde{Q}^b_j) = \frac{F-N}{F}$ (note that the anomaly-free R-charge of matter fields does not depend on a, while the gaugino has always R-charge equal to 1). Therefore, the group of continuous global symmetries at the quantum level is $G_F = SU(F)_L \times SU(F)_R \times U(1)_B \times U(1)_R$. Notice that for $F = 0$, namely for pure SYM, there do not exist an axial current and in turn the R-symmetry is inevitably anomalous. This difference will play a crucial role later on.

What we are interested in is the gauge coupling running, namely the β function

$$\beta = \mu \frac{\partial g}{\partial \mu} = -\frac{b_1}{16\pi^2} g^3 + \mathcal{O}(g^5). \tag{9.36}$$

The one-loop coefficient b_1 can be easily computed from the field content of the classical Lagrangian and reads $b_1 = 3N - F$. The question we would like to answer is whether holomorphy can tell us something about higher-loop (and non-perturbative) corrections.

Let us consider pure SYM, first, whose action is

$$\mathcal{L} = \frac{1}{16\pi i} \int d^2\theta \; \tau \, \mathrm{Tr} \, W^\alpha W_\alpha + h.c., \tag{9.37}$$

with τ the complexified gauge coupling, $\tau = \frac{\theta_{\rm YM}}{2\pi} + \frac{4\pi i}{g^2}$. Notice that τ appears holomorphically in the action above, but the gauge fields are not canonically normalized (to go to a basis where gauge fields are canonically normalized one should shift $V \to gV$, as we did in Chapter 5 when constructing matter-coupled actions).

One thing that will prove useful for what we want to do, is to trade the dynamical generated scale Λ for a complex parameter. For $G = SU(N)$ the one-loop running of the gauge coupling is

$$\frac{1}{g^2(\mu)} = -\frac{3N}{8\pi^2} \log\left(\frac{|\Lambda|}{\mu}\right), \qquad |\Lambda| = \mu_0 \, e^{-\frac{8\pi^2}{3N g^2(\mu_0)}}, \tag{9.38}$$

where $|\Lambda|$ what we previously called Λ. We can then define a holomorphic scale Λ as

$$\Lambda = |\Lambda| e^{i\frac{\theta_{\rm YM}}{3N}} = \mu \, e^{\frac{2\pi i \tau}{3N}}, \tag{9.39}$$

in terms of which the one-loop complexified gauge coupling reads

$$\tau_{\text{1-loop}} = \frac{3N}{2\pi i} \log \frac{\Lambda}{\mu}. \tag{9.40}$$

What about higher order corrections? Suppose we integrate down to a scale μ, then

$$W_{\rm eff} = \frac{\tau(\Lambda;\mu)}{16\pi i} \, \mathrm{Tr} \, W^\alpha W_\alpha. \tag{9.41}$$

Since the physics is periodic under $\theta_{\rm YM} \to \theta_{\rm YM}+2\pi$, the following rescaling

$$\Lambda \to e^{\frac{2\pi i}{3N}}\Lambda \tag{9.42}$$

is a symmetry of the theory, under which $\tau \to \tau+1$. The most general form for τ transforming this way under (9.42) is

$$\tau(\Lambda;\mu) = \frac{3N}{2\pi i}\log\frac{\Lambda}{\mu} + f(\Lambda;\mu), \tag{9.43}$$

with f a function of Λ (and not of $\overline{\Lambda}$!) having the following properties:

- Under (9.42) τ should transform as $\tau \to \tau+1$. This is already accounted for by the one-loop contribution (9.40), so the function f should be invariant under (9.42).
- In the limit $\Lambda \to 0$, which is a classical limit, f must vanish since we should get back the one-loop result.

The first requirement implies that f should be a function of Λ but not of $\log\Lambda$. By the second requirement, this implies that f should admit a positive Taylor expansion in Λ. The upshot is that the effective coupling (9.43) should have the following form

$$\tau(\Lambda;\mu) = \frac{3N}{2\pi i}\log\frac{\Lambda}{\mu} + \sum_{n=1}^{\infty} a_n \left(\frac{\Lambda}{\mu}\right)^{3Nn}. \tag{9.44}$$

Recalling that the instanton action is

$$e^{-S_{\rm inst}} = \left(\frac{\Lambda}{\mu}\right)^{3N}, \tag{9.45}$$

we conclude that the function f receives only non-perturbative corrections and these corrections come from n-instantons contributions. The upshot is that τ is one-loop exact, in perturbation theory.

The one-loop exactness of the SYM gauge coupling can be equivalently proven as follows. The θ-term is a topological term so it does not get renormalized perturbatively. Therefore the β-function, $\beta = \beta(\tau)$ can only involve $\operatorname{Im}\tau$. If β should be a holomorphic function of τ this implies that it can only be a imaginary constant (a holomorphic function $f(z)$, which is independent of $\operatorname{Re} z$, is an imaginary constant). Therefore

$$\beta(\tau) \equiv \mu\frac{d}{d\mu}\tau = ia, \tag{9.46}$$

which implies

$$\mu \frac{d}{d\mu} \theta_{\rm YM} = 0, \quad \mu \frac{d}{d\mu} g = -\frac{a}{8\pi} g^3. \tag{9.47}$$

So we see that, indeed, the gauge coupling does not receive corrections beyond one-loop, in perturbation theory (for the theory at hand $a = 3N/2\pi$).

All what we said above applies identically to SQCD (again, working in the basis where gauge fields are not canonically normalized and the complexified gauge coupling enters holomorphically in the action), the only difference being that the one-loop coefficient of the β-function is now $3N - F$, with F being the number of flavors.

Remarkably, in some specific cases one can show that also non-perturbative corrections are absent. One such instances is pure SYM, and the argument goes as follows. As already noticed, the R-symmetry of pure SYM is anomalous

$$\partial_\mu j^\mu_R = 0 \xrightarrow[\text{corrections}]{\text{quantum}} \partial_\mu j^\mu_R = -\frac{2N}{32\pi^2} F^a_{\mu\nu} \widetilde{F}^{\mu\nu}_a. \tag{9.48}$$

The $U(1)_R$, however, is not fully broken. This can be seen as follows. A R-symmetry transformation with parameter α, under which the gaugino transforms as

$$\lambda \to e^{i\alpha} \lambda \tag{9.49}$$

is equivalent to a shift of the θ-angle

$$\theta_{\rm YM} \to \theta_{\rm YM} - 2N\alpha, \tag{9.50}$$

recall eqs. (9.24) and (9.26). The point is that the transformation $\theta_{\rm YM} \to \theta_{\rm YM} + 2\pi k$ where $k \in \mathbb{Z}$, is a symmetry of the theory. So, whenever the $U(1)_R$ parameter α equals $\pi k/N$, the theory is unchanged also at the quantum level. This implies that a discrete subgroup of the original continuous abelian symmetry is preserved,

$$U(1)_R \longrightarrow \mathbb{Z}_{2N}. \tag{9.51}$$

Treating the complexified gauge coupling τ as a spurion field and assigning to it transformation properties so to compensate for the shift, we can define a spurious symmetry acting as

$$\lambda \to e^{i\alpha} \lambda, \quad \tau \to \tau + \frac{N\alpha}{\pi}. \tag{9.52}$$

This constrains the coefficients a_n in the expansion (9.44). Indeed, under the spurious symmetry the holomorphic scale $\Lambda = \mu\, e^{\frac{2\pi i \tau}{3N}}$ transforms as

$$\Lambda \to e^{i\frac{2N\alpha}{3N}}\Lambda. \tag{9.53}$$

Hence we have

$$\tau(\Lambda;\mu) \to \frac{N\alpha}{\pi} + \frac{3N}{2\pi i}\log\frac{\Lambda}{\mu} + \sum_{n=1}^{\infty} a_n \left(\frac{\Lambda}{\mu}\right)^{3Nn} e^{2iN\alpha n}. \tag{9.54}$$

Since whenever $n \neq 0$ then $e^{2iN\alpha n} \neq 0$, it follows that to match the spurious symmetry (9.52) we need to have

$$a_n = 0 \quad \forall n > 0. \tag{9.55}$$

Hence in pure SYM also non-perturbative corrections to the gauge coupling are absent! Actually, the same conclusion holds for any supersymmetric gauge theory with an anomalous global symmetry. For example, in SQCD the R-symmetry is not anomalous, and one would not get any constraint on the coefficients a_n, since $\theta_{\rm YM}$ does not shift under an R-symmetry rotation. However, in SQCD there is an axial anomaly, under which $\theta_{\rm YM}$ does shift, the anomaly coefficient being now $A = F$. Repeating the same argument as above, one then concludes that also in SQCD the (holomorphic — recall the comment after eq. (9.37)) gauge coupling, which is one-loop exact in perturbation theory, does not receive non-perturbative corrections, either.

If we collect all what we have learned so far we might have the feeling that something wrong is going on. There are three apparently incompatible results regarding the running of the SQCD gauge coupling.

- Due to holomorphy, the gauge coupling runs only at one-loop in perturbation theory, and the full perturbative β-function hence reads

$$\beta = -\frac{g^3}{16\pi^2}(3N - F), \tag{9.56}$$

- There exists a well known result in the literature which claims that the exact, all-loops β-function of SQCD is

$$\beta = -\frac{g^3}{16\pi^2}\frac{\left[3N - \sum_{i=1}^{F}(1-\gamma_i)\right]}{1 - Ng^2/8\pi^2} \tag{9.57}$$

where $\gamma_i = d\log Z_i(\mu)/d\log\mu$ are matter fields anomalous dimensions. This result gets contribution at all loops and is in clear contradiction with the previous result. Eq. (9.57) is sometime called NSVZ β-function (after Novikov, Shifman, Vainshtein and Zakharov).

- Another well known fact about the β-function of SQCD (and, in general, of any gauge theory) is that its one and two-loop coefficients are universal, in the sense that are renormalization scheme independent. This can be easily proven as follows. Changing renormalization scheme amounts to define a new coupling g' which is related to g as

$$g' = g + ag^3 + \mathcal{O}(g^5). \tag{9.58}$$

Suppose that the β-function for g is

$$\beta_g = b_1 g^3 + b_2 g^5 + \mathcal{O}(g^7). \tag{9.59}$$

We get for the β-function for g'

$$\beta_{g'} = \beta_g \frac{\partial g'}{\partial g} = \beta_g \left(1 + 3ag^2 + \mathcal{O}(g^4)\right) = b_1 g^3 + (b_2 + 3ab_1)g^5 + \mathcal{O}(g^7). \tag{9.60}$$

We can invert the relation between g and g' as

$$g = g' - ag'^3 + \mathcal{O}(g'^5), \tag{9.61}$$

and get finally

$$\beta_{g'} = b_1 g'^3 - 3ab_1 g'^5 + (b_2 + 3ab_1)\, g'^5 + \mathcal{O}(g'^7) = b_1 g'^3 + b_2 g'^5 + \mathcal{O}(g'^7), \tag{9.62}$$

which shows that the first two coefficients of the β-function are universal. Given the universality of the β-function up to two loops, the discrepancy between the two expressions (9.56) and (9.57), which only agree at one loop, cannot just be a matter of renormalization scheme.

How can we reconcile this apparent contradiction? The answer turns out to be surprisingly simple. Let us first consider pure SYM whose action is

$$\mathcal{L} = \frac{1}{16\pi i}\int d^2\theta\, \tau\, \mathrm{Tr}\, W^\alpha W_\alpha + h.c.. \tag{9.63}$$

As we already noticed, if one integrates in superspace one gets a space-time action where gauge fields are not canonically normalized

$$\mathcal{L} = -\frac{1}{4g^2}\, \mathrm{Tr}\, F_{\mu\nu}F^{\mu\nu} + \cdots . \tag{9.64}$$

Let us call the gauge coupling defined in this frame *holomorphic* gauge coupling g_h, defined via the complexified gauge coupling as $\tau = 4\pi i/g_h^2$. In order to get a Lagrangian in terms of canonically normalized fields one should rescale the vector superfield V as $V \to gV$. In other words, we should perform the change of variables $V_h = g_p V_p$. In terms of this physical gauge coupling g_p the Lagrangian reads

$$\mathcal{L} = \frac{1}{4}\int d^2\theta \left(\frac{1}{g_p^2} - i\frac{\theta_{\rm YM}}{8\pi^2}\right) \operatorname{Tr} W^\alpha(g_p V_p) W_\alpha(g_p V_p) + h.c., \tag{9.65}$$

Notice that the Lagrangian above is not holomorphic in the physical coupling since g_p is real as $g_p V_p$ should also be real. The crucial point now is that the two Lagrangians (9.63) and (9.65) are *not* equivalent under the change of variables $V_h = g_p V_p$ in the path integral, since there is a rescaling anomaly (there is an anomalous Jacobian in passing from V_h to V_p), that is $\mathcal{D}(g_p V_p) \neq \mathcal{D} V_p$. In particular, one can show that

$$\begin{aligned}&\mathcal{D}(g_p V_p)\\ &\quad = \mathcal{D}V_p \exp\left[-\frac{i}{4}\int d^2\theta \left(\frac{2T(Adj)}{8\pi^2}\log g_p\right) \operatorname{Tr} W^\alpha(g_p V_p) W_\alpha(g_p V_p) + h.c.\right].\end{aligned} \tag{9.66}$$

Hence we get for the partition function

$$\begin{aligned} Z &= \int \mathcal{D}V_h \exp\left[\frac{i}{4}\int d^2\theta \frac{1}{g_h^2} \operatorname{Tr} W^\alpha(V_h) W_\alpha(V_h) + h.c.\right] = [V_h = g_p V_p] \\ &= \int \mathcal{D}(g_p V_p) \exp\left[\frac{i}{4}\int d^2\theta \frac{1}{g_h^2} \operatorname{Tr} W^\alpha(g_p V_p) W_\alpha(g_p V_p) + h.c.\right] \\ &= \int \mathcal{D}V_p \exp\left[\frac{i}{4}\int d^2\theta \left(\frac{1}{g_h^2} - \frac{2T(Adj)}{8\pi^2}\log g_p\right)\right. \\ &\qquad\qquad \left. \times \operatorname{Tr} W^\alpha(g_p V_p) W_\alpha(g_p V_p) + h.c.\right], \end{aligned} \tag{9.67}$$

which implies

$$\frac{1}{g_p^2} = \operatorname{Re}\left(\frac{1}{g_h^2}\right) - \frac{2T(Adj)}{8\pi^2}\log g_p = \operatorname{Re}\left(\frac{1}{g_h^2}\right) - \frac{2N}{8\pi^2}\log g_p. \tag{9.68}$$

where in the last equality we used the fact that for $SU(N)$ the Dynkin index for the adjoint $T(Adj) = N$. Differentiating with respect to $\log\mu$, and using the expression (9.56) for the holomorphic gauge coupling β-function, one gets for the physical gauge coupling g_p precisely the NSVZ β-function (9.57), where in both expressions we have set $F = 0$.

One can repeat an identical reasoning for SQCD where the relation between the physical and the holomorphic gauge couplings reads

$$\frac{1}{g_p^2} = \mathrm{Re}\left(\frac{1}{g_h^2}\right) - \frac{2T(Adj)}{8\pi^2}\log g_p - \sum_i \frac{T(r_i)}{8\pi^2}\log Z_i. \tag{9.69}$$

Differentiating with respect to $\log\mu$ (using again $T(Adj) = N$ and taking matter to be in the fundamental, for which $T(r) = 1/2$), one gets for the physical gauge coupling exactly the expression (9.57).

We now see why there is no contradiction with the two-loops universality of the β-function. The point is simply that the relation between the holomorphic and the physical gauge coupling is not analytic. In other words, one cannot be Taylor-expanded in the other, because of the log-term (it is a *singular* change of renormalization scheme: the so-called holomorphic scheme is not related continuously to any other physical renormalization scheme). Furthermore, we now also understand where higher-loop contributions to the physical β-function come from. This is just because of wave function renormalization (both of the vector superfield as well as of matter superfields): the physical gauge coupling differs from the holomorphic gauge coupling by effects coming from wave-function renormalization, which get contribution at all loops. Consistently, the physical β-function can be expressed exactly in terms of anomalous dimension of fields, once the one-loop coefficient (which agrees with that of the holomorphic β-function) has been calculated.

One can repeat the same kind of reasoning for gauge theories with extended supersymmetry which, after all, are just (very) special cases of $\mathcal{N} = 1$ theories. In doing so one immediately gets the result we anticipated when discussing non-renormalization theorems in section 6.3. Let us start from $\mathcal{N} = 2$ pure SYM. Using $\mathcal{N} = 1$ language we have a vector and a chiral multiplet, the latter transforming in the adjoint of the gauge group. As we have already seen, due to $\mathcal{N} = 2$ supersymmetry, the kinetic terms of V and Φ are both changed according to the holomorphic gauge coupling. Hence, going to canonical normalization for all fields we must rescale them the same way, $V_h = g_p V_p$ and $\Phi_h = g_p \Phi_p$. The crucial point is that the anomalous Jacobian for V cancels exactly that from Φ! In other words, $\mathcal{D}(g_p V_p)\mathcal{D}(g_p \Phi_p) = \mathcal{D}V_p\, \mathcal{D}\Phi_p$, implying that the holomorphic and physical gauge couplings coincide (basically, the second and third terms in (9.69) cancel each other). Adding matter nothing changes since, as we have already noticed, kinetic terms for hypermultiplets do not renormalize. Hence, $\mathcal{N}=2$ gauge theories, with and without matter, are (perturbatively) one-loop finite. Applying this result to $\mathcal{N}=4$ SYM we conclude that the latter is tree-level exact, since the $\mathcal{N}=4$ one-loop β-function vanishes. It's that simple!

9.6 Holomorphic decoupling

Holomorphy helps also in getting effective superpotentials when one has to integrate out some massive modes and study the theory at scales lower than the corresponding mass scale.

Let us consider a model of two chiral superfields, L and Φ, interacting via the following superpotential

$$W=\frac{1}{2}M\Phi^2+\frac{\lambda}{2}L^2\Phi, \tag{9.70}$$

This is a Wess-Zumino-like model, so this superpotential does not suffer from quantum corrections, neither perturbatively nor non-perturbatively. The spectrum of the theory is that of a massless chiral superfield and a massive one. If we want to study the system at energies $\mu<M$, we have to integrate Φ out. In order to do so we can use holomorphicity arguments, and proceed as we did when proving the exactness of the WZ superpotential. Let us first promote the couplings to spurion fields and, consequently, enlarge the global symmetries as follows

$$\begin{array}{cccc} & U(1)_a & U(1)_b & U(1)_R \\ L & 0 & 1 & 1 \\ \Phi & 1 & 0 & 0 \\ M & -2 & 0 & 2 \\ \lambda & -1 & -2 & 0 \end{array} \tag{9.71}$$

The low energy effective superpotential, in which Φ should not enter, should be a dimension-three function of λ, M and L respecting the above symmetries. In particular it should have $U(1)_{a,b}$ charge 0 and $U(1)_R$ charge 2. The only possible answer is

$$W_{\text{eff}}=a\frac{\lambda^2L^4}{M}, \tag{9.72}$$

where a is an undetermined constant of order one.

The same result can be obtained by using the ordinary integrating out procedure. At scales well below M, the chiral superfield Φ is frozen at its

VEV (we do not have enough energy to make it fluctuate). Therefore, we can integrate the field out by solving its equation of motion, which is just an algebraic one, involving only the F-term, since the kinetic term (the D-term) is trivially zero, that is

$$-\frac{1}{4}\overline{D}^2\overline{\Phi} + \frac{\partial W}{\partial \Phi} = 0 \longrightarrow \frac{\partial W}{\partial \Phi} = M\Phi + \frac{\lambda}{2}L^2 = 0 \longrightarrow \Phi = -\frac{\lambda}{2M}L^2. \tag{9.73}$$

Substituting back into the superpotential we get

$$W_{\text{eff}} = -\frac{1}{8}\frac{\lambda^2 L^4}{M}, \tag{9.74}$$

which is the same as (9.72) (with the undetermined coefficient being fixed). Notice that the superpotential we have just obtained is the effective superpotential one generates in perturbation theory, in the limit of small λ, see Figure 9.3.

Let us emphasize that in this case, differently from the WZ model discussed in the previous section, we have allowed for negative powers of M. In other words, we have not required any smoothness in the $M \to 0$ limit. The reason is that the effective theory we are considering is valid only at energies lower than M, which is a UV cut-off for the theory. Hence, we can accept (and actually do expect) singularities as we send M to zero: new massless degrees of freedom are expected to arise when $M \to 0$. They are those associated to Φ, the superfield we have integrated out.

As a final instructive example, let us consider a perturbation of the previous model. The superpotential we would like to analyze is

$$W = \frac{1}{2}M\Phi^2 + \frac{\lambda}{2}L^2\Phi + \frac{\epsilon}{6}\Phi^3. \tag{9.75}$$

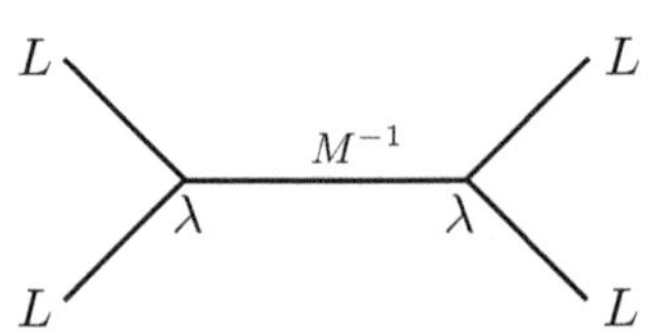

Fig. 9.3 The tree level (super)graph which produces the effective superpotential (9.72) in the weak coupling limit.

Again, if we want to study the system at energies $\mu < M$, we have to integrate the massive field out. The equation of motion for Φ gives

$$\Phi = -\frac{M}{\epsilon}\left(1 \mp \sqrt{1 - \frac{\epsilon\lambda L^2}{M^2}}\right). \tag{9.76}$$

We have now two possible solutions, hence two different vacua sitting at Φ_+ and Φ_-, respectively. Consistently, as we send ϵ to zero one of the two vacua approaches the one of the unperturbed model while the second is pushed all the way to infinity

$$\begin{aligned}
\Phi_+ &= -\frac{M}{\epsilon}\left[1 + \left(1 - \frac{\epsilon\lambda L^2}{2M^2} + \mathcal{O}(\epsilon^2)\right)\right] \underset{\epsilon\to 0}{\longrightarrow} \infty \\
\Phi_- &= -\frac{M}{\epsilon}\left[1 - \left(1 - \frac{\epsilon\lambda L^2}{2M^2} + \mathcal{O}(\epsilon^2)\right)\right] \underset{\epsilon\to 0}{\longrightarrow} -\frac{\lambda}{2M}L^2.
\end{aligned}$$

Substituting (9.76) into (9.75) we get for the effective superpotential

$$W_{\text{eff}} = \frac{M^3}{3\epsilon^2}\left[1 - \frac{3}{2}\frac{\epsilon\lambda L^2}{M^2} \mp \left(1 - \frac{\epsilon\lambda L^2}{M^2}\right)\sqrt{1 - \frac{\epsilon\lambda L^2}{M^2}}\right]. \tag{9.77}$$

There are again singularities, both in parameter space as well as in field L space, now. Comparing to the unperturbed case one can suspect that at these points extra massless degrees of freedom show up. Indeed, computing the (effective) mass for the field we have integrated out, we get

$$M_{\text{eff}} = \frac{\partial^2 W}{\partial\Phi^2} = M + \epsilon\Phi = \text{(on the solution)} = \pm M\sqrt{1 - \frac{\epsilon\lambda L^2}{M^2}}. \tag{9.78}$$

The field Φ becomes massless at $\langle L\rangle = \pm M/\sqrt{\epsilon\lambda}$, precisely the two singularities of the effective superpotential (9.77). In the limit $\epsilon \to 0$, keeping M fixed, one recovers, again, the result of the unperturbed theory.

As in previous example, the same result could have been obtained just using holomorphicity arguments. Promoting also the coupling ϵ to a spurion field with charges

$$\begin{array}{cccc}
 & U(1)_a & U(1)_b & U(1)_R \\
\epsilon & -3 & 0 & 2
\end{array} \tag{9.79}$$

and repeating the same argument as in the previous example, one could conclude that the effective superpotential should have the following structure

$$W_{\text{eff}} = \frac{M^3}{\epsilon^2} f\left(\frac{\epsilon \lambda L^2}{M^2}\right), \tag{9.80}$$

which has precisely the structure of the exact expression (9.77). Taking various limits one can actually fix also the form of the function f and get the expression (9.77) (modulo an overall numerical coefficient, as before).

This way of integrating out in supersymmetric theories, which preserves holomorphy, is called *holomorphic decoupling*. We will heavily use holomorphic decoupling when studying the quantum properties of SQCD in our next chapter. For instance, using this technique it is possible to get the effective superpotential for SQCD with an arbitrary number of flavors once the exact expression for a given number of flavors is known. Everything amounts to integrate flavors in and out (more later).

9.7 Exercises

Exercise 9.1. Using holomorphy and (spurious) symmetries, show that the superpotential

$$W = \mu_1 \Phi + \mu_2 \Phi^2 + \cdots + \mu_n \Phi^n, \tag{9.81}$$

is not renormalized at any order in perturbation theory and non-perturbatively.

Exercise 9.2. Starting from the expression (9.68) and differentiating with respect to $\log \mu$, show that one gets the NSVZ β-function (9.57) with $F = 0$.

References

[1] S. Coleman, *Aspects of Symmetry: Selected Erice Lectures — The uses of Instantons*, Chapter 7, Cambridge University Press (1985).

[2] M. Shifman, *Advanced topics in quantum field theory. A lecture course*, Chapter 5, Cambridge University Press (2012).

[3] M. A. Shifman and A. I. Vainshtein, *Instantons versus supersymmetry: Fifteen years later*, In *Shifman, M.A.: ITEP lectures on particle physics and field theory, vol. 2* 485–647 [hep-th/9902018].

[4] J. Terning, *Modern supersymmetry: Dynamics and duality*, Chapters 7 and 8, Oxford University Press (2006).

[5] P. Argyres, *An Introduction to Global Supersymmetry*, Sections 2.3.2 and 2.3.3, http://www.physics.uc.edu/%7Eargyres/661/susy2001.pdf
P. Argyres, *Introduction to Supersymmetry*, Sections 7 and 14, http://www.physics.uc.edu/%7Eargyres/661/susy1996.pdf

[6] N. Dorey, T. J. Hollowood, V. V. Khoze and M. P. Mattis, *The Calculus of many instantons*, *Phys. Rept.* **371** (2002), 231–459 [arXiv:hep-th/0206063 [hep-th]].

[7] N. Seiberg, Naturalness Versus Supersymmetric Non-renormalization Theorems, *Phys. Lett. B* **318** (1993) 469 [arXiv:hep-ph/9309335].

[8] N. Seiberg, Exact Results On The Space Of Vacua Of Four-Dimensional Susy Gauge Theories, *Phys. Rev. D* **49** (1994) 6857 [arXiv:hep-th/9402044].

[9] N. Seiberg, *The Power of holomorphy: Exact results in 4-D SUSY field theories*, arXiv:hep-th/9408013.

[10] V. A. Novikov, M. A. Shifman, A. I. Vainshtein and V. I. Zakharov, Exact Gell-Mann-Low Function of Supersymmetric Yang-Mills Theories from Instanton Calculus, *Nucl. Phys. B* **229** (1983), 381.

[11] N. Arkani-Hamed and H. Murayama, Holomorphy, rescaling anomalies and exact beta functions in supersymmetric gauge theories, *JHEP* **0006** (2000) 030 [arXiv:hep-th/9707133].

Chapter 10

Supersymmetric Gauge Dynamics: Minimal Supersymmetry

Given a quantum field theory, the very first questions one should answer regard the way its symmetries are realized in its vacua, and what the dynamics around such vacua is:

- Given a QFT with gauge group G and global symmetry group $G_{\rm F}$, how are these realized in the vacuum?
- Which phases may enjoy such a theory?
- Are there tools to give not only qualitative but also quantitative answers to these questions?

It is very difficult to fully or even partially answer these questions, in general. However, as we will discuss in this and subsequent chapters, for supersymmetric theories this *is* possible, sometime. Before entering into any details, there are a number of general remarks we want to make regarding the low energy behavior of field theories, more specifically asymptotically free gauge theories. This will help to better appreciate what comes next.

10.1 Confinement and mass gap in QCD, YM and SYM

It is often said that asymptotically free gauge theories enjoy many interesting and fascinating phenomena at low energy, *i.e.* confinement, the generation of a mass gap, chiral symmetry breaking, etc... This is certainly true, but it turns out that such phenomena may be realized very differently, for different theories. Below we are going to consider three specific theories, namely QCD, YM and $\mathcal{N}=1$ SYM, which are all UV-free and all said to be confining, and show how different the IR dynamics of these theories actually is.

QCD, the theory of strong interactions. At high energy QCD is a weakly coupled theory, a $SU(3)$ gauge theory of weakly interacting quarks and gluons. It grows, through renormalization effects, to become strong in low energy processes. So strong so to bind quarks into nucleons. The strong coupling scale of QCD is

$$\Lambda_{\rm QCD} \sim 300\,{\rm MeV}. \tag{10.1}$$

Note that as compared to protons and neutrons (whose mass is order 1 GeV), constituent quarks are relatively light (the u and d quarks are order of a few MeV; the s quark is order 100 MeV). Most of the mass of nucleons comes from quark kinetic energy and the interactions binding quarks together.

The reason why we cannot see free quarks, we usually say, is confinement: quarks are bound into nucleons and cannot escape. In fact, this statement is not completely correct: if we send an electron deep into a proton, we can make the quark escape!

If the electron is energetic enough, a large amount of energy, in the form of chromoelectric field, appears in the region between the escaping quark and the rest of the proton. When the field becomes strong enough, of order $\Lambda_{\rm QCD}^4 \sim (300\ {\rm MeV})^4$, flux lines can break and produce $q - \overline{q}$ pairs (this is a familiar phenomenon also in electromagnetism: electric fields beyond a certain magnitude cannot survive; strong fields with energy density bigger that $m_e^4 \sim (1\ {\rm Mev})^4$ decay by producing $e^+ - e^-$ pairs). The $\overline{q}$ quark binds to the escaping quark while the q quark binds to the other two quarks in the proton. Therefore, the original quark does escape, the force between it and the remaining proton constituent drops to zero. Just, the escaping quark is not alone, it is bound into a meson. This should be better called charge screening, rather than confinement.

Can we have confinement in a more strict sense? Suppose that quarks were much more massive, say $m_q \sim 1$ TeV. Now the proton mass would be of the order of the TeV. The dynamics drastically changes, now. Repeating the previous experiment, when the chromoelectric field becomes order $\Lambda_{\rm SYM}^4$,

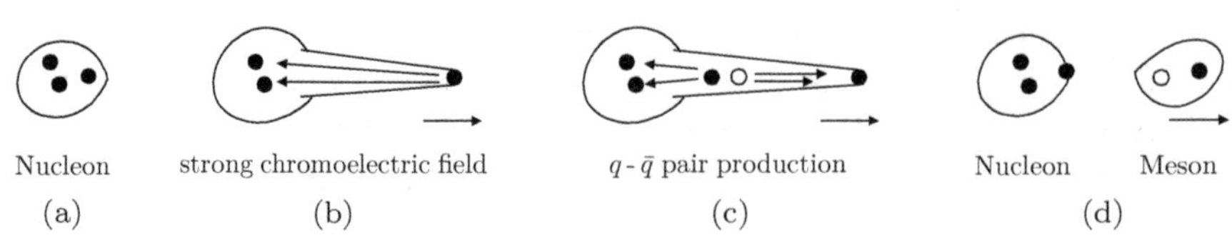

Fig. 10.1 Charge screening: the way QCD confines.

there is not enough energy now to produce $q-\overline{q}$ pairs. The force between the escaping quark and the proton goes to a constant: a tube of chromoelectric flux of thickness $\sim \Lambda_{\rm QCD}^{-1}$ and tension (energy per unit length) of order $\Lambda_{\rm QCD}^2$ connects the two. Not only the quark is confined, it is the flux itself which is confined. This is certainly a more precise definition of confinement: it holds regardless of quarks, in the sense that it holds also in the limit $m_q \to \infty$, namely when the quarks disappear from the spectrum (they become chromoelectric static sources and play no role in the dynamics). It is a property of the pure glue. Strict confinement would be a property of QCD only if the quarks were very massive, more precisely in the limit $F/N \ll 1$, where F is the number of light quarks and N the number of colors. Real-life quarks are light enough to let the chromoelectric flux tube break. Hence, actual QCD does not confine in the strict sense.

Let us discuss the structure of QCD vacua in more detail, looking at how the global symmetry group of QCD is realized in the vacuum. In what follows, we consider only the three light quarks, u, d and s and forget the other ones (which are dynamically less important). So we have $F = 3$ flavors. Moreover, we will first put ourselves in the limit where the light quarks are massless, which is approximately true for u, d and s quarks constituting ordinary matter (protons and neutrons). Only later we will consider the effect of the small quark masses. In this massless limit the QCD Lagrangian reads

$$\mathcal{L} = -\frac{1}{4}\,{\rm Tr}\,F_{\mu\nu}F^{\mu\nu} + \sum_i \overline{q}_{\rm L}^i i \not{D} q_{\rm L}^i + \sum_i \overline{q}_{\rm R}^i i \not{D} q_{\rm R}^i, \quad i = 1,2,3. \tag{10.2}$$

Quark quantum numbers under the global symmetry group are

$$\begin{array}{ccccc} & SU(3)_{\rm L} & SU(3)_{\rm R} & U(1)_A & U(1)_{\rm B} \\ q_{\rm L} & 3 & 1 & 1 & 1 \\ q_{\rm R} & 1 & \overline{3} & 1 & -1 \end{array} \tag{10.3}$$

As well known, there is an axial anomaly, in the sense that the $U(1)_A$ symmetry is broken to $\mathbb{Z}_{2F}$ at the quantum level. Therefore, the continuos global symmetries at the quantum level are just

$$G_{\rm F} = SU(3)_{\rm L} \times SU(3)_{\rm R} \times U(1)_{\rm B}. \tag{10.4}$$

As we have already discussed, the theory undergoes confinement (or better charge screening) and quarks and gluons are bounded into color singlet states. But what about $G_{\rm F}$? Experimental and theoretical considerations plus several numerical simulations on the lattice lead to a definite picture of the realization of the global symmetry at low energy. It is believed that at low energy only a subgroup survives

$$SU(3)_{\rm D}\times U(1)_{\rm B}, \tag{10.5}$$

under which hadrons are classified: $SU(3)_{\rm D}$ gives the flavor quantum numbers and $U(1)_{\rm B}$ is the baryon charge. The remaining generators must be broken, somehow. The intuitive picture is as follows. Due to confinement, at strong coupling quarks and anti-quarks are bound into pairs, and the vacuum is filled by a condensate of these color singlet quark bilinears

$$\langle q_{\rm L}^i q_{\rm R}^j\rangle = \Delta\,\delta^{ij}, \tag{10.6}$$

where $\Delta\sim\Lambda^3_{\rm QCD}$. This condensate is invariant under a diagonal $SU(3)$ subgroup of the original $SU(3)_{\rm R}\times SU(3)_{\rm L}$ group and is then responsible for the spontaneous breaking of the chiral symmetry of the original symmetry group $G_{\rm F}$

$$SU(3)_{\rm L}\times SU(3)_{\rm R}\times U(1)_{\rm B}\to SU(3)_{\rm D}\times U(1)_{\rm B}. \tag{10.7}$$

Eight global symmetries are broken by the quark condensates and hence we would expect eight Goldstone bosons. The latter are indeed observed experimentally, and correspond to the eight pseudoscalar mesons, the pions, $\pi^{0,\pm}, K^{0,1}, \overline{K}^{0,+}, \eta$

$$\begin{aligned} &\pi^+ = u\bar d, \quad \pi^- = d\bar u, \quad \pi^0 = d\bar d - u\bar u, \quad \eta = u\bar u + d\bar d - 2s\bar s\\ &K^0 = s\bar d, \quad K^- = \bar u s, \quad \overline{K}^0 = \bar s d, \quad \overline{K}^+ = \bar s u \end{aligned} \tag{10.8}$$

Let us first notice that if $U(1)_A$ were not anomalous, we would have had a ninth meson, the so-called η' meson, which would have corresponded to a shift in the phase of the condensate (10.6). The condensate breaks spontaneously the $\mathbb{Z}_{2F}$ symmetry down to $\mathbb{Z}_2$, but this does not give rise to any massless particle. The η' has a periodic potential with F minima, each of them being $\mathbb{Z}_2$ invariant, and related one another by $\mathbb{Z}_F$ rotations. These minima are not isolated, though, since they are connected via $SU(F)_{\rm L}\times SU(F)_{\rm R}$ rotations. This means that there is a moduli space of vacua. In Figure 10.2 a qualitative picture of QCD vacuum structure is reported.

Via a $SU(3)$ rotation acting separately on $q_{\rm L}$ and $q_{\rm R}$, the condensate (10.6) can be put in the form

$$\langle q_{\rm L}^i q_{\rm R}^j\rangle = \Delta\,U^{ij} \tag{10.9}$$

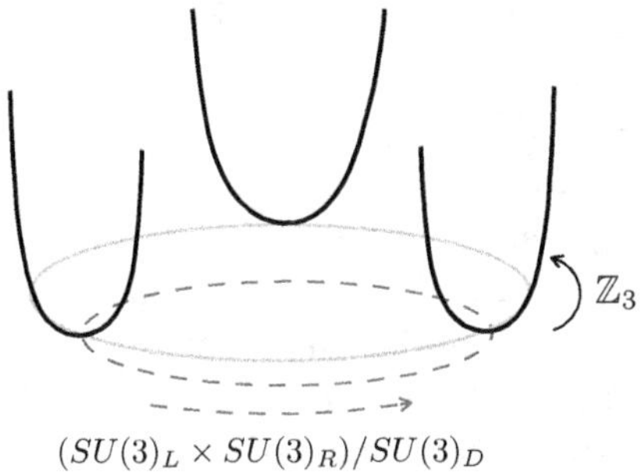

Fig. 10.2 The vacuum structure of 3-flavor massless QCD. The η' particle is massive. The three vacua associated to the periodic potential along the η' direction are rotated one another by the broken $\mathbb{Z}_3$ generators, but are not isolated since there are flat directions connecting them, associated to eight massless excitations (the pions) which parametrize the coset $(SU(3)_\text{L} \times SU(F)_\text{R})/SU(3)_\text{D}$.

where U^{ij} is a $SU(3)$ matrix on which a $SU(3)_\text{L} \times SU(3)_\text{R}$ rotation acts as

$$U \to A_\text{L}^\dagger U A_\text{R}, \tag{10.10}$$

which shows that there exists a $SU(3)_\text{D}$ rotation ($A_\text{L} = A_\text{R}$) under which the matrix U is invariant. So the moduli space of vacua is a $SU(3)$ manifold.

The quantum fluctuations of the entries of this matrix represent the massless excitations around the vacua of massless QCD, the pions. An effective Lagrangian for such excitations can be written in terms of $U(x)$ and its derivatives. This Lagrangian should be invariant under the full global symmetry group G_F, hence non-derivative terms are not allowed (the only G_F-invariant function would be $U^\dagger U = 1$), which is simply saying that the pions are massless in the massless QCD limit we are considering. The structure of the effective Lagrangian hence reads

$$\mathcal{L}_\text{eff} = f_\pi^2 \left(\partial_\mu U^\dagger \partial^\mu U\right) + \kappa\, \partial_\mu U^\dagger \partial^\mu U \partial_\nu U^\dagger \partial^\nu U + \cdots , \tag{10.11}$$

where $\kappa \sim 1/M^2$, with M being some intrinsic mass scale of the theory, and traces on flavor indices are understood. At low momenta only the first term contributes and we then get a definite prediction for pion scattering amplitudes, in terms of a single parameter f_π.

In fact, quarks are not exactly massless and therefore the above picture is only approximate. In reality, the $SU(3)_\text{L} \times SU(3)_\text{R}$ symmetry is only approximate since quark masses correspond to (weak) G_F-breaking terms. This has the effect to make the pions be only pseudo-Goldstone bosons.

Hence, one would expect them to be massive, though pretty light, and this is indeed what we observe in Nature.

In principle, one should have gotten the chiral Lagrangian (10.11) from the UV Lagrangian (10.2) by integrating out high momentum modes. This is difficult (next to impossible, in fact). However, we know in advance the expression (10.11) to be right, since that is the most general Lagrangian one can write describing pion dynamics and respecting the original symmetries of the problem. Combining the expression (10.11) with weak G_F-breaking terms induced by actual quark masses, one gets a Lagrangian which, experimentally, does a good job.

Summarizing, we see that combining symmetry arguments, lattice simulations, experimental observations and some physical reasoning, we can reach a rather reasonable understanding of the low energy dynamics of QCD. This is all very nice but one would like to gain, possibly, a theoretical (*i.e.* more microscopic) understanding of QCD phenomena. As of today, this is still an open question for QCD. And, more generally, it is so for any generic gauge theory. As we will later see, supersymmetry lets one have more analytical tools to answer this kind of questions, having sometime the possibility to derive strong coupling phenomena like confinement, chiral symmetry breaking and the generation of a mass gap, from first principles.

YM theory, gauge interactions without matter fields. Let us consider a YM theory with gauge group $SU(N)$. This is again a UV-free theory, the one-loop β-function being now

$$\beta_g = \frac{g^3}{16\pi^2}\left(-\frac{11}{3}N\right). \tag{10.12}$$

There are two claims about this theory (based mostly on lattice calculations and on theoretical reasoning in comparing YM with what we know, experimentally, about QCD).

(1) The theory has a *mass gap*, *i.e.* there are no massless fields in the spectrum. Rather, there is a discrete set of states with masses of order Λ, the scale where the one-loop gauge coupling diverges (higher loop and non-perturbative effects do not change the actual value of Λ in any sensible way)

$$\Lambda = \mu\, e^{-\frac{8\pi^2}{g^2(\mu)b_1}} \quad \text{where } b_1 = \frac{11}{3}N. \tag{10.13}$$

The low energy spectrum consists of glueballs. These are sort of gluons bound states which however do not consist of a fixed number of gluons (gluon number is not a conserved quantum number in strong interactions), but rather of a shifting mass of chromoelectric flux lines. Unlike gluons, for which a mass term is forbidden (because they have only two polarizations), glueballs include scalars and vectors with three polarizations (as well as higher spin particle states), for which a mass term is allowed. Such mass should clearly be of the order of the dynamical scale, $m \sim \Lambda$, so not to contradict perturbation theory.

The low energy spectrum is very different from QCD. In QCD there is a mass gap just because quarks are massive. If u, d and s quarks were massless, we would not have had a mass gap in QCD since pions would have been exact Goldstone bosons and hence massless. Here instead there is a genuine mass gap.

(2) The theory undergoes *confinement* (now in the strict sense). The chromoelectric flux is confined, it cannot spread out in space over regions larger than about Λ^{-1} in radius. How can we see confinement, namely the presence of strings which contain the chromoelectric flux? Let us add some heavy quarks to the theory and let us see whether these quarks are confined, as it was the case for very massive QCD. The Lagrangian would read

$$\mathcal{L} = -\frac{1}{4}\,\mathrm{Tr}\, F^{\mu\nu}F_{\mu\nu} + i\overline{\psi}\not{D}\psi - M\overline{\psi}\psi, \quad M \gg \Lambda. \tag{10.14}$$

In the limit $M \to \infty$ the test particles become chromoelectrostatic sources, and play no role in the dynamics.

If confinement occurs, we would expect a linear potential between the two quarks. Indeed, in an unconfined theory, the electric flux is uniformly distributed over a sphere surrounding a charge, and falls-off as $1/r^2$. In a confining theory with flux tubes, the flux tube has a fixed cross-sectional area $\sim \Lambda^{-2}$. Thus, for any sphere of radius $r \gg R \equiv \Lambda^{-1}$, the flux is zero except in a region of area Λ^{-2}, see Figure 10.3.

Hence, the electric field in that region has a magnitude which is r-independent, which implies that the force it generates on a test charge is also r-independent, and so the potential V between charges would grow linearly in r. The force goes to a constant, it never drops to zero, see Figure 10.4.

Let us take a closer look to the potential

$$V(r) = T_{\mathrm{R}}\, r. \tag{10.15}$$

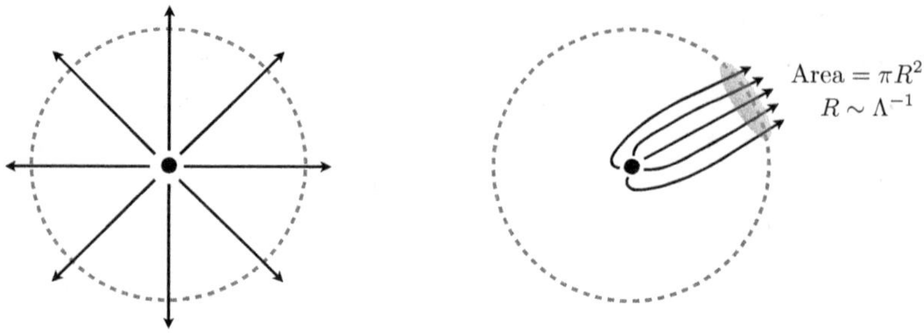

Fig. 10.3 Gauss law for unconfined (left) and confined (right) flux.

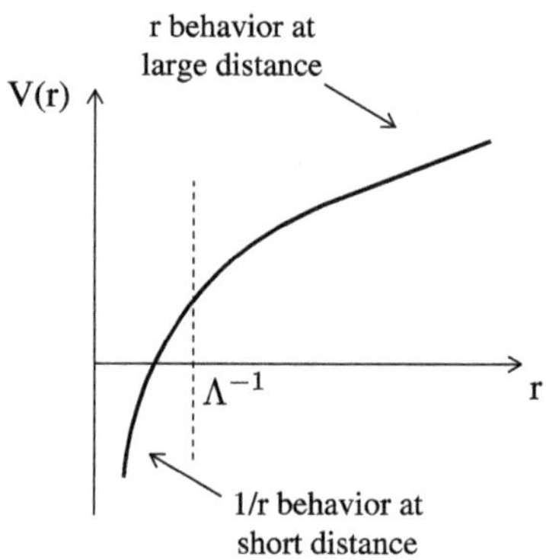

Fig. 10.4 The potential between two test charges in YM theory.

The proportionality coefficient has dimension of an energy per unit length, and it is the so-called string tension. On general ground, one would expect the string tension to depend in some way on the gauge group representation the test charges transform. This is pretty obvious since, e.g., for the singlet representation $T_{\rm R}$ is clearly zero, while for actual quarks, which transform in the fundamental representation, it is not. In fact, as we are going to show below, the string tension does not depend on the representation itself, but on what is called the *N-ality* of the representation.

Let us consider a (gauge) group G. Its center, $Z(G)$ is defined as the part of G which commutes with all generators. For $G = SU(N)$, we have that

$$Z(G) = \left\{ U^{\alpha}_{\overline{\beta}} = e^{2\pi i k/N} \delta^{\alpha}_{\overline{\beta}}; \quad k = 0, 1, \ldots, N-1 \bmod N \right\}, \tag{10.16}$$

where α an index in the fundamental and $\bar{\beta}$ in the anti-fundamental of the gauge group. Hence, in this case, $Z(G) = \mathbb{Z}_N$. Let us now consider some representation R. An element ρ of this representation is labeled by n upper indices α_i and $\bar{n}$ lower indices $\bar{\beta}_i$, each upper index transforming in the fundamental and each lower index transforming in the anti-fundamental representation. If one acts with the center of the group on ρ one gets

$$Z(G) \; : \; \rho \to e^{2\pi i k(n-\bar{n})/N} \rho. \tag{10.17}$$

The coefficient $n - \bar{n}$ is called the N-ality of the representation ρ. If ρ has N-ality p, then the complex conjugate representation ρ^C has N-ality $-p$, which is nothing but $N-p$, since from eq. (10.17) it follows that the N-ality is defined modulo N. For instance, while the adjoint representation and the trivial representation have $p = 0$, the fundamental representation has $p = 1$ and the anti-fundamental has $p = N - 1$.

Clearly, representations break into equivalence classes under the center of the gauge group. It turns out that the string tension T_{R} is not a function of the representation but actually of the N-ality. The basic reason for this is that gluon number is not a conserved quantity in YM theories, while N-ality is, as we now show. Let us consider heavy test particles transforming either in the anti-symmetric representation or in the symmetric representation of the gauge group, ψ_{S} and ψ_{A}, respectively. Each of them will have its own string tension, T_{S} and T_{A} (but same N-ality, $p = 2$).

Suppose that $T_{\mathrm{S}} > T_{\mathrm{A}}$. Since gluon number is not a conserved quantity, we can add a gluon A_μ coming from the chromoelectric flux tube next to ψ_{S}. The charge of the bound state $\psi_{\mathrm{S}} A_\mu$ is Symmetric $\otimes$ Adj $= \oplus$ Representations, where all representations entering the sum have the same N-ality (the same as the symmetric representation, in fact, since the N-ality is an additive quantity, and that of the adjoint representation is zero). For example, choosing $G = SU(3)$ we have

$$6 \otimes 8 = \overline{3} + 6 + \overline{15} + 24, \tag{10.18}$$

where the first representation on the r.h.s. is the anti-symmetric representation. Since we have assumed that $T_{\mathrm{S}} > T_{\mathrm{A}}$ it is energetically favored to pop a gluon out of the vacuum and put it near to ψ_{S} (and another one near to $\overline{\psi_{\mathrm{S}}}$) since this has an energy cost (of order Λ) which is lower than the energy gain, proportional to $(T_{\mathrm{S}} - T_{\mathrm{A}})r$, which for sufficiently large r always wins. In other words, in YM theory the representation of a chromoelectric source is not a conserved quantum number; only the N-ality is. Therefore, for all

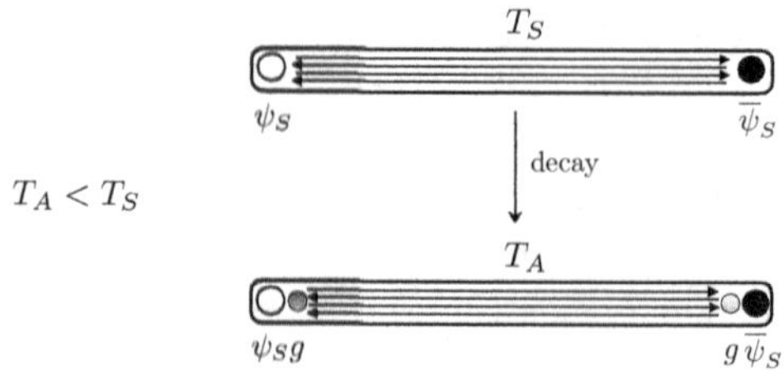

Fig. 10.5 The string tensions corresponding to the antisymmetric and symmetric representations. All strings have the same N-ality, $p = 2$. The flux tube in the symmetric representation, which is more energetic, decays into that of the anti-symmetric one by popping-up a gluon out of the vacuum.

representations with the same N-ality, there is only one stable configuration of strings, the one with lowest tension, as shown in Figure 10.5. In summary, the tension of *stable* flux tubes are labeled by p, the N-ality, not by R, the representation.

Notice that charge conjugation symmetry ensures that $T_p = T_{N-p}$. Therefore, there are order $N/2$ stable flux tube configurations for $SU(N)$. For $SU(3)$ there is only one single confining string, that with N-ality $p = 1$, since $T_0 = 0$ and $T_2 = T_1$. Multiple flux tubes only arise for larger gauge groups. For instance, for $G = SU(4)$, there can exist two different string tensions, with N-ality $p = 1$ and $p = 2$, respectively).

All what we said let us also understand how to classify gauge singlets bound states. While gluons are not confined by flux tubes, since $T_{\text{Adj}} = T_0 = 0$, any heavy quark ψ with N-ality $\neq 0$ will experience a linear potential and a constant force which will confine it to an antiquark $\overline{\psi}$ (these are the mesons) or, more generally, to some combination of quarks and anti-quarks with opposite N-ality. For instance, such combination can be made of $N - 1$ quarks and the bound state is called a baryon.

SYM, supersymmetric gauge interactions without matter fields. We will study this theory in detail later. Here, we just want to emphasize the similarities and differences with respect to YM theories and QCD.

Similarly to YM, SYM enjoys strict confinement, a mass gap and no pions. Similarly to QCD, it has a sort of chiral symmetry breaking and an anomaly, which makes the corresponding η'-like particle being massive. Finally, it differs from both since it has multiple *isolated* vacua.

We choose again, for definiteness, the gauge group to be $SU(N)$, but qualitatively identical statements can be done for any other choice of (non-abelian) gauge group. The structure of the (on-shell) SYM Lagrangian

$$\mathcal{L}_{\mathrm{SYM}} = -\operatorname{Tr}\left[\frac{1}{4}F_{\mu\nu}F^{\mu\nu} + i\bar{\lambda}\not{D}\lambda\right]. \tag{10.19}$$

First notice that gauginos cannot break flux tubes since they transform in the adjoint representation, which is in the same N-ality class of the singlet representation. So gauginos behave very differently from QCD quarks, in this respect. Basically, the presence of these fermion fields does not change the confining behavior of pure YM glue, since gauginos cannot break flux tubes. That is why SYM enjoys strict confinement, differently from QCD.

On the other hand, the $U(1)_R$ symmetry resembles the axial symmetry of QCD, since it is anomalous and it is broken to $\mathbb{Z}_{2N}$ at the quantum level (recall gauginos have R-charge equal to one). Like QCD, SYM enjoys chiral symmetry breaking, since it turns out (and we will be able to prove) that gaugino bilinears acquire a non-vanishing VEV in the vacuum. More precisely, we have

$$\langle\lambda\lambda\rangle \sim \Lambda^3 e^{2\pi ik/N}, \quad k = 0, 1, \ldots, N-1, \tag{10.20}$$

which breaks $\mathbb{Z}_{2N} \to \mathbb{Z}_2$. Hence there are N isolated vacua, each of them $\mathbb{Z}_2$ symmetric, related by $\mathbb{Z}_N$ rotations, as shown in Figure 10.6. The η' is the phase of the condensate (10.20) (similarly to QCD), but the vacua are isolated, so there is a dynamical mass gap (unlike QCD and like YM).

10.1.1 *Intermezzo: Wilson loops as order parameters for confinement*

We have discussed the different ways in which confinement is realized in YM, QCD and SYM by probing the theories with external heavy sources. In particular, YM and SYM are believed to enjoy strict confinement and the potential between static sources grows linearly at large distance. QCD,

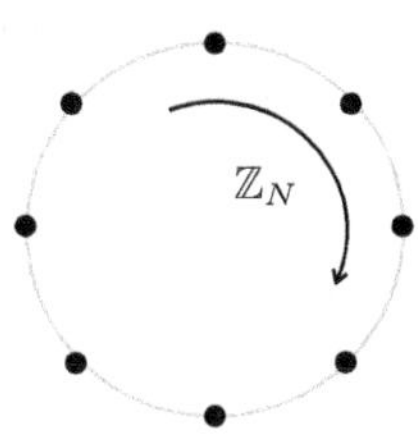

Fig. 10.6 The vacuum structure of pure $\mathcal{N} = 1$ SYM. The N vacua are isolated, and related by $\mathbb{Z}_N$ rotations (compare with Figure 10.2).

instead, enjoys a milder version of confinement, namely charge screening. In this case, the potential between static sources goes to a constant asymptotically since the force drops to zero at large distance, being the charge screened.

Interestingly, similar conclusions can be reached without resorting to external sources but rather computing expectation values of specific line operators, known as *Wilson loops*, as we now review.

In the same vein as in General Relativity the Levi-Civita connection tell us how to parallel transport vectors around a manifold, in a gauge theory the gauge connection A_μ tells how the internal degrees of freedom of a charged particle in representation R of the gauge group (*i.e.* the vector ω_i, with $i = 1, \ldots, \dim R$, describing its color degrees of freedom) change as the particle moves along some given path γ. The way the vector ω_i rotates along γ depends on the initial and final points of γ, as well as on γ itself.

If we take the path to be closed, this tells how the vector differs from its starting value (hence measuring the holonomy). In this case one can form a gauge invariant object, the Wilson loop, defined as

$$W_{\rm R}(\gamma) = {\rm Tr}_{\rm R}\, \mathcal{P} \exp\left(i \int_\gamma A_\mu dx^\mu \right), \tag{10.21}$$

where $\mathcal{P}$ is the path-ordering operator (meaning that when expanding the exponential the matrices $A_\mu = A_\mu^a T_a$ are ordered so that those at earlier times are placed to the left). The Wilson loop is a gauge invariant but non-local operator.

What does this have to do with confinement? It turns out that the VEV of the Wilson loop operator is an order parameter for confinement! Let us see how this comes. The VEV of the Wilson loop reads

$$\langle W_{\rm R}(\gamma) \rangle = \int \mathcal{D}A\, {\rm Tr}_{\rm R}\, \mathcal{P} \exp\left(i \int_\gamma A_\mu dx^\mu \right) e^{iS} \tag{10.22}$$

where S is the action. Let us take as γ the closed path in Figure 10.7 and compute the VEV of the corresponding Wilson loop in the fundamental representation. A physical interpretation of such a path can be thought of as having put a quark q and an anti-quark $\overline{q}$ at some distance L and then compute the interaction potential as a function of time. Equivalently, one can think of creating a $q/\overline{q}$ pair out of the vacuum at some time in the past. These then propagate for a time T before they annihilate back in the vacuum.

Let us work in Euclidean space. If we take the time T to be very large, eventually $T \to \infty$, the path integral projects the system onto lowest energy

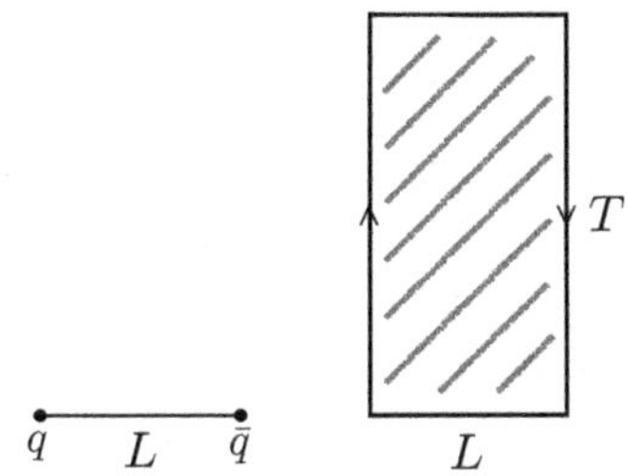

Fig. 10.7 A rectangular Wilson loop in the fundamental representation of the gauge group G.

states. Before q and $\bar{q}$ appear and after they disappear this energy is zero, while in between this is $V(L)$, the potential energy between the two quarks. Hence in this limit we get

$$\lim_{T\to\infty}\langle W_f(\gamma)\rangle \sim e^{-V(L)T} = \begin{cases} e^{-\sigma LT} = e^{-\sigma A[\gamma]} & \text{Confinement} \\ e^{-V_0 T} = e^{-\frac{V_0}{2}P[\gamma]} & \text{Charge screening} \end{cases} \tag{10.23}$$

where we have considered the cases of a confining potential $V \sim \sigma L$ (as for YM or SYM) and a charge screening one $V \sim V_0$ (as for QCD), respectively. A is the area enclosed by γ and P its perimeter. The second equality in the second equation holds because in the limit $T \to \infty$ we get $P = 2T + 2L \simeq 2T$.

So we learn that in a confining theory, for representations with non-vanishing N-ality (as for instance the fundamental representation), the Wilson loop follows the area law, while for unconfined theories (including those enjoying charge screening, as QCD!) it follows the perimeter law, no matter the representation.

In the limit in which one takes the loop infinitely large, one could suspect, naively, that, regardless the phase the theory enjoys, both VEVs in eqs. (10.23) vanish, since both the area and the perimeter become infinite. This is not correct. Infinities can be re-absorbed by counterterms. These however are local, so for a line operator as the Wilson loop one can regulate the perimeter but not the area. The upshot is then that for $\gamma \to \infty$, in a confining theory the VEV of the Wilson loop operator vanishes while for a theory where the charge is screened it goes to a constant

$$\begin{aligned} \langle W_f(\gamma)\rangle &= 0 \quad \text{YM, SYM} \\ \langle W_f(\gamma)\rangle &\neq 0 \quad \text{QCD} \end{aligned} \tag{10.24}$$

In this sense, the VEV of the Wilson loop operator can be regarded as an order parameter for (strict) confinement.

Eq. (10.24) may suggest the existence of a symmetry which is preserved or spontaneously broken depending whether the theory confines (YM,SYM) or enjoys charge screening (QCD). Well, yes and no.

Yes, because there is indeed a symmetry. Note, however, that this symmetry is not an ordinary symmetry under which local operators are charged, but rather a one-form symmetry under which line operators, as the Wilson loop, are charged. Higher-form symmetries can only be abelian and in this case the relevant symmetry is discrete, related to the centre of the gauge group, $\mathbb{Z}_N^{(1)}$ for $G = SU(N)$ (the upper index refers to the symmetry being a one-form symmetry). According to eq. (10.24), YM and SYM do preserve such symmetry.

No, because the non-vanishing VEV for the Wilson loop in QCD does not correspond to spontaneous symmetry breaking. This is because QCD breaks the one-form symmetry explicitly, so there is not a symmetry to start with. A way to understand why this is the case is as follows. In QCD, besides the Wilson loop, one can construct a different gauge invariant line operator, a Wilson line with a quark and an antiquark (which are part of the theory) at its two ends. Locally, this dressed line operator has the same charge as the Wilson loop but this charge is not a well defined quantum number. Indeed, the sphere measuring the flux and hence the charge of the Wilson line can be continuously deformed into a sphere not linking the defect anymore, hence measuring now a vanishing charge, as shown in Figure 10.8. This is clearly inconsistent (in other words, the charge operator

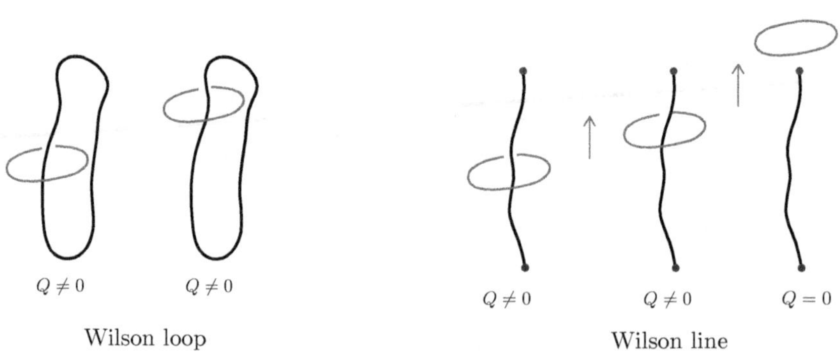

Fig. 10.8 Ill defined one-form symmetry in QCD. The sphere measuring the one-form charge (the light grey circle in the figure) cannot unwind a Wilson loop operator while it can unwind a Wilson line, hence measuring a different charge upon continuous deformations.

is not topological, this being a necessary condition for the symmetry to act faifthfully on charged operators).

A theory for which the Wilson loop could have a non-vanishing VEV but where, unlike QCD, this would indicate a spontaneous breaking of the one-form symmetry would be for example $SU(N)$ YM coupled to adjoint scalars. Being the scalars in the adjoint, whose N-ality is 0, (open) Wilson lines are now not well-defined operators since they cannot be made gauge invariant (there are no local charged operators they can end on). Therefore, the one-form symmetry is now a well-defined symmetry of the theory. If a Higgs-like potential exists and vacua are at non-vanishing scalar field VEV the gauge group would be broken as $SU(N) \to U(1)^{N-1}$ and the massless spectrum would be made just of $N-1$ photons. The potential between test particles would go as $1/r$, so Coulomb-like. Computing the VEV of the Wilson loop of Figure 10.7 one would get a non-vanishing result

$$\lim_{\gamma\to\infty} \langle W_f(\gamma)\rangle \sim \lim_{T,L\to\infty} e^{-V(L)T} = \lim_{T,L\to\infty} e^{-T/L} \neq 0. \tag{10.25}$$

Since the one-form symmetry is now a symmetry of the theory, a non-vanishing VEV for the Wilson loop would indicate the spontaneous breaking of the one-form symmetry. Quite interestingly, it turns out that in the IR $\mathbb{Z}_N^{(1)}$ is enhanced to $\left[U(1)^{(1)}\right]^{N-1}$ and therefore the non-vanishing VEV (10.25) breaks a *continuous* one-form symmetry, the $N-1$ massless photons being the corresponding Goldstone bosons! More precisely, the enhancement is to $\left[U(1)^{(1)} \times U(1)^{(1)}\right]^{N-1}$, the second abelian factor being related to a magnetic one-form symmetry we will say more about in Section 12.5, but which does not play any crucial role in the present discussion. In any event, the upshot is that for adjoint YM a vanishing or non-vanishing VEV for the Wilson loop operator would indeed correspond to a symmetry being preserved (confining phase) or spontaneously broken (Coulomb phase), respectively.

10.2 Phases of gauge theories: examples

After this detour on "the meaning of confinement" and the different ways in which confinement is realized in a sample of UV-free gauge theories, we would like to consider, in more general terms, which kind of phases a generic gauge theory can enjoy. There are basically three qualitatively different such phases:

- *Higgs* phase: the gauge group G is spontaneously broken, all vector bosons obtain a mass.

- *Coulomb* phase: vector bosons remain massless and mediate $1/r$ long range interactions. This phase can be either interacting (this is sometime what is referred to as actual Coulomb phase) or free, meaning that asymptotic states do not interact, at low enough energy.
- *Wilson* or confining phase: color sources, like quarks, gluons, etc..., are bound into color singlets. As discussed before, this can be realized as charge screening or strict confinement, depending on the details of the theory under consideration.

It is worth notice that the Coulomb phase is not specific to abelian gauge theories, as QED. For example, a non-abelian gauge theory with enough matter content may become IR-free, giving a long-range potential between color charges $V(r) \sim a(r) \times 1/r$, with $a(r)$ a coefficient decreasing logarithmically with r. And, as we will see in the following, also interacting non-abelian Coulomb phases can exist.

There can of course be intermediate situations, where for instance the original gauge group is Higgsed down to a subgroup H, which then confines, or is in a Coulomb phase (this is what happens in the SM of electroweak interactions or in the adjoint YM example discussed in previous section). In these cases the phase of the gauge theory is defined by what happens to H in the vacua, regardless of the fate of the original gauge group G.

Below we consider two examples which will hopefully clarify the meaning of some of above statements, but also point out some subtleties one could encounter when dealing with concrete models.

10.2.1 *Coulomb phase and free phase*

Let us consider SQED, whose structure we have discussed in Chapter 5. The scalar potential of SQED reads

$$V = m^2|\phi_-|^2 + m^2|\phi_+|^2 + \frac{1}{2}e^2\left(|\phi_+|^2 - |\phi_-|^2\right)^2, \tag{10.26}$$

where ϕ_- and ϕ_+ are the scalar fields belonging to the two chiral superfields Φ_- and Φ_+ with electric charge ± 1 respectively, and a superpotential mass term $W = m\Phi_-\Phi_+$ has been also added. Let us consider massive and massless cases separately.

- $m \neq 0$. In this case the vacuum is at $\langle\phi_-\rangle = \langle\phi_+\rangle = 0$. Heavy static probe charges would experience a potential

$$V \sim \frac{\alpha(r)}{r}, \quad \alpha(r) \sim \frac{1}{\log r}. \tag{10.27}$$

However, the logarithmic fall-off is frozen at distance $r = m^{-1}$: for larger distances α stops running. Hence, the asymptotic potential reads

$$V(r) \sim \frac{\alpha_*}{r}, \quad \alpha_* = \alpha(r = m^{-1}), \tag{10.28}$$

which simply says that massive SQED is in a (interacting) Coulomb phase.

- $m = 0$. In this case the potential gets contributions from D-terms only. Now there are more vacua, actually a moduli space of vacua. Besides the origin of field space, also any $\langle \phi_- \rangle = \langle \phi_+ \rangle \neq 0$ satisfies the D-equations. One can parameterize the supersymmetric vacua in terms of the gauge invariant combination $u = \langle \phi_- \phi_+ \rangle$. We have then two options. When $u \neq 0$ we are in a Higgs phase, the gauge group $U(1)$ is broken and the photon becomes massive (the theory is described by a massive vector multiplet and a massless chiral multiplet). When instead $u = 0$ the gauge group remains unbroken. Still, we are in a different phase with respect to the massive case. The basic difference is that the coupling $\alpha(r)$ does not stop running, now, since $m = 0$, and hence it ends-up vanishing at large enough distances. In other words, the potential again reads

$$V(r) \sim \frac{\alpha(r)}{r}, \quad \alpha(r) \sim \frac{1}{\log r} \tag{10.29}$$

but now $\alpha = 0$ for $r \to \infty$. This is called a *free* Coulomb phase. At low energy (large enough distances) the theory is a theory of free massless particles.

Let us emphasize again that both the interacting and the free Coulomb phases are not specific to abelian gauge theories, and can be enjoyed also by non-abelian theories. We will see examples of this phenomenon soon.

10.2.2 *Continuously connected phases*

Another important aspect we want to emphasize is that there can exist instances where there is no gauge-invariant distinction between different phases. Let us show this non-trivial fact with a simple example.

Let us consider a $SU(2)$ gauge theory with a $SU(2)$ scalar doublet ϕ (a Higgs field), a $SU(2)$ singlet $e_{\rm R}$ and a $SU(2)$ doublet $L = (\nu_{\rm L}, e_{\rm L})$, with interaction Lagrangian

$$\mathcal{L}_{\rm int} = \overline{L}\phi e_{\rm R} + h.c.. \tag{10.30}$$

This is nothing but a one-family electroweak (EW) theory model.

As it happens in standard EW theory, this theory can be realized in the Higgs phase, where the field ϕ gets a non-vanishing VEV

$$\langle\phi\rangle = \frac{1}{\sqrt{2}}\begin{pmatrix}0\\ v\end{pmatrix}. \tag{10.31}$$

In this phase all three gauge bosons get a mass, the neutrino ν_{L} remains massless while the electron gets a mass $m_e = v/\sqrt{2}$.

Suppose instead that the theory were realized in a different phase, a confinement phase. In such phase one would not observe massless gauge bosons (as above) while fermions and Higgs bosons would bind into $SU(2)$ singlet combinations

$$E_{\mathrm{L}} = \phi^{\dagger} L, \quad N_{\mathrm{L}} = \epsilon_{ab}\phi_a L_b, \quad e_{\mathrm{R}} \tag{10.32}$$

in terms of which the interaction Lagrangian becomes

$$\mathcal{L}_{\mathrm{int}} = m\overline{E}_{\mathrm{L}} e_{\mathrm{R}} + h.c., \tag{10.33}$$

where $m \sim \Lambda$. So we see that E_{L} and e_{R} pair-up and become massive while N_{L} remains massless. The spectrum in this phase is the same as that of the Higgs phase: there is no gauge-invariant distinction between the two phases! Consistently, also a Wilson loop computation cannot tell in which phase the theory is. In a Higgs phase gauge bosons are massive and cannot mediate long range forces, so the potential between static test charges goes to a constant asymptotically. Confinement in presence of massless (charged) matter is realized as charge screening and also in this case the force drops to zero asymptotically. Hence, in both case the VEV of a Wilson loop follows a perimeter law, and it cannot distinguish between the two phases. This said, there are of course quantitative differences between the two phases: for instance, E_{L} is a composite field and its pair production would be suppressed by a form factor which is not observed in real world EW theory. Still, we need experiments to discern between the two phases and understand which one is actually realized in Nature.

The general lesson we want to convey is that by adjusting some parameter of a gauge theory, sometime one can move continuously from one type of phase to another. In this non-abelian example, there is no invariant distinction between Higgs and Wilson phase. Theories of this kind are said to enjoy *complementarity*. We will encounter more situations of this sort in the following.

10.3 $\mathcal{N}$=1 SQCD: perturbative analysis

In what follows we will consider SQCD and its quantum behavior (including non-perturbative effects) and try to answer the basic questions about its dynamical properties in the most analytical possible way. Let us first summarize what we have already learned about the classical and quantum — though perturbative only — behavior of SQCD.

SQCD is a renormalizable supersymmetric gauge theory with gauge group $SU(N)$, F flavors $(Q, \widetilde{Q})$ and no superpotential. Interaction terms are present and come from D-terms. The group of continuous global symmetries at the quantum level is $G_{\rm F} = SU(F)_{\rm L} \times SU(F)_{\rm R} \times U(1)_{\rm B} \times U(1)_R$ with the following charge assignment

	$SU(F)_{\rm L}$	$SU(F)_{\rm R}$	$U(1)_{\rm B}$	$U(1)_R$
Q^i_a	F	$\bullet$	1	$\frac{F-N}{F}$
$\widetilde{Q}^b_j$	$\bullet$	$\overline{F}$	-1	$\frac{F-N}{F}$

As already emphasized, for pure SYM the R-symmetry is anomalous, and only a $\mathbb{Z}_{2N}$ subgroup of $U(1)_R$ survives at the quantum level.

From the analysis in Chapter 5 we have learned that, classically, there is a huge moduli space of supersymmetric vacua, described by the D-term equations

$$D^A = Q^{\dagger\,b}_{\ i} (T^A)^c_b Q^i_c - \widetilde{Q}^{\ b}_i (T^A)^c_b \widetilde{Q^\dagger}^{\ i}_c = 0, \tag{10.34}$$

where $A = 1, 2, \ldots, N^2 - 1$ is an index in the adjoint representation of $SU(N)$.

Up to flavor and global gauge rotations, a solution of the above equations can be found for both $F < N$ and $F \geq N$. For $F < N$ one can show that on the moduli space (10.34) the matrices Q and $\widetilde{Q}$ can be put, at most, in the following form

$$Q = \begin{pmatrix} v_1 & 0 & \ldots & 0 & 0 & \ldots & 0 \\ 0 & v_2 & \ldots & 0 & 0 & \ldots & 0 \\ \ldots & \ldots & \ldots & \ldots & 0 & \ldots & 0 \\ 0 & 0 & \ldots & v_F & 0 & \ldots & 0 \end{pmatrix} = \widetilde{Q}^T \tag{10.35}$$

Hence, at a generic point of the moduli space the gauge group is broken to $SU(N-F)$. The (classical) moduli space can be parameterized in terms of mesons fields

$$M^i_{\ j} = Q^i_a \widetilde{Q}^a_j \tag{10.36}$$

without any classical constraint between them, since the meson matrix has maximal rank.

For $F \geq N$ the matrices Q and $\widetilde{Q}$ can also be brought to a diagonal form on the moduli space

$$Q = \begin{pmatrix} v_1 & 0 & \dots & 0 \\ 0 & v_2 & \dots & 0 \\ \dots & \dots & \dots & \dots \\ 0 & 0 & \dots & v_N \\ 0 & 0 & 0 & 0 \\ \dots & \dots & \dots & \dots \\ 0 & 0 & 0 & 0 \end{pmatrix}, \quad \widetilde{Q}^T = \begin{pmatrix} \tilde{v}_1 & 0 & \dots & 0 \\ 0 & \tilde{v}_2 & \dots & 0 \\ \dots & \dots & \dots & \dots \\ 0 & 0 & \dots & \tilde{v}_N \\ 0 & 0 & 0 & 0 \\ \dots & \dots & \dots & \dots \\ 0 & 0 & 0 & 0 \end{pmatrix} \tag{10.37}$$

where $|v_i|^2 - |\tilde{v}_i|^2 = a$, with a a i-independent number. At a generic point of the moduli space the gauge group is now completely broken. The moduli space is efficiently described in terms of mesons and baryons but there exist classical constraints between them, now. The mesons are again defined as in eq. (10.36) but the meson matrix does not have maximal rank anymore. Baryons are gauge invariant single trace operators made out of N fields Q respectively N fields $\widetilde{Q}$, with fully anti-symmetrized indices and read

$$\begin{aligned} B_{i_1 \dots i_{F-N}} &= \epsilon_{i_1 i_2 \dots i_{F-N} j_1 \dots j_N} \epsilon^{a_1 a_2 \dots a_N} \, Q^{j_1}_{a_1} Q^{j_2}_{a_2} \dots Q^{j_N}_{a_N} \\ \widetilde{B}^{i_1 \dots i_{F-N}} &= \epsilon^{i_1 i_2 \dots i_{F-N} j_1 \dots j_N} \epsilon_{a_1 a_2 \dots a_N} \, \widetilde{Q}^{a_1}_{j_1} \widetilde{Q}^{a_2}_{j_2} \dots \widetilde{Q}^{a_N}_{jN}, \end{aligned} \tag{10.38}$$

where a_i are gauge indices and i_l, j_l are flavor indices.

Finally, as far as quantum correction are concerned, we know the exact (perturbative) expression for the gauge coupling which, in the holomorphic scheme, reads

$$\tau = \frac{\theta_{\rm YM}}{2\pi} + i\frac{4\pi}{g^2(\mu)} = \frac{b_1}{2\pi i}\log\frac{\Lambda}{\mu}, \quad b_1 = 3N - F \quad \text{and} \quad \Lambda = \mu\, e^{2\pi i \tau / b_1}. \tag{10.39}$$

10.4 $\mathcal{N}$=1 SQCD: non-perturbative dynamics

Our goal is to understand how the above picture is modified once non-perturbative corrections are taken into account and get, if possible, a quantum exact description of the vacua of the theory and of the low energy dynamics around them. Given the expression of the one-loop coefficient of the β-function (10.39), the region of the parameter space where to gain such an understanding is harder it is obviously $F < 3N$.

A generic prediction for a UV-free theory with a classical moduli space, as SQCD is for $F < 3N$, is that quantum corrections are expected to modify the perturbative analysis only near the origin of field space. Indeed, for large value of scalar field VEVs, the gauge group gets broken (and the gauge coupling hence stops running) for small values of the gauge coupling constant

$$e^{-\frac{8\pi^2}{g^2(\langle Q\rangle)}+i\theta_{\rm YM}} = \left(\frac{\Lambda}{\langle Q\rangle}\right)^{3N-F} \longrightarrow 0 \quad \text{for } \langle Q\rangle \to \infty. \qquad (10.40)$$

This implies that for large field VEVs the gauge coupling freezes at a value g_* where (semi)-classical analysis works properly. The smaller the field VEV the more important are quantum corrections. Hence, generically, we expect non-perturbative dynamics to modify the perturbative answer mostly near the origin of field space.

This said, it also turns out that for any fixed value of N, several non-perturbative dynamical properties change with the number of flavors, F. Hence, in what follows, we will consider qualitative different cases separately.

10.4.1 *Pure SYM: gaugino condensation*

We have already discussed this case, at a qualitative level. Let us first recall that this is the only case in which there does not exist an anomaly-free

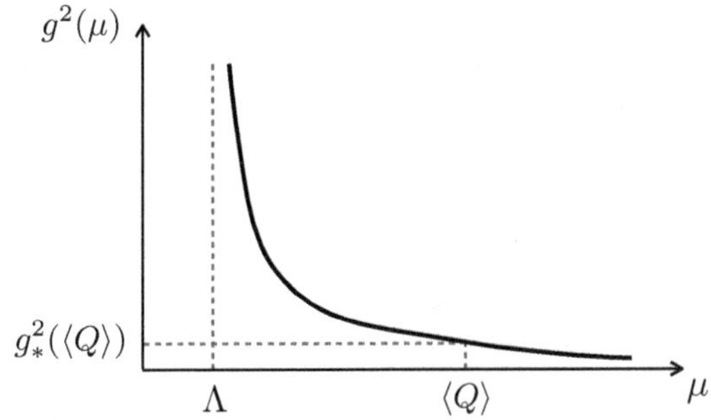

Fig. 10.9 The gauge coupling running of a UV-free theory. The large $\langle Q\rangle$ region is a weakly coupled region where classical analysis is correct, since the value at which the gauge coupling stops running, $g = g_*$, is small.

R-symmetry. At the quantum level, only a discrete $\mathbb{Z}_{2N}$ R-symmetry survives. Promoting τ to a spurion field and using holomorphy arguments, it is easy to see what the structure of the non-perturbative generated superpotential should be. Let us first notice that the operator $e^{2\pi i\tau/N}$ has R-charge $R = 2$. Indeed, due to the transformation properties of $\theta_{\rm YM}$ under R-symmetry transformations, $\theta_{\rm YM} \to \theta_{\rm YM} + 2N\alpha$, we have

$$e^{2\pi i\tau/N} \to e^{2i\alpha} e^{2\pi i\tau/N}. \tag{10.41}$$

Because of confinement, *assuming* a mass gap, the effective Lagrangian should depend only on τ, and hence $W_{\rm eff}$, if any, should also depend only on τ. Imposing R-symmetry, by dimensional analysis the only possible term reads

$$W_{\rm eff} = c\,\mu^3 e^{2\pi i\tau/N} = c\,\Lambda^3. \tag{10.42}$$

where c is an undetermined coefficient (which in principle could also be zero, of course). This innocent-looking constant superpotential contribution contains one crucial physical information. Given the presence of a massless strong interacting fermion field (the gaugino), one could wonder whether in SYM theory gauginos undergo pair condensation, as it is believed to happen in QCD, where quark bilinears condense. Looking at the SYM Lagrangian

$$\mathcal{L} = \frac{1}{32\pi}{\rm Im}\left[\int d^2\theta\,\tau\,{\rm Tr}\,W^\alpha W_\alpha\right], \tag{10.43}$$

we see that $\lambda^\alpha\lambda_\alpha$ is the scalar component of $W^\alpha W_\alpha$ and (minus) F_τ acts as a source for it (recall we are thinking of τ as a spurion superfield, $\tau = \tau + \sqrt{2}\theta\psi_\tau - \theta\theta F_\tau$). Therefore, in order to compute the gaugino condensate one should just differentiate the logarithm of the partition function $Z = \int \mathcal{D}V e^{i\int\mathcal{L}}$ with respect to F_τ. Under the assumption of a mass gap, the low energy effective action depends only on τ, since gauge fields have been integrated out, and it coincides with the effective superpotential (10.42), giving for the gaugino condensate

$$\langle\lambda\lambda\rangle = -16\pi\frac{\partial}{\partial F_\tau}\log Z = -16\pi i\frac{\partial}{\partial F_\tau}\int d^2\theta W_{\rm eff}(\tau) = 16\pi i\frac{\partial}{\partial\tau}W_{\rm eff}(\tau) \tag{10.44}$$

where in doing the second step we have used the fact that

$$W_{\rm eff} = w_{\rm eff}(\tau) + \sqrt{2}\frac{\partial W_{\rm eff}}{\partial\tau}\theta\psi_\tau - \theta\theta\left(\frac{\partial W_{\rm eff}}{\partial\tau}F_\tau + \frac{1}{2}\frac{\partial^2 W_{\rm eff}}{\partial\tau^2}\psi_\tau^2\right).$$

Plugging (10.42) in eq. (10.44) we get

$$\langle\lambda\lambda\rangle = -\frac{32\pi^2}{N} c\,\mu^3 e^{2\pi i\tau/N} \equiv a\,\Lambda^3, \tag{10.45}$$

which means that if $c \neq 0$ gauginos do condense in SYM. Since gauginos have $R = 1$, this implies that in the vacuum the $\mathbb{Z}_{2N}$ symmetry is broken to $\mathbb{Z}_2$ and that there are in fact N distinct (and isolated) vacua. All these vacua appear explicitly in the above formula since the transformation

$$\theta_{\rm YM} \to \theta_{\rm YM} + 2\pi k, \tag{10.46}$$

which is a symmetry of the theory, sweeps out N distinct values of the gaugino condensate

$$\langle\lambda\lambda\rangle \to e^{2i\alpha}\langle\lambda\lambda\rangle, \quad \theta_{\rm YM} \to \theta_{\rm YM} + 2N\alpha \simeq \theta_{\rm YM} + 2\pi k \tag{10.47}$$

where $k = 0, 1, \ldots, 2N-1$, and $k = i$ and $k = i + N$ give the same value of the gaugino condensate. In other words, we can label the N vacua with N distinct phases of the gaugino condensate $\left(0, 2\pi\frac{1}{N}, 2\pi\frac{2}{N}, \ldots, 2\pi\frac{N-1}{N}\right)$, recall Figure 10.6.

This ends our discussion of pure SYM. It should be stressed that to have a definitive picture we should find independent ways to compute the constant c in eq. (10.42), since if it were zero, then all our conclusions would have been wrong (in particular, there would not be any gaugino condensate, and hence we would have had a unique vacuum preserving the full $\mathbb{Z}_{2N}$ symmetry). We will come back to this important point later.

10.4.2 *$F < N$: the ADS superpotential*

For $F < N$ the classical analysis tells that there is a moduli space of complex dimension F^2, parameterized by meson field VEVs. The question, again, is whether an effective superpotential is generated due to strong coupling dynamics. This can be answered using again holomorphy.

First, notice that the complexified gauge coupling, which we tread by Λ^{3N-F}, is not charged under $SU(F)_{\rm L} \times SU(F)_{\rm R}$, while the only single trace operators made of the meson matrix and invariant under the non-abelian flavor symmetry is $\det M$. The quantum numbers under the abelian symmetries of $\det M$ and Λ^{3N-F} are

	$U(1)_{\rm B}$	$U(1)_A$	$U(1)_R$
$\det M$	0	$2F$	$2(F-N)$
Λ^{3N-F}	0	$2F$	0

The form of W_{eff} can be fixed requiring it to be invariant under the non-abelian global symmetry, $SU(F)_{\text{L}} \times SU(F)_{\text{R}}$, with R-charge two, $U(1)_A$ and $U(1)_{\text{B}}$ charges equal zero and dimension three. From the table above it follows that W_{eff} should be made by (a sum of) terms like

$$W_{\text{eff}} \sim \Lambda^{(3N-F)n} \,(\det M)^p \,, \tag{10.48}$$

where n and p are integer numbers subject to the constraints

$$\begin{cases} U(1)_A\colon\, 0 = 2nF + 2pF \\ U(1)_R\colon\, 2 = 2p(F-N) \end{cases} \quad \longrightarrow \quad \begin{cases} n = -p \\ p = -1/(N-F) \end{cases} \tag{10.49}$$

So we see that the only superpotential term which can be generated should have the following form

$$W_{\text{eff}} = c_{N,F} \left(\frac{\Lambda^{3N-F}}{\det M} \right)^{\frac{1}{N-F}} \,, \tag{10.50}$$

where, again, the overall constant, which generically will be some function of N and F, is undetermined. Notice that in the range we are considering, $F < N$, both $3N - F$ and $N - F$ are positive, so in the classical limit, $\Lambda \to 0$, W_{eff} vanishes, as needed. The expression (10.50) is the celebrated Affleck-Dine-Seiberg (ADS) superpotential.

In what follows we would like to analyze several properties of the ADS superpotential, trying to understand where it may come from, physically, and eventually determine the coefficient $c_{N,F}$.

Let us consider again the classical moduli space. At a generic point of the moduli space the $SU(N)$ gauge group is broken to $SU(N-F)$. Suppose for simplicity that all scalar field VEVs are equal, $v_i = v$, recall (10.35). Clearly the theory behaves differently at energies higher or lower than v. At energies higher than v the gauge coupling running is that of SQCD with gauge group $SU(N)$ and F massless flavors. At energies lower than v all matter fields become massive (and should be integrated out) and the theory behaves as pure SYM with gauge group $SU(N-F)$. Hence the gauge coupling runs differently at energies larger or smaller than v and, accordingly, the dynamical generated scale is also different. More precisely we have

$$\begin{aligned} E > v \quad \frac{4\pi}{g^2(\mu)} &= \frac{3N-F}{2\pi} \log \frac{\mu}{\Lambda} \\ E < v \quad \frac{4\pi}{g_{\text{L}}^2(\mu)} &= \frac{3(N-F)}{2\pi} \log \frac{\mu}{\Lambda_{\text{L}}}, \end{aligned} \tag{10.51}$$

where $\Lambda_{\rm L}$ is the strong coupling scale of the low energy effective theory. If supersymmetry is preserved, the two above equations should match at $E = v$. This is known as *scale matching* (that there are no threshold factors reflects a choice of subtraction scheme, on which threshold factors depend; this is the correct matching in, e.g. the $\overline{DR}$ scheme). Hence we get

$$\Lambda_{\rm L}^{3(N-F)} = \Lambda^{3N-F}\frac{1}{v^{2F}} = \frac{\Lambda^{3N-F}}{\det M} \quad \longrightarrow \quad \Lambda_{\rm L}^{3} = \left(\frac{\Lambda^{3N-F}}{\det M}\right)^{\frac{1}{N-F}}. \tag{10.52}$$

This implies that

$$W_{\rm eff} = c_{N,F}\left(\frac{\Lambda^{3N-F}}{\det M}\right)^{\frac{1}{N-F}} = c_{N,F}\Lambda_{\rm L}^{3}, \tag{10.53}$$

which means that

$$c_{N,F} = c_{N-F,0}. \tag{10.54}$$

Besides getting a relation between c's for different theories (recall these are (N, F)-dependent constants, in general), we also get from the above analysis some physical intuition for how the ADS superpotential is generated. One can think of $W_{\rm eff}$ being generated by gaugino condensation of the left over $SU(N-F)$ gauge group (recall that gaugino condensation is in one-to-one correspondence with the very existence of an effective superpotential for pure SYM theory: the two are fully equivalent statements).

Let us now start from SQCD with a given number of flavors and suppose to give a mass m to the F-th flavor. At high enough energy this does not matter much. But below the scale m the theory behaves as SQCD with one flavor less, as far as the gauge coupling running is concerned. More precisely, we have

$$\begin{aligned} E > m \quad & \frac{4\pi}{g^2(\mu)} = \frac{3N-F}{2\pi}\log\frac{\mu}{\Lambda_F} \\ E < m \quad & \frac{4\pi}{g_{\rm L}^2(\mu)} = \frac{3N-(F-1)}{2\pi}\log\frac{\mu}{\Lambda_{L,F-1}}, \end{aligned} \tag{10.55}$$

where Λ_F refers to the strong coupling scale of SQCD with F flavors and $\Lambda_{L,F-1}$ is the strong coupling scale of the low energy effective theory, SQCD with $F-1$ flavors. Matching the scale at $E = m$ we obtain the following relation between Λ_F and $\Lambda_{L,F-1}$

$$\Lambda_{L,F-1}^{3N-F+1} = m\,\Lambda_F^{3N-F}. \tag{10.56}$$

Let us now use holomorphic decoupling to connect the theories above and below the scale m. The superpotential of SQCD with $F-1$ massless flavors and one massive one reads

$$W_{\text{eff}} = c_{N,F}\left(\frac{\Lambda_F^{3N-F}}{\det M}\right)^{\frac{1}{N-F}} + mQ^{\mathrm{F}}\widetilde{Q}_{\mathrm{F}}. \tag{10.57}$$

At low enough energy we can trade the equations of motion of mesons involving the massive flavor by their F-term equations. The F-term equation for $M^{\mathrm{F}}{}_i$ for $i \neq F$ implies $M^{\mathrm{F}}{}_i = 0$, and similarly for $M^i{}_{\mathrm{F}}$. So the meson matrix can be put into the form

$$M = \begin{pmatrix} \widetilde{M} & 0 \\ 0 & t \end{pmatrix}, \quad t \equiv M^{\mathrm{F}}{}_{\mathrm{F}}, \tag{10.58}$$

where $\widetilde{M}$ is the meson matrix made out of $F-1$ flavors. The F-term equation for t gives

$$0 = -\frac{c_{N,F}}{N-F}\left(\frac{\Lambda_F^{3N-F}}{\det \widetilde{M}}\right)^{\frac{1}{N-F}}\left(\frac{1}{t}\right)^{1+\frac{1}{N-F}} + m \tag{10.59}$$

which implies

$$t = \left[\frac{N-F}{c_{N,F}}\, m\left(\frac{\Lambda_F^{3N-F}}{\det \widetilde{M}}\right)^{\frac{1}{F-N}}\right]^{\frac{F-N}{N-F+1}}. \tag{10.60}$$

Plugging this back into eq. (10.57) one gets

$$W_{\text{eff}} = (N-F+1)\left(\frac{c_{N,F}}{N-F}\right)^{\frac{N-F}{N-F+1}}\left(\frac{m\Lambda_F^{3N-F}}{\det \widetilde{M}}\right)^{\frac{1}{N-F+1}}. \tag{10.61}$$

We can now use eq. (10.56) and get for the effective superpotential of SQCD with $F-1$ flavors, which is what the theory reduces to at low enough energies,

$$\begin{aligned} W_{\text{eff}} &= (N-F+1)\left(\frac{c_{N,F}}{N-F}\right)^{\frac{N-F}{N-F+1}}\left(\frac{\Lambda_{L,F-1}^{3N-F+1}}{\det \widetilde{M}}\right)^{\frac{1}{N-F+1}} \\ &= c_{N,F-1}\left(\frac{\Lambda_{L,F-1}^{3N-F+1}}{\det \widetilde{M}}\right)^{\frac{1}{N-F+1}}, \end{aligned} \tag{10.62}$$

giving finally the following relation

$$c_{N,F-1} = (N - F + 1) \left(\frac{c_{N,F}}{N - F} \right)^{\frac{N-F}{N-F+1}} . \tag{10.63}$$

Combining this result with the relation we found before, eq. (10.54), one concludes that all coefficients are related one another as

$$c_{N,F} = (N - F)\, c^{\frac{1}{N-F}} , \tag{10.64}$$

with a unique common coefficient c to be determined. This result tells that if the ADS superpotential can be computed exactly for a given value of F (hence fixing c), then we know its expression for any other value!

Let us consider the case $F = N - 1$, which is the extreme case in the window $F < N$. In this case

$$c_{N,N-1} = c. \tag{10.65}$$

Interestingly, for $F = N - 1$ the gauge group is fully broken, so there is no left-over strong IR dynamics. In other words, any term appearing in the effective action should be visible in a weak-coupling analysis. Even more interesting, the ADS superpotential for $F = N - 1$ is proportional to Λ^{2N+1} which is nothing but how one-instanton effects contribute to gauge theory amplitudes (recall that for $F = N - 1$ $b_1 = 2N + 1$, and $e^{-S_{inst}} \sim \Lambda^{b_1}$), suggesting that in this case the ADS superpotential is generated by instantons. At weak coupling, a reliable one-instanton calculation can indeed be done and gives $c = 1$. Via eq. (10.64) this result hence fixes uniquely $c_{N,F}$ for arbitrary values of N and F as

$$c_{N,F} = N - F, \tag{10.66}$$

giving finally for the ADS superpotential the following exact expression

$$W_{\text{ADS}} = (N - F) \left(\frac{\Lambda^{3N-F}}{\det M} \right)^{\frac{1}{N-F}} . \tag{10.67}$$

Notice that this also fixes the coefficient of the effective superpotential of pure SYM theory (10.42), which is

$$W_{\text{eff}} = N\, \Lambda^3, \tag{10.68}$$

implying, via eq. (10.44), that gauginos do condense!

Let us finally see how does the ADS superpotential affect the moduli space of vacua. From the expression (10.67) we can compute the potential,

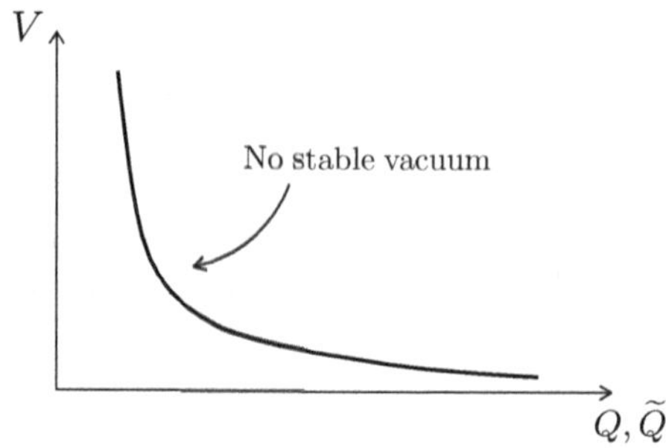

Fig. 10.10 The runaway behavior of the quantum corrected potential of $SU(N)$ SQCD with $F < N$.

which is expected not to be flat anymore, since the effective superpotential W_{ADS} depends on scalar fields (through the meson matrix). The potential

$$V_{\text{ADS}} = \sum_i \left|\frac{\partial W_{\text{ADS}}}{\partial Q_i}\right|^2 + \left|\frac{\partial W_{\text{ADS}}}{\partial \widetilde{Q}^i}\right|^2 \tag{10.69}$$

is minimized at infinity in field space, namely for $Q = \widetilde{Q} \to \infty$, where it reaches zero, see Figure 10.10. This can be easily seen noticing that $\det M \sim M^F$, which implies that $V_{\text{ADS}} \sim |M|^{-\frac{2N}{N-F}}$, which is indeed minimized at infinity. This means that the theory does not admit any stable vacuum at finite distance in field space: the (huge) classical moduli space is completely lifted at the quantum level!

This seemingly strange behavior actually makes sense, if one thinks about it for a while. For large field VEVs, eventually for $v \to \infty$, we recover pure SYM which has indeed supersymmetric vacua (that is, zero energy states). This is part of the space of D-term solutions of SQCD; any other configuration would have higher energy and would hence be driven to the supersymmetric one. Let us suppose this picture were wrong and that SQCD had a similar behavior as QCD: confinement and chiral symmetry breaking. Then we would have expected a quark condensate to develop $\langle \psi_{Q_i} \psi_{\widetilde{Q}^j} \rangle \neq 0$. Such condensate, differently from a gaugino condensate (which we certainly have), would break supersymmetry, since it is nothing but a F-term for the meson matrix M_i^j. Hence this configuration would have $E > 0$ and thus any configuration with $E = 0$ would be preferred. The latter are all configurations like (10.35) which, by sending v_i all the way to infinity, reduce to SYM, which admits supersymmetry preserving vacua. The ADS superpotential simply shows this.

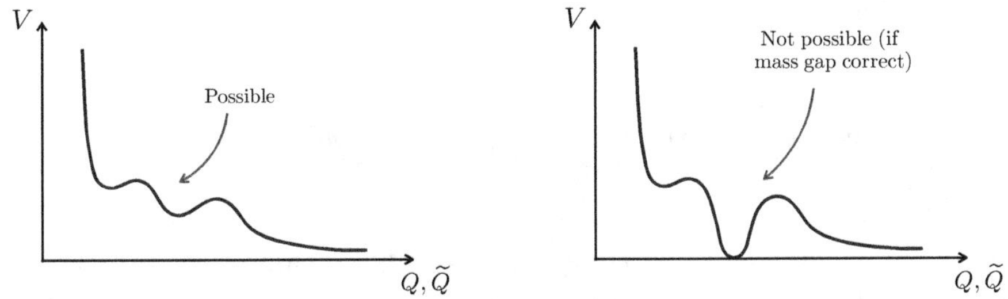

Fig. 10.11 The effect of a non-canonical Kähler potential on the ADS potential. The picture on the right cannot hold if the assumption of mass gap for pure SYM is correct.

There is a caveat in all this discussion. In our analysis we have not included wave-function renormalization effects. The latter could give rise, in general, to non-canonical Kähler potential terms, which could produce wiggles or even local minima in the potential. However, at best this could give rise to metastable vacua (which our holomorphic analysis cannot see), but it would not lift the absolute supersymmetric minima at infinity, a region where the Kähler potential is nearly canonical in the UV-variables Q and $\widetilde{Q}$. On the other hand, no supersymmetric minima can arise at finite distance in field space. These would correspond to singularities of the Kähler metric, implying that at those specific points in field space extra massless degrees of freedom show-up. This cannot be, if the assumption of mass gap for pure SYM (to which the theory reduces at low enough energy, at generic points on the classical moduli space) is correct.

10.4.3 *Integrating in and out: the linearity principle*

The superpotential of pure SYM is sometime written as

$$W_{\mathrm{VY}} = NS\left(1 - \log\frac{S}{\Lambda^3}\right) \tag{10.70}$$

where $S = -\frac{1}{32\pi^2}\operatorname{Tr} W^\alpha W_\alpha$ is the so-called glueball superfield and the subscript VY stands for Veneziano-Yankielowicz. Let us first notice that integrating S out (recall we are supposing pure SYM has a mass gap) we get

$$\frac{\partial W_{\mathrm{VY}}}{\partial S} = N\left(1 - \log\frac{S}{\Lambda^3}\right) + NS\left(-\frac{1}{S}\right) = 0, \tag{10.71}$$

which implies

$$\langle S \rangle = \Lambda^3. \tag{10.72}$$

Plugging this back into the VY superpotential gives

$$W_{\rm VY} = N\Lambda^3, \tag{10.73}$$

which is nothing but the effective superpotential of pure SYM we have previously derived, eq. (10.68). From this view point the two descriptions seem to be equivalent, at least as far as low enough energies are concerned: the effective superpotential (10.68) can be obtained from the VY superpotential by integrating S out.

Analogously, the ADS superpotential is sometime written as

$$W_{\rm TVY} = (N-F)\, S \left[1 - \frac{1}{N-F} \log \left(\frac{S^{N-F} \, \det M}{\Lambda^{3N-F}} \right)\right], \tag{10.74}$$

where TVY stands for Taylor-Veneziano-Yankielowicz. Integrating S out one now recovers the ADS superpotential. Instead, adding a mass term for all matter fields and integrating M out, from the TVY superpotential one gets, consistently, the VY superpotential.

The fact is that one can also revert the procedure, and obtain the VY and TVY superpotentials starting from the expressions (10.68) and (10.67) and integrating the glueball superfield S *in*. At first sight this could look strange since one does not expect the Wilsonian RG to be invertible, of course. The TVY or VY superpotentials include one more dynamical field with respect to (10.67)–(10.68), namely the glueball superfield S, so we expect them to contain some dynamical information more. As we are going to discuss below, this intuition is not correct: the two descriptions are completely equivalent.

Let us try to be as general as possible and consider a supersymmetric gauge theory admitting also a tree-level superpotential $W_{\rm tree}$. Given a set of chiral superfields Φ_i, the generic form of such superpotential is

$$W_{\rm tree} = \sum_r \lambda_r X_r(\Phi_i), \tag{10.75}$$

where λ_r are coupling constants and X_r gauge invariant combinations of the chiral superfields Φ_i. In general, one would expect the non-perturbative generated superpotential $W_{\rm non-pert}$ to be a (holomorphic) function of the couplings λ_r, the gauge invariant operators X_r, and of the dynamical generated scales Λ_s (we are supposing, to be as most general as possible, the gauge group not to be simple, hence we allow for several dynamical scales). In fact, as shown by Intriligator, Leigh and Seiberg, $W_{\text{non-pert}}$ does not depend on the couplings λ_r. This fact implies that the full effective superpotential (which includes both the tree level and the non-perturbative contributions) is linear in the couplings. This is sometime referred to as *linearity principle.* The upshot is that, in general, we have

$$W_{\text{eff}} = \sum_r \lambda_r X_r + W_{\text{non-pert}}(X_r, \Lambda_s). \tag{10.76}$$

Let us focus on the dependence on, say, λ_1. At low enough energy (where the superpotential piece dominates — let us suppose for now that X_1 is massive) we can integrate out the field X_1 by solving its F-term equation, which, because of eq. (10.76), reads

$$\lambda_1 = -\frac{\partial}{\partial X_1} W_{\text{non-pert}}. \tag{10.77}$$

The above equation is the same as a Legendre transform. In other words, the coupling λ_r and the gauge invariant operator X_r behave as Legendre dual variables. Solving for X_1 in terms of λ_1 and all other variables, and substituting in eq. (10.76), one obtains an effective superpotential with a complicated dependence on λ_1 but where X_1 has been integrated out. Repeating the same reasoning for all X_r one can integrate out all fields and end-up with an effective superpotential written in terms of couplings only

$$W_{\text{eff}}(\lambda_r, \Lambda_s) = \left[\sum_r \lambda_r X_r + W_{\text{non-pert}}(X_r, \Lambda_s)\right]_{X_r(\lambda,\Lambda)}. \tag{10.78}$$

The point is that the Legendre transform is invertible. Therefore, as we can integrate out a field, we can also integrate it back in, by reversing the procedure

$$\langle X_r \rangle = \frac{\partial}{\partial \lambda_r} W_{\text{eff}}(\lambda_r, \Lambda_s). \tag{10.79}$$

The reason why the two descriptions, one in terms of the fields, one in terms of the dual couplings, are equivalent is because we have not considered D-terms. D-terms contain the dynamics (e.g. the kinetic term). Hence, if we ignore D-terms, namely if we only focus on holomorphic terms as we are doing here, integrating out or in a field is an operation which does not make us loose or gain information. As far as the holomorphic part of the effective action is concerned, a field and its dual coupling are fully equivalent.

What about the dynamical scales Λ_s? Can one introduce canonical pairs for them, too? The answer is yes, and this is where the physical equivalence between ADS and TVY superpotentials we claimed about becomes explicit.

Let us start by considering pure SYM. One can write the gauge kinetic term as a contribution to the tree level superpotential in the sense of eq. (10.75)

$$W_{\text{tree}} = \frac{\tau(\mu)}{16\pi i} \operatorname{Tr} W^{\alpha} W_{\alpha} = 3N \log\left(\frac{\Lambda}{\mu}\right) S, \tag{10.80}$$

where S is a X-like field and $3N \log(\Lambda/\mu)$ the dual coupling. In other words, one can think of S and $\log \Lambda$ as Legendre dual variables. From this view point, the SYM superpotential (10.68) is an expression of the type (10.78), where the field S has been integrated out and the dependence on the dual coupling is hence non-linear. Indeed (10.68) can be re-written as

$$W_{\text{eff}} = N\Lambda^3 = N\mu^3 e^{\frac{1}{N} 3N \log \frac{\Lambda}{\mu}}, \tag{10.81}$$

which explicitly shows that this is the case. Using now eq. (10.79) applied to this dual pair, one gets

$$\langle S \rangle = \frac{1}{3N} \Lambda \frac{\partial}{\partial \Lambda} W_{\text{eff}} = \Lambda^3. \tag{10.82}$$

Therefore

$$W_{\text{non-pert}}(S) = W_{\text{eff}} - W_{\text{tree}} = NS - 3N \log\left(\frac{\Lambda}{\mu}\right) S = NS - NS \log \frac{S}{\mu^3}, \tag{10.83}$$

which is correctly expressed, according to the linearity principle, in terms of S only, and not the coupling, $\log \Lambda$. We can now add the two contributions, the one above and (10.80) and get for the effective superpotential an expression in the form (10.76)

$$W_{\text{eff}} = W_{\text{non-pert}} + W_{\text{tree}} = NS\left(1 - \log \frac{S}{\Lambda^3}\right) \tag{10.84}$$

which is nothing but the VY superpotential! The same reasoning can be applied to a theory with flavor and/or with multiple dynamical scales. The upshot is one and the same: integrating in (TVY) or out (ADS) fields holomorphically, are operations which one can do at no cost. The two descriptions are physically equivalent.

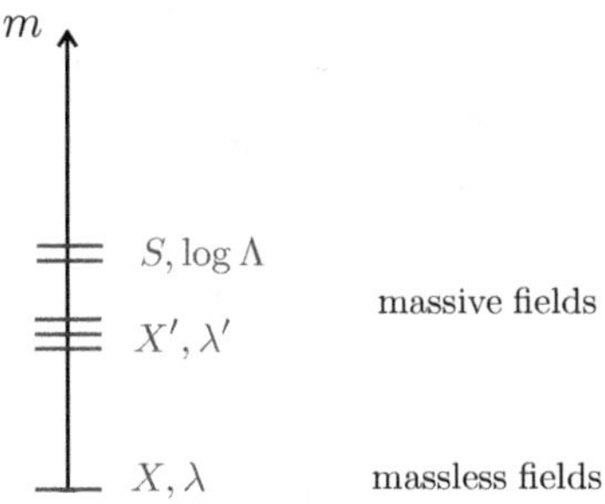

Fig. 10.12 The mass spectrum of a generic theory.

To sum-up, the relation between couplings and dual field variables for the most generic situation is

Couplings	$b_1^1 \log \frac{\Lambda_1}{\mu}$	$b_1^2 \log \frac{\Lambda_1}{\mu}$	...	λ_1	λ_2	...
Fields	S_1	S_2	...	X_1	X_2	...

Suppose now that the mass spectrum of such generic theory is as in Figure 10.12: the X's are a set of massless fields, the X''s are massive ones, and the S's are glueball superfields (which are all massive because of mass gap of the pure glue theory, *i.e.* all Λ's $\neq 0$).

The most Wilsonian thing to do would be to describe the effective superpotential in terms of fields X, and couplings λ' and Λ

$$W_{\text{eff}} = W_{\text{eff}}(X, \lambda', \Lambda). \tag{10.85}$$

In this sense, the ADS superpotential is more Wilsonian than the TVY. Seemingly, for pure SYM the most Wilsonian thing to do is to express the effective superpotential as a function of the coupling only (since the glueball superfield is massive), namely as $W_{\text{eff}} = N\Lambda^3$. However, since (as far as the holomorphic part of the effective action is concerned) integrating in and out fields are equivalent operations, one can very well choose to write down the effective superpotential by integrating X fields out and X' and S fields in (or anything in between these two extreme cases)

$$W_{\text{eff}} = W_{\text{eff}}(\lambda, X', S), \tag{10.86}$$

getting an equivalent way of describing the low energy effective theory superpotential. This said, one should bare in mind that as far as the massless fields X, there is no actual energy range for which integrating them out makes real physical sense, and this would be indicated by the Kähler potential of the effective theory being ill-defined (in other words, there is no energy range in which the kinetic term of such massless fields is negligible, since the energy is always bigger or equal than the field mass, which is vanishing). On the contrary, in presence of a mass gap, that is in the absence of X-like fields, the two descriptions, one in terms of couplings the other in terms of fields, are equivalent, since now no singularities are expected in the Kähler potential. And this is a more and more exact equivalence the lower the energy.

10.4.4 $F \geq N$: *persistent moduli space*

Let us now go back to our analysis of the IR dynamics of SQCD with gauge group $SU(N)$ and F flavors. What about the case $F \geq N$? As we are going to see, things change drastically. For one thing, a properly defined effective superpotential cannot be generated. Using couplings and fields we have (mesons, baryons and the dynamical scale Λ), there is no way of constructing an object respecting all symmetries, with the correct dimension, and being vanishing in the classical limit. The consequence is that for $F \geq N$ the classical moduli space is not lifted. This does not mean that nothing interesting happens. For instance, the moduli space can be *deformed* by strong dynamics effects. Moreover, the perturbative analysis does not tell us what the low energy effective theory looks like; as we will see instead, in some cases we will be able to make very non-trivial statements about the way light degrees of freedom interact, and in turn about the phase the theory enjoys.

In what follows, we will consider qualitatively different cases separately. Let us start analyzing the case $F = N$. It is easy to see that in this case all gauge invariant operators have R-charge $R = 0$, so one cannot construct an effective superpotential with $R = 2$. However, as we are going to show, something does happen due to strong dynamics.

Besides the mesons, there are now two baryons

$$B = \epsilon^{a_1 a_2 \ldots a_N} \, Q^1_{a_1} Q^2_{a_2} \ldots Q^N_{a_N}$$
$$\widetilde{B} = \epsilon_{a_1 a_2 \ldots a_N} \, \widetilde{Q}^{a_1}_1 \widetilde{Q}^{a_2}_2 \ldots \widetilde{Q}^{a_N}_N \, .$$

The classical moduli space is parameterized by VEVs of mesons and baryons. There is, however, a classical constraint between them

$$\det M - B\widetilde{B} = 0 \qquad (10.87)$$

(this comes because for $N = F$ we have that $\det Q = B$ and $\det \widetilde{Q} = \widetilde{B}$ and the determinant of the product is the product of the determinants). One can ask whether this classical constraint is modified at the quantum level. In general, one could expect the quantum version of the above classical constraint to be

$$\det M - B\widetilde{B} = a\,\Lambda^{2N}, \tag{10.88}$$

where a is a (undetermined for now) dimensionless and charge-less constant. One can get easily convinced that this is the only possible modification compatible with all physical requirements. First it has the correct physical dimension, namely the same as the l.h.s, $\Delta = 2N$. Second, it correctly vanishes in the classical limit, $\Lambda \to 0$. Third, it has vanishing R-charge and $U(1)_A$ charge $2N$, as the l.h.s has. Finally, it is suggestive that the power that Λ enters in eq. (10.88) is the one-loop coefficient of the β-function and is exactly that associated with a one instanton correction, since for $N = F$ we have for the instanton action

$$e^{-S_{\rm inst}} \sim e^{-\frac{8\pi^2}{g^2}+i\theta_{\rm YM}} \sim \Lambda^{2N}. \tag{10.89}$$

So, there are no reasons not to allow for a modification as (10.88), modulo the constant a which can very well be vanishing, after all. So, given that in principle a modification like (10.88) is allowed, everything boils down to determine whether the constant a is vanishing or has a finite value.

The constraint (10.88) can be implemented, formally, by means of a Lagrange multiplier, allowing a superpotential

$$W = A\left(\det M - B\widetilde{B} - a\Lambda^{2N}\right) \tag{10.90}$$

where A is the Lagrange multiplier, whose equation of motion is by construction the constraint (10.88). The interesting thing is that one can use holomorphic decoupling to fix the constant a. Adding a mass term for the N-th flavor, the low energy theory reduces to SQCD with $F = N - 1$. Imposing that after having integrated out the N-th flavor one obtains an effective superpotential which matches the ADS superpotential for $F = N - 1$, fixes $a = 1$, that is

$$\det M - B\widetilde{B} = \Lambda^{2N}. \tag{10.91}$$

So the quantum constraint is there, after all. Actually, it is *necessary* for it to be there in order to be consistent with what we already know about the quantum properties of SQCD with $F < N$!

Several comments are in order at this point.

$F = N$ SQCD is the first case we meet where a moduli space of supersymmetric vacua persists at the quantum level. Still, the quantum moduli space is different from the classical one. The moduli space (10.87) is singular. It has a singular submanifold reflecting the fact that on this submanifold additional massless degrees of freedom arise. This is the submanifold where not only (10.87) is satisfied, but also $d(\det M - B\widetilde{B}) = 0$, which makes the tangent space singular and therefore good local coordinates not being well-defined. This happens whenever baryon VEVs vanish, $B = \widetilde{B} = 0$, and the meson matrix has rank $k \leq N-2$ since

$$d(\det M - B\widetilde{B}) = \text{minor}\{M^i_j\}\, dM^i_j - Bd\widetilde{B} - \widetilde{B}dB, \tag{10.92}$$

and for this to be zero each term should vanish separately. Note that when only one of the two baryon operators vanishes ($B = 0, \widetilde{B} \neq 0$ or viceversa) the rank of M can be as large as $N-1$ since $\det M = \det Q \det \widetilde{Q}$. In this case $\text{minor}\{M^i_j\} \neq 0$ implying that also $d(\det M - B\widetilde{B}) \neq 0$ and the metric is hence not singular.

On the submanifold where (10.92) vanishes a $SU(N-k)$ gauge group remains unbroken, and corresponding gluons (as well as some otherwise massive matter fields) remain massless. The quantum moduli space (10.91) is instead smooth. Basically, when $B = \widetilde{B} = 0$ the rank of the meson matrix is not diminished since its determinant does not vanish, now: everywhere on the quantum moduli space the gauge group is fully broken.

Classically, the origin is part of the space of vacua. Hence, chiral symmetry can be unbroken. At the quantum level, instead, the origin is excised (all the classical singular submanifold, in fact), so in any allowed vacuum chiral symmetry is broken (like in QCD). Moreover, being the moduli space non-singular, means there are no massless degrees of freedom other than mesons and baryons. But the latter are massless, since are moduli. Hence in SQCD with $N = F$ there is no mass gap (as for massless QCD). Note that because of supersymmetry, this means that there are also massless composite fermions.

Obviously, the chiral symmetry breaking pattern is not unique. Different points on the moduli space display different patterns. At a generic point, where all gauge invariant operators get a VEV, all global symmetries are broken. But there are submanifolds of enhanced global symmetry. For instance, along the *mesonic branch*, defined as

$$M^i_j = \Lambda^2 \delta^i_j, \quad B = \widetilde{B} = 0, \tag{10.93}$$

we have that

$$SU(F)_{\rm L} \times SU(F)_{\rm R} \times U(1)_{\rm B} \times U(1)_R \longrightarrow SU(F)_{\rm D} \times U(1)_{\rm B} \times U(1)_R, \tag{10.94}$$

a chiral symmetry breaking pattern very much similar to QCD. Along the *baryonic branch*, which is defined as

$$M^i_j = 0, \quad B = -\widetilde{B} = \Lambda^N, \tag{10.95}$$

we have instead

$$SU(F)_{\rm L} \times SU(F)_{\rm R} \times U(1)_{\rm B} \times U(1)_R \longrightarrow SU(F)_{\rm L} \times SU(F)_{\rm R} \times U(1)_R, \tag{10.96}$$

which is very different from QCD (the full non-abelian chiral symmetry is preserved).

Which phases does the theory enjoy? The classical singular submanifold is excised. Therefore, the gauge group is always (fully) broken and the theory is hence in a Higgs phase. Still, near the origin the theory can be better thought to be in a confined phase, since the effective theory is smooth in terms of mesons and baryons, and, moreover, we are in the strongly coupled region of field space, where an inherently perturbative Higgs description is not fully appropriate. In fact, there is no order parameter which can distinguish between the two phases; there is no phase transition between them (this is similar to the prototype example of one-family EW theory we discussed already). Notice that the Wilson loop is not a useful order parameter here since it follows the perimeter law, no matter where one sits on the moduli space: we do not have strict confinement but just charge screening, as in QCD, since we have (light) matter transforming in the fundamental representation of the gauge group, and therefore flux lines can (and do) break. The qualitative difference between classical and quantum moduli spaces, and their interpretation is depicted in Figure 10.13.

A non-trivial consistency check of this picture comes from computing 't Hooft anomalies in the UV and in the IR. Let us consider, for instance, the mesonic branch. The charges under the unbroken global symmetries, $SU(F)_{\rm D} \times U(1)_{\rm B} \times U(1)_R$ of the UV (fundamental) and IR (composite)

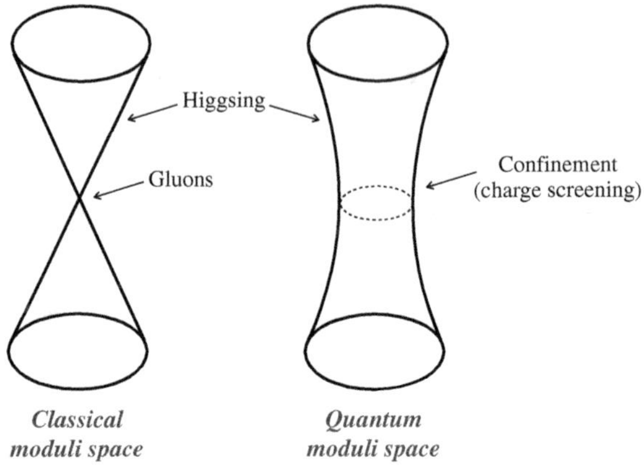

Fig. 10.13 Classical picture (left): at the origin the full gauge symmetry is recovered and chiral symmetry is not broken. Quantum picture (right): the (singular) origin has been replaced by a circle of theories where chiral symmetry is broken (rather than Higgs phase, this resembles more closely the physics of a confining vacuum).

degrees of freedom are as follows

	$SU(F)_{\rm D}$	$U(1)_{\rm B}$	$U(1)_R$
ψ_Q	F	1	-1
$\psi_{\widetilde{Q}}$	$\overline{F}$	-1	-1
λ	$\bullet$	0	1
ψ_M	Adj	0	-1
$\psi_{\rm B}$	$\bullet$	F	-1
$\psi_{\widetilde{B}}$	$\bullet$	$-F$	-1

where we have used the constraint (10.91) to eliminate the fermionic partner of $\operatorname{Tr} M$, so that ψ_M transforms in the Adjoint of $SU(F)_{\rm D}$. We can now compute several triangular anomalies and see whether computations done in terms of UV and IR degrees of freedom agree. We get

$$
\begin{array}{lcc}
 & UV & IR \\
SU(F)^2_{\rm D}\, U(1)_R & 2N\frac{1}{2}(-1) = -N & F(-1) = -F \\
U(1)^2_{\rm B}\, U(1)_R & -2NF & -2F^2 \\
U(1)^3_R & -2NF + N^2 - 1 & -(F^2-1) - 1 - 1 = -F^2 - 1
\end{array}
\tag{10.97}
$$

Since (crucially!) $F = N$ we see that 't Hooft anomaly matching holds. A similar computation can be done for the baryonic branch finding again perfect agreement between the UV and IR 't Hooft anomalies. This rather non-trivial agreement ensures that our low energy effective description in terms of mesons and baryons, subject to the constraint (10.91), is most likely correct.

Let us move on and consider the next case, $F = N + 1$. The moduli space is again described by mesons and baryons. We have $N + 1$ baryons of type B and $N + 1$ baryons of type $\widetilde{B}$ now

$$B_i = \epsilon_{ij_1 \ldots j_N} \, \epsilon^{a_1 a_2 \ldots a_N} \, Q^{j_1}_{a_1} Q^{j_2}_{a_2} \ldots Q^{j_N}_{a_N}$$
$$\widetilde{B}^i = \epsilon^{ij_1 \ldots j_N} \, \epsilon_{a_1 a_2 \ldots a_N} \, \widetilde{Q}^{a_1}_{j_1} \widetilde{Q}^{a_2}_{j_2} \ldots \widetilde{Q}^{a_N}_{jN}.$$

As we are going to show, differently from the previous case, the classical moduli space not only is unlifted, but is quantum exact, also. In other words, there are no quantum modifications to it.

This result can be proved using holomorphic decoupling. The rationale goes as follows. This system can be described, formally, by the following superpotential

$$W_{\text{eff}} = \frac{a}{\Lambda^{2N-1}} \left(\det M - B_i \, M^i_{\ j} \, \widetilde{B}^j \right), \tag{10.98}$$

where $i = 1, 2, \ldots, N+1$ is a flavor index, $2N - 1$ is the one-loop β-function coefficient and a, as usual, is for now an undetermined coefficient. The above superpotential has all correct symmetry properties, including the R-charge, which is indeed equal to 2. Notice, though, that since the rank of the meson matrix $k \leq N$, then $\det M = 0$, classically. So the above equation should be really thought of as a quantum equation, valid off-shell, so to say.

Let us now add a mass m to the F-th flavor. This gives

$$W_{\text{eff}} = \frac{a}{\Lambda^{2N-1}} \left(\det M - B_i \, M^i_{\ j} \, \widetilde{B}^j \right) - m M^{\text{N+1}}_{\ \text{N+1}}. \tag{10.99}$$

Integrating out massive modes, which tantamounts to impose the F-flatness conditions for $M^{\text{N+1}}_{\ \ i}, M^i_{\ \text{N+1}}, B_i$ and $\widetilde{B}^i$ for $i < N + 1$, reduces the meson matrix and the baryons to

$$M = \begin{pmatrix} \widehat{M}^i_{\ j} & 0 \\ 0 & t \end{pmatrix}, \quad B = \begin{pmatrix} 0_i \\ \widehat{B} \end{pmatrix}, \quad \widetilde{B} = \begin{pmatrix} 0^i \\ \widehat{\widetilde{B}} \end{pmatrix} \tag{10.100}$$

where now $i, j = 1, \ldots, N$, and $t = M^{\mathrm{N+1}}_{\mathrm{N+1}}$. The F-flatness condition for t reads

$$\frac{a}{\Lambda^{2N-1}}\left(\det \widehat{M} - \widehat{B}\widehat{\widetilde{B}}\right) - m = 0 \tag{10.101}$$

which implies

$$\det \widehat{M} - \widehat{B}\widehat{\widetilde{B}} = \frac{1}{a}\, m\, \Lambda^{2N-1} = \frac{1}{a}\, \Lambda_{\mathrm{L}}^{2N}, \tag{10.102}$$

where in the last step we have used the relation (10.56). This shows that the ansatz (10.98) is correct, since upon holomorphic decoupling we get exactly the quantum constraint of $F = N$ SQCD (and a gets fixed to one).

From eq. (10.98), by differentiating with respect to $M^i_{\ j}$, B_i and $\widetilde{B}^i$ we get the moduli space equations (*i.e.* the classical, still quantum exact, constraints between baryons and mesons)

$$\begin{cases} M \cdot \widetilde{B} = B \cdot M = 0 \\ \det M \cdot \left(M^{-1}\right)_i^{\ j} - B_i \widetilde{B}^j = 0 \end{cases} \tag{10.103}$$

where $\det M \cdot \left(M^{-1}\right)_i^{\ j} \equiv \text{minor}\,\{M\}_i^{\ j}$ (recall that above equations are on-shell, and on-shell $\det M$ itself vanishes but the minor $\{M\}_i^{\ j}$ does not!).

As a non-trivial check of this whole picture one can verify, choosing any preferred point in the space of vacua, that 't Hooft anomalies match (and hence that our effective description holds).

Now that we know eq. (10.98) is correct, let us try to understand what does it tell us about the vacuum structure of SQCD with $F = N + 1$. First, unlike $F = N$ SQCD, the origin of field space, where all VEVs vanish, is part of the moduli space. In such vacuum chiral symmetry is unbroken. This is an instance of a theory displaying confinement (actually charge screening) without chiral symmetry breaking. Theories with such a property, like $F = N + 1$ SQCD at the origin of field space, are said to be *s-confining*.

Classically, the singularities at the origin are interpreted as extra massless gluons (and matter fields), since the theory gets unhiggsed for vanishing values of matter field VEVs. At the quantum level, the physical interpretation is different, since because the theory is UV-free, the region around the origin is the more quantum one. Singularities are more naturally associated with additional massless mesons and baryons which pop-up since eqs. (10.103) are trivially realized at the origin and therefore do not provide any actual constraint between meson and baryon components. In other words, at the origin the number of mesonic and baryonic massless degrees of freedom is larger than the dimension of the moduli space. This can be checked, again, by 't Hooft anomaly matching.

On the other hand, as $F = N$ SQCD, this theory exhibits *complementarity*, in the sense that one can move smoothly from a confining phase (near the origin) to a Higgs phase (at large field VEVs) without any order parameter being able to distinguish between them (again, the Wilson loop follows the perimeter law in both phases).

One could try to go further, and apply the same logic to $F = N + 2$ (and on). On general ground one would expect $M^i_{\ j}$, B^{ij}, $\widetilde{B}_{ij}$ (baryons have now two free flavor indices) to be the dynamical degrees of freedom in the IR, and could then try to construct an effective (off-shell) superpotential of the kind of (10.98). This, however, does not work. Looking at the charges of the various gauge invariant operators and dynamical scale Λ one can easily see that an effective superpotential with R-charge equal to 2, correct physical dimensions and symmetries, cannot be constructed. Indeed, the only $SU(F)_{\rm L} \times SU(F)_{\rm R}$ invariant superpotential one could construct should be the obvious generalization of (10.98), that is

$$W_{\rm eff} \sim \det M - B_{il}\, M^i_{\ j} M^l_{\ m}\, \widetilde{B}^{jm}, \tag{10.104}$$

which, to start with, does not have $R = 2$ but actually $R = 4$ (things get worse the larger the number of flavors). Even 't Hooft anomaly matching condition can be proven not to work. For instance, choosing for simplicity the origin of field space where meson and baryons are unconstrained, one can easily see that 't Hooft anomalies do not match. More generally, it turns out that increasing F, 't Hooft anomaly coefficients computed using (unconstrained) IR degrees of freedom increase much faster than those computed using UV degrees of freedom, and only for $F = N + 1$ they match.

In fact, things turn out to be rather different. As we will show, the correct degrees of freedom to describe the low energy effective dynamics of $F = N + 2$ SQCD are those of an IR-free theory (!) described by a supersymmetric gauge theory with gauge group $SU(2)$, F chiral superfields q transforming in the fundamental of $SU(2)$, F chiral superfields $\tilde{q}$ transforming in the anti-fundamental (hence F flavors) and F^2 singlet chiral superfields Φ, plus a cubic tree level superpotential coupling $q, \tilde{q}$ and Φ. How that can be?

Two pieces of information are needed in order to understand this apparently weird result and, more generally, to understand what is going on for $F \geq N+2$. Both are due to Seiberg and in the following we will review them in turn.

10.4.5 *Conformal window*

A first proposal is that SQCD in the range $\frac{3}{2}N < F < 3N$ flows to an interacting IR fixed point (meaning it does not confine!). In other words, even if the theory is UV-free and hence the gauge coupling g increases through the IR, at low energy g reaches a constant RG-fixed value. Let us try to see how such claim comes about. The SQCD β-function for the physical gauge coupling (which hence takes into account wave-function renormalization effects) is

$$\beta(g) = -\frac{g^3}{16\pi^2}\frac{3N - F\left[1-\gamma(g^2)\right]}{1-Ng^2/8\pi^2}, \tag{10.105}$$

where γ is the anomalous dimension of matter fields and can be computed in perturbation theory to be

$$\gamma(g^2) = -\frac{g^2}{8\pi^2}\frac{N^2-1}{N} + \mathcal{O}(g^4). \tag{10.106}$$

Expanding formula (10.105) in powers of g^2 we get

$$\beta(g) = -\frac{g^3}{16\pi^2}\left[3N - F + \left(3N^2 - 2FN + \frac{F}{N}\right)\frac{g^2}{8\pi^2} + \mathcal{O}(g^4)\right]. \tag{10.107}$$

From the above expression it is clear that there can exist values of F and N such that the one-loop contribution is negative but the two-loops contribution is positive. This suggests that in principle there could be a non-trivial fixed point, a value of the gauge coupling $g = g_*$, for which $\beta(g_*) = 0$.

Let us consider F slightly smaller than $3N$. Defining

$$\epsilon = 3 - \frac{F}{N} \ll 1 \tag{10.108}$$

we can re-write the β-function as

$$\beta(g) = -\frac{g^3}{16\pi^2}\left[\epsilon N - \left[3(N^2-1) + \mathcal{O}(\epsilon)\right]\frac{g^2}{8\pi^2} + \mathcal{O}(g^4)\right]. \tag{10.109}$$

The first term inside the parenthesis is positive while the second is negative and hence we see we have a solution $\beta(g) = 0$ at

$$g_*^2 = \frac{8\pi^2}{3}\frac{N}{N^2-1}\,\epsilon, \tag{10.110}$$

up to $\mathcal{O}(\epsilon^2)$ corrections. This is called Banks-Zaks (BZ) fixed point. Seiberg argued that an IR fixed point like the one above exists not only for F so near to $3N$ but actually for any F in the range $\frac{3}{2}N < F < 3N$, the so-called *conformal window*. According to this proposal, the IR dynamics of SQCD in the range $\frac{3}{2}N < F < 3N$ is described by an interacting superconformal theory: quarks and gluons are *not* confined but appear as interacting massless particles, the Coulomb-like potential being

$$V(r) \sim \frac{g_*^2}{r}. \tag{10.111}$$

Hence, according to this proposal, SQCD in the conformal window enjoys a non-abelian Coulomb phase.

Let us try to understand why the conformal window is bounded from below and from above. The possibility of making exact computations in a SCFT shows that for $F < \frac{3}{2}N$ the theory should be in a different phase. In a SCFT the dimension of a field satisfies the following relation

$$\Delta \geq \frac{3}{2}|R|, \tag{10.112}$$

where R is the field R-charge (recall that in a SCFT the generator of the R-symmetry enters the algebra, and hence an R-symmetry is always present). The equality holds for chiral (or anti-chiral) operators. Since M is a chiral operator, this implies that

$$\Delta(M) = \frac{3}{2}R(M) = \frac{3}{2}R(Q\widetilde{Q}) = 3\frac{F-N}{F} \equiv 2 + \gamma_*. \tag{10.113}$$

This means that the anomalous dimension of the meson matrix at the IR fixed point is $\gamma_* = 1 - 3N/F$.

Now, the lowest component of the meson matrix M is a scalar operator. In four space-time dimensions the dimension of a scalar must satisfy

$$\Delta \geq 1. \tag{10.114}$$

Indeed, when $\Delta < 1$ the operator, which is in a unitary representation of the superconformal algebra, would include a negative norm state which cannot exist in a unitary theory. This implies that $F = \frac{3}{2}N$ is a lower bound since there $\Delta(M) = 1$ and lower values of F make no sense: for $F < \frac{3}{2}N$ the theory should be in a different phase. A clue to what such phase could be is that at $F = \frac{3}{2}N$ the field M becomes free. Indeed, for $F = \frac{3}{2}N$ we get that $\Delta(M) = 1$ which is possible only for free, non-interacting scalar operators. Perhaps it is the whole theory of mesons and baryons which becomes free, somehow. We will make this intuition more precise later.

As for the upper bound, let us notice that for $F \geq 3N$ SQCD is not asymptotically free anymore, since the β-function changes sign (for $F = 3N$ the one-loop β function is 0 but one can show that the two-loop contribution is positive). The spectrum at large distance consists of elementary quarks and gluons interacting through a potential

$$V \sim \frac{g^2}{r} \quad \text{with } g^2 \sim \frac{1}{\log(r\Lambda)}, \tag{10.115}$$

which implies that SQCD is in a non-abelian free phase. It is interesting to notice that for $F = 3N$ the anomalous dimension of M is actually zero, consistent with the fact that from that value on, the IR dimension of gauge invariant operators is not renormalized since the theory becomes IR-free. A summary of the IR behavior of SQCD for $F > \frac{3}{2}N$ is reported in Figure 10.14.

10.4.6 $\mathcal{N} = 1$ *electric-magnetic duality*

The second proposal put forward by Seiberg regards the existence of a sort of electro-magnetic duality. The IR physics of SQCD for $F > N + 1$ has an equivalent description in terms of another supersymmetric gauge theory, known as the *magnetic dual* theory. This equivalence, however, is just an IR equivalence. SQCD, sometime called electric theory in this context,

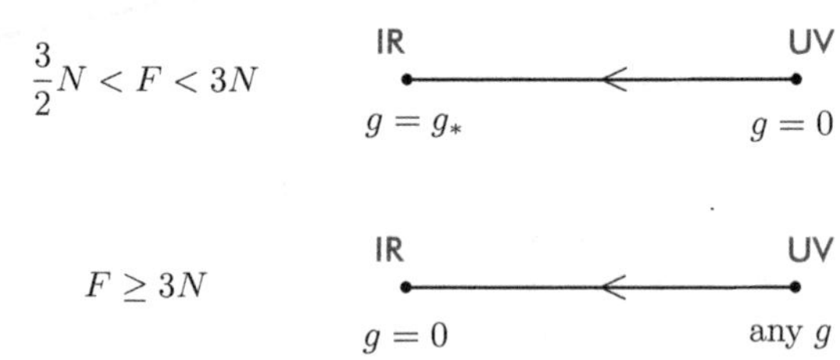

Fig. 10.14 The IR behavior of SQCD in the window $\frac{3}{2}N < F < 3N$, where the theory flows to an IR fixed point with $g = g_*$, and for $F \geq 3N$, where $g_* = 0$ and the theory is in a non-abelian IR-free phase.

and its magnetic dual are very different in the UV as well as along the RG-flow. They just provide two equivalent ways to describe the dynamics around the space of vacua. We speak in this case of an IR duality (it should be said that perturbing SQCD by suitable operators, e.g. by quartic operators, one can sometime promote this IR duality to a full duality, valid along the whole RG; discussing such instances, however, is beyond our scope).

In order to understand this claim (and its implications), and define such dual theory more precisely, we first need to do a step back. In trying to extend to higher values of F the reasoning about SQCD with $F = N+1$, one should consider the following gauge invariant operators

$$M^i_{\ j}, \quad B_{i_1 i_2 \ldots i_{F-N}}, \quad \widetilde{B}^{i_1 i_2 \ldots i_{F-N}}. \tag{10.116}$$

The baryons have $\widetilde{N} = F - N$ free indices so, at least group-theoretically, one could think about them as if they were bound states of $\widetilde{N}$ components, some new quark-like fields q and $\tilde{q}$ of some supersymmetric gauge theory theory with gauge group $SU(\widetilde{N}) = SU(F-N)$ for which q and $\tilde{q}$ transform in the $\widetilde{N}$ and $\overline{\widetilde{N}}$ representations, respectively. Then the SQCD baryons would have a dual description as

$$B_{i_1 i_2 \ldots i_{\widetilde{N}}} \sim \epsilon_{a_1 a_2 \ldots a_{\widetilde{N}}} q^{a_1}_{i_1} q^{a_2}_{i_2} \ldots q^{a_{\widetilde{N}}}_{i_{\widetilde{N}}} \tag{10.117}$$

and similarly for $\widetilde{B}$. Recall that in terms of the original matter fields Q and $\widetilde{Q}$, the baryons are composite fields made out of N components.

Seiberg made this naive idea concrete (and physical), putting forward the following proposal: SQCD with gauge group $SU(N)$ and $F > N+1$ flavors can be equivalently described, in the IR, by a different SQCD-like theory with gauge group $SU(F-N)$ and F flavors plus an additional chiral superfield Φ which is a gauge singlet and which transforms in the fundamental representation of $SU(F)_{\rm L}$ and in the anti-fundamental representation of $SU(F)_{\rm R}$, and which interacts with q and $\tilde{q}$ via a cubic superpotential

$$W = h\, q_i\, \Phi^i_j\, \tilde{q}^j. \tag{10.118}$$

As bizarre this proposal may look like, let us try to understand it better. Let us first consider the Seiberg dual theory (which from now on we dub mSQCD, where 'm' stands for magnetic) without superpotential term, and let us focus on the SQCD conformal window, $\frac{3}{2}N < F < 3N$, first.

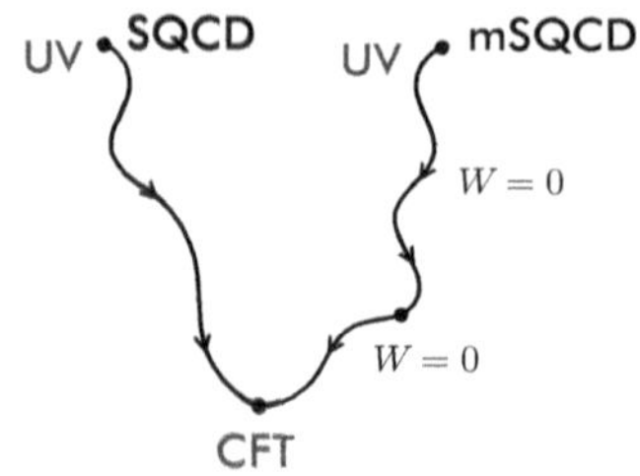

Fig. 10.15 How the RG goes for SQCD and mSQCD in the conformal window. In the far IR, the two theories reach the same fixed point.

For $W = 0$ the field Φ is completely decoupled and mSQCD is just SQCD with gauge group $SU(F-N)$ and F flavors. So, in the conformal window mSQCD is itself UV-free since its one-loop β-function coefficient is $b_1 = 2F - 3N$ and hence positive. In fact, as one can easily check, the SQCD conformal window is a conformal window also for mSQCD! Hence mSQCD (without the singlet Φ) flows to an IR fixed point for $\frac{3}{2}N < F < 3N$. At such fixed point the superpotential coupling, that we now switch-on, is relevant, since

$$\Delta(W) = \Delta(\Phi) + \Delta(q) + \Delta(\tilde{q}) = 1 + \frac{3}{2}N/F + \frac{3}{2}N/F < 3, \qquad (10.119)$$

where in the second step we have used the relation (10.117) (and the analogue one relating $\widetilde{B}$ and the $\tilde{q}$'s) and the values of baryon R-charges. The claim is that the relevant perturbation (10.118) drives the theory to some new fixed point (where the β-functions of the dual gauge coupling and of the coupling h both vanish!) which is actually the same fixed point of SQCD. This idea is summarized in Figure 10.15.

How does mSQCD look like for $F \leq \frac{3}{2}N$? Since for mSQCD $b_1 = 2F - 3N$, for $F = \frac{3}{2}N$ the β-function vanishes and for lower values of F it changes its sign and the theory becomes IR-free. Hence, the bound $F = \frac{3}{2}N$ has the same role that the bound $F = 3N$ has for SQCD (not surprisingly, one can apply the BZ fixed point argument to mSQCD for F slightly larger than $\frac{3}{2}N$ and find the existence of a perturbative fixed point). This explains why, if Seiberg duality is correct, the IR dynamics of SQCD in the range $N + 1 < F \leq \frac{3}{2}N$ differs from the behavior in the conformal window, something we had already some indications, when studying the lower bound in F of the SQCD conformal window. Indeed, we can now make our former intuition precise: using a clever set of variables (*i.e.* the magnetic dual variables), one concludes that for $N+1 < F \leq \frac{3}{2}N$ SQCD IR dynamics is described by a theory of freely interacting (combinations of) meson and baryon fields. These can be described in terms of free dual quarks interacting with a Coulomb-like potential

$$V_{\rm m} \sim \frac{g_{\rm m}^2}{r} \quad \text{with } g_{\rm m}^2 \sim \frac{1}{\log(r\Lambda_{\rm m})}, \tag{10.120}$$

where $g_{\rm m}$ is the mSQCD gauge coupling and $\Lambda_{\rm m}$ the mSQCD strong coupling scale (which in this regime of parameters is a UV cut-off, since the theory is IR-free). This phase of SQCD is dubbed *free magnetic phase*, a theory of freely interacting (dual) quarks. The fact that the IR dynamics of SQCD for $N+1 < F \leq \frac{3}{2}N$, where the theory is confining, can be described this way is a rather powerful statement: since mSQCD is IR-free, in terms of magnetic dual variables the Kähler potential is canonical (up to subleading $1/\Lambda_{\rm m}^2$ corrections), meaning that we know the full effective IR Lagrangian of SQCD for $N+1 < F \leq \frac{3}{2}N$, at low enough energies!

As for the conformal window, which variables to use depends on F. The larger F, the nearer to IR-freedom SQCD is, and the more UV-free mSQCD is. In other words, the conformal window IR-fixed point is at smaller and smaller value of the electric gauge coupling the nearer F is to $3N$, and eventually becomes 0 for $F = 3N$. For mSQCD things are reversed. The IR-fixed point arises at weaker coupling the nearer F is to $\frac{3}{2}N$, and for $F = \frac{3}{2}N$ we have that $g_*^{\rm m} = 0$. Therefore, the magnetic description is the simplest to describe SQCD non-abelian Coulomb phase for F near to $\frac{3}{2}N$; the electric description is instead the most appropriate one when F is near to $3N$.

For $F \geq 3N$ the magnetic theory does not reach anymore an IR interacting fixed point. The value $F = 3N$ plays for mSQCD the same role the value $F = \frac{3}{2}N$ plays for SQCD. Indeed, the mSQCD meson matrix, $U = q\tilde{q}$ has $\Delta = 1$ for $F = 3N$, and becomes a free field, while for larger values of F it would get a dimension lower than one, which is not acceptable. For $F \geq 3N$ the theory should enter in a new phase. This is something we knew already: in this region we are in the SQCD IR-free phase.

Can we provide some consistency checks for the validity of this proposed duality?

Let us first note that two basic necessary requirements for its validity are met: the two theories have the same global symmetry group as well as the same number of IR degrees of freedom. In order to see this, let us first make the duality map precise. The mapping between chiral operators of

SQCD and mSQCD (at the IR fixed point) is

$$M \longleftrightarrow \Phi \quad : \quad {\Phi^i}_j = \frac{1}{\mu}{M^i}_j \tag{10.121}$$

$$B \longleftrightarrow b \quad : \quad b^{j_1 j_2 \dots j_N} = c\,\epsilon^{i_1 i_2 \dots i_{F-N} j_1 j_2 \dots j_N}\, B_{i_1 i_2 \dots i_{F-N}} \tag{10.122}$$

and similarly for $\tilde{b}$ and $\widetilde{B}$, with b and $\tilde{b}$ being the baryons of mSQCD. The scale μ relating SQCD mesons with the mSQCD gauge singlet Φ appears for the following reason. In SQCD mesons are composite fields and their dimension in the UV, where SQCD is free, is $\Delta = 2$. On the other hand, Φ is an elementary field in mSQCD and its dimension in the UV is $\Delta = 1$. Hence the scale μ needs to be introduced to match Φ to M in the UV. Clearly, upon RG-flow both fields acquire an anomalous dimension and should flow to one and the same operator in the IR, if the duality is correct. Applying formula (10.112), which for chiral operators is an equality, one easily sees that this is indeed what happens, since $R(M) = R(\Phi)$. Eq. (10.122) is obtained from (10.117) multiplying the latter by $\epsilon^{i_1 i_2 \dots i_{F-N} j_1 j_2 \dots j_N}$, while the scale c is there for similar reasons as μ and has mass dimension $F - 2N$. Its precise value, which turns out to be a function of μ in fact, will be fixed later.

From its very definition, it follows that the magnetic theory has a global symmetry group which is nothing but the one of SQCD, $G_{\mathrm{F}} = SU(F)_{\mathrm{L}} \times SU(F)_{\mathrm{R}} \times U(1)_{\mathrm{B}} \times U(1)_R$ and, using the map (10.121)(10.122), one can read-off the following charges for the elementary fields

	$SU(F)_{\mathrm{L}}$	$SU(F)_{\mathrm{R}}$	$U(1)_{\mathrm{B}}$	$U(1)_R$
q_i^a	$\overline{F}$	$\bullet$	$\frac{N}{F-N}$	$\frac{N}{F}$
$\tilde{q}_b^j$	$\bullet$	F	$-\frac{N}{F-N}$	$\frac{N}{F}$
Φ	F	$\overline{F}$	0	$2\frac{F-N}{F}$
$\tilde{\lambda}$	$\bullet$	$\bullet$	0	1

while the superpotential (10.118) has $R = 2$.

We can now use global symmetries to see that SQCD and mSQCD have the same number of IR degrees of freedom. Basically, there is a one-to-one map between gauge invariant operators, and these operators have the same global symmetries (which counts physically distinct degrees of freedom). Indeed, the meson matrix M enjoys the same symmetries as the mSQCD singlet Φ, and the SQCD baryons $B, \widetilde{B}$ the same as the mSQCD baryons $b, \tilde{b}$. One might feel uncomfortable since the mesons of the magnetic dual theory, $U_i^j = q_i \tilde{q}^j$, seem not to match with anything in the electric theory.

This is where the superpotential (10.118) comes into play. Recall that the supposed equivalence between SQCD and mSQCD is just an IR equivalence. The F-equations for Φ fix the dual meson to vanish on the moduli space, since $F_\Phi = hq\tilde{q} = hU = 0$. Hence, in the IR the two theories do have the same number of degrees of freedom (and we now see the importance of the superpotential term (10.118)!).

Given these necessary requirements, we now want to present further checks for the validity of Seiberg's proposal.

- The first such checks comes from 't Hooft anomaly matching. There are several global symmetries and related 't Hooft anomalies one can cmpute. Let us choose, for instance, those associated to $SU(F)^2_{\mathrm{L}}\, U(1)_{\mathrm{B}}$, $U(1)^2_{\mathrm{B}}\, U(1)_R$ and $U(1)^3_R$. The computation, using SQCD and mSQCD degrees of freedom, respectively, gives the following results

SQCD

$$SU(F)^2_{\mathrm{L}}\, U(1)_{\mathrm{B}} \qquad \tfrac{1}{2}N(+1) = \tfrac{1}{2}N$$

$$U(1)^2_{\mathrm{B}}\, U(1)_R \qquad 2NF(+1)(-\tfrac{N}{F}) = -2N^2$$

$$U(1)^3_R \qquad \left(-\tfrac{N}{F}\right)^3 2NF + N^2 - 1 = -2\tfrac{N^4}{F^2} + N^2 - 1$$

mSQCD

$$SU(F)^2_{\mathrm{L}}\, U(1)_{\mathrm{B}} \qquad \tfrac{1}{2}(F-N)\tfrac{N}{F-N} = \tfrac{1}{2}N$$

$$U(1)^2_{\mathrm{B}}\, U(1)_R \qquad 2(F-N)F\left(\tfrac{N}{F-N}\right)^2\left(\tfrac{N-F}{F}\right) = -2N^2$$

$$U(1)^3_R \qquad \left(\tfrac{N-F}{F}\right)^3 2(F-N)F + F^2\left(\tfrac{F-2N}{F}\right)^3 + \\ + (F-N)^2 - 1 = -2\tfrac{N^4}{F^2} + N^2 - 1$$

This shows that 't Hooft anomalies indeed match between SQCD and its IR-equivalent mSQCD description.

Note that for the matching to work it turns out that the presence of dual gauginos is crucial (in the above computation, they contribute to the mSQCD $U(1)^3_R$ 't Hooft anomaly). This explicitly shows that the description of SQCD baryons in terms of some sort of dual quarks is not just a mere group representation theory accident. There is a truly dynamical dual gauge group, under which dual quarks are charged, and dual vector superfields (which include dual gauginos) which interact with them. This means that SQCD, in the range $N+1 < F < 3/2N$ (a range in which the theory confines), enjoys an emergent gauge symmetry in the IR!

- The duality relation *is* a duality, which means that acting twice with the duality map one recovers the original theory (as far as IR physics!). Let us start from SQCD with N colors and F flavors and act with the duality map twice

$$\text{SQCD} : SU(N),\, F\,,\, W = 0$$
$$\downarrow \text{duality}$$
$$\text{mSQCD} : SU(F-N),\, F\,,\, W = \frac{1}{\mu} q_i M^i{}_j \tilde{q}^j = q_i \Phi^i{}_j \tilde{q}^j$$
$$\downarrow \text{duality}$$
$$\text{mmSQCD} : SU(N),\, F\,,\, W = \frac{1}{\mu} q_i M^i{}_j \tilde{q}^j + \frac{1}{\tilde{\mu}} d^i U_i{}^j \tilde{d}_j = q_i \Phi^i{}_j \tilde{q}^j + d^i \Psi_i{}^j \tilde{d}_j$$

where $U_i{}^j = q_i \tilde{q}^j$ is the meson matrix of mSQCD, $\Psi_i{}^j$ is the gauge singlet chiral superfield dual to U and belonging to the magnetic dual of mSQCD and, for the ease of notation, we have put $h = 1$ in eq. (10.118). Choosing $\tilde{\mu} = -\mu$, we can rewrite the superpotential of mmSQCD as

$$W = \frac{1}{\mu} \mathrm{Tr}\left[UM - dU\tilde{d}\right]. \tag{10.123}$$

The fields U and M are hence massive and can be integrated out (recall we claim the IR equivalence of Seiberg-dual theories, not the equivalence at all scales). This implies

$$\frac{\partial W}{\partial U} = 0 \to M^i{}_j = d^i \tilde{d}_j, \quad \frac{\partial W}{\partial M} = 0 \to U = 0 \tag{10.124}$$

showing that the dual of the dual quarks are nothing but the original quark superfields Q and $\widetilde{Q}$, and that $U = 0$ (hence $W = 0$) in the IR. Summarizing, after integrating out heavy fields, we are left with SQCD with gauge group $SU(N)$, F flavors and no superpotential, exactly the theory we have started with! In passing, let us note that in order to make the duality working we have to set $\tilde{\mu} = -\mu$, a mass scale which is not fixed by the duality itself.

- The duality is preserved under mass perturbations, namely upon holomorphic decoupling. Let us again consider SQCD with gauge group $SU(N)$ and F flavors and let us add a mass term to the F-th flavor,

$W = mM^{\mathrm{F}}_{\mathrm{F}}$. This corresponds to $SU(N)$ SQCD with $F-1$ massless flavors and one massive one. In the dual magnetic theory this gives a superpotential

$$W = \frac{1}{\mu} q_i M^i_{\ j} \tilde{q}^j + m\, M^{\mathrm{F}}_{\mathrm{F}}. \tag{10.125}$$

The F-flatness conditions for $M^{\mathrm{F}}_{\mathrm{F}}$ and q_{F} and $\tilde{q}^{\mathrm{F}}$ are

$$q^a_{\mathrm{F}}\, \tilde{q}^{\mathrm{F}}_a + \mu\, m = 0, \quad (M \cdot \tilde{q}_a)^{\mathrm{F}} = (q^a \cdot M)_{\mathrm{F}} = 0, \tag{10.126}$$

where a is a $SU(F-N)$ gauge index. The first equation induces a VEV for the dual quarks with flavor index F, which breaks the gauge group down to $SU(F-N-1)$. The other two equations imply that the F-th row and column of the SQCD meson matrix M vanish. We hence end-up with $SU(F-N-1)$ SQCD with $F-1$ flavors, a gauge singlet M which is a $(F-1)\times(F-1)$ matrix, while the superpotential (10.125) reduces to eq. (10.118) where now i, j run from 1 to $F-1$ only. This is the correct Seiberg dual of SQCD with $F-1$ flavors.

This analysis shows that a mass term in the electric theory corresponds to higgsing in the magnetic dual theory, according to the table below

SQCD		mSQCD
$SU(N), F$	$\longleftrightarrow$ dual	$SU(F-N), F$
$\downarrow$ mass		$\downarrow$ higgsing
$SU(N), F-1$	$\longleftrightarrow$ dual	$SU(F-N-1), F-1$

The converse is also true (though slightly harder to prove): a mass term in mSQCD corresponds to higgsing in SQCD.

- Let us use holomorphic decoupling to go from the last value of F where we have the duality, $F = N+2$, to $F = N+1$. If Seiberg duality is correct, we should recover the description of SQCD with $F = N+1$ flavors we discussed previously. Let us consider mSQCD when $F = N+2$. The magnetic gauge group is $SU(2)$. Upon holomorphic decoupling, an analysis identical to the one we did above produces a cubic superpotential at low energy as

$$W \sim q_i M^i_j \tilde{q}^j \quad i, j = 1, \ldots, N+1 \tag{10.127}$$

where q_i are nothing but the baryons B^i of SQCD with $F = N+1$ and $\tilde{q}_j$ the baryons $\widetilde{B}_j$, just take eq. (10.117) with $\widetilde{N} = F - N = 1$. At the same time, the VEVs for $q_{\rm N+2}$ and $\tilde{q}^{\rm N+2}$ break the $SU(2)$ gauge symmetry completely. From mSQCD view point this is a situation similar to SQCD with $F = N-1$ where the full breaking of gauge symmetry group allowed an exact instanton calculation providing the $\sim$ det$M$ contribution to the effective superpotential. The same happens here and the final answer one gets for the low energy effective superpotential is

$$W_{\rm eff} \sim \left(q_i M^i_j \tilde{q}^j - \det M\right), \tag{10.128}$$

which is precisely the effective superpotential of SQCD with $F = N+1$! This shows that by holomorphic decoupling we can actually connect the description of the IR dynamics of SQCD for *any* number of flavors, from $F = 0$ to any larger values of F, at fixed N.

Let us finally notice, in passing, that even for $F = N+1$ we can sort of speak of a magnetic dual theory. Just it is trivial, since there is no magnetic dual gauge group.

- There is yet an important relation between the three a priori different mass scales entering the duality: the electric dynamical scale $\Lambda_{\rm el}$, the magnetic scale $\Lambda_{\rm m}$, and the matching scale μ. This reads

$$\Lambda_{\rm el}^{3N-F}\Lambda_{\rm m}^{3(F-N)-F} = (-1)^{F-N}\mu^F. \tag{10.129}$$

That this relation is there, can be seen in different ways. First, one can check that the relation is duality invariant, as it should. Indeed, applying the duality map (recall that $\tilde{\mu} = -\mu$, while $\Lambda_{\rm el}$ and $\Lambda_{\rm m}$ get interchanged by the duality) one gets

$$\Lambda_{\rm m}^{3(F-N)-F}\Lambda_{\rm el}^{3N-F} = (-1)^N\tilde{\mu}^F = (-1)^{N-F}\mu^F, \tag{10.130}$$

which is identical to (10.129). One can also verify the consistency of the relation (10.129) upon higgsing and/or holomorphic decoupling.

Finally, by matching $F = N+2$ SCQD to $F = N+1$ SQCD via holomorphic decoupling, one can also fix the value of c in eq. (10.122) to be

$$c = 1/\sqrt{-(-\mu)^{N-F}\Lambda_{\rm el}^{3N-F}} = 1/\sqrt{-\mu^N\Lambda_{\rm m}^{3N-2F}}. \tag{10.131}$$

Eq. (10.129) shows that as the electric theory becomes stronger (*i.e.* $\Lambda_{\rm el}$ increases), the magnetic theory becomes weaker (*i.e.* $\Lambda_{\rm m}$ decreases). By using the relation between dynamical scales and gauge couplings, this can be translated into a relation between gauge coupling constants, and gives an inverse relation between them

$$g_{\rm el}^2 \sim g_{\rm m}^{-2}, \tag{10.132}$$

showing that large values of the electric gauge coupling $g_{\rm el}$ correspond to small values of the magnetic one, and viceversa. This is why Seiberg duality can be thought as a sort of electric-magnetic duality.

Depending on where in the (F, N) space one sits, the meaning of the dynamical scales changes. In the conformal window both SQCD and mSQCD are UV-free. Both theories have a non-trivial RG-flow and, upon non-perturbative effects, driven by $\Lambda_{\rm el}$ and $\Lambda_{\rm m}$, reach an IR fixed point (which is one and the same, in fact). In the free-magnetic phase, mSQCD is IR-free and SQCD is UV-free. Therefore, in this regime $\Lambda_{\rm m}$ should be better thought of as a UV cut-off for the magnetic theory, which is an effective theory. In this regime SQCD can be thought of as the (or better, a possible) UV-completion of mSQCD. The electric free phase can be thought of in a similar way, with the role of SQCD and mSQCD reversed. Within this interpretation it is natural to tune the free parameter μ to make the two theories have one single non-perturbative scale, the scale at which non-perturbative SQCD effects come into play and the scale below which the magnetic effective description takes over. From relation (10.129) one sees that this is obtained by equating, up to an overall phase, the matching scale μ with $\Lambda_{\rm el}$ and $\Lambda_{\rm m}$

$$\mu = \Lambda_{\rm el} = \Lambda_{\rm m} (\equiv \Lambda). \tag{10.133}$$

Using the above relation for $F = N + 2$ and adding a mass term for the F-th flavor, upon holomorphic decoupling one gets the expression (10.128) including the correct power of Λ, that is

$$W_{\rm eff} = \frac{1}{\Lambda^{2N-1}} \left(B_i \, M^i_j \, \widetilde{B}^j - \det M \right). \tag{10.134}$$

where we have already used the fact that the dual quarks are nothing but the baryon themselves, in this case.

Figure 10.16 contains a qualitative description of the three different regimes we have just discussed.

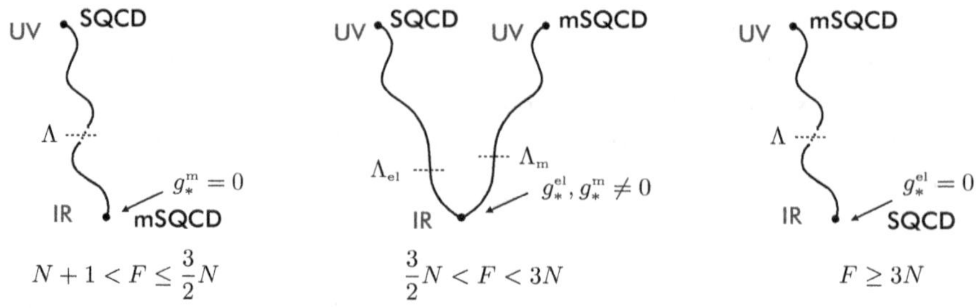

Fig. 10.16 The three qualitative different phases of SQCD with $F > N + 1$. In the magnetic-free and electric-free phases we have chosen (for convenience) the arbitrary matching scale μ in such a way to make Λ_{el} and Λ_m being identified, eq. (10.133).

10.5 The phase diagram of $\mathcal{N}=1$ SQCD

After this long tour on quantum properties of SQCD, it is time to wrap-up and summarize its phase diagram.

SQCD with $F = 0$ (*i.e.* pure SYM) enjoys strict confinement, displays N isolated supersymmetric vacua and a mass gap.

For $0 < F < N$ the theory doesn't exist by its own. The classical moduli space is completely lifted and a runaway potential, with no absolute minima at finite distance in field space, is generated.

For $F = N, N+1$ a moduli space persists at the quantum level and SQCD enjoys confinement with charge screening (the asymptotic states are gauge singlets but flux lines can break) and no mass gap. Asymptotic states are mesons and baryons. The theory displays complementarity, as any theory where there are scalars transforming in the fundamental representation of the gauge group: there is no invariant distinction between Higgs phase, which is the more appropriate description for large field VEVs, and confinement phase, which takes over near the origin of field space. The potential between static test charges goes to a constant asymptotically since in the Higgs phase gauge bosons are massive and there are no long-range forces. As already observed, this holds also in the confining description, since we actually have charge screening and Wilson loops do not follow the area law in this case. At the origin of field space, $F = N + 1$ SQCD s-confines.

For $N + 2 \leq F \leq \frac{3}{2}N$ we are still in a confinement phase, but the theory is in the so-called free magnetic phase and can be described at large enough distance in terms of freely interacting dual quarks and gluons. What is amusing here is that while asymptotic massless states are composite of elementary electric degrees of freedom (*i.e.* mesons and baryons), they are charged with respect to a magnetic gauge group whose dynamics is not visible in the electric description and which is generated, non-perturbatively, by the theory itself.

For $F > \frac{3}{2}N$ SQCD does not confine anymore, not even in the weak sense: asymptotic states are quarks and gluons (and their superpartners). The potential between asymptotic states, though, differs if $F \geq 3N$ or $F < 3N$. In the former case the theory is IR-free and it is described by freely interacting particles. Hence the potential vanishes, at large enough distance. For $\frac{3}{2}N < F < 3N$, instead, the theory (which is still UV-free) is in a non-abelian Coulomb phase. Charged particles are not confined but actually belong to a SCFT, and interact by a $1/r$ potential with fixed, non-vanishing coupling.

A diagram summarizing the gross features of the quantum dynamics of SQCD is reported below.

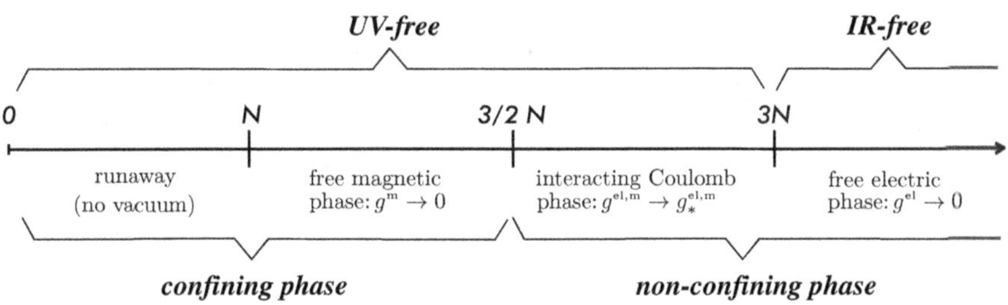

Our focus in this chapter has been on SQCD with gauge group $SU(N)$. This should be regarded as a prototype for many other $\mathcal{N} = 1$ supersymmetric gauge theories whose dynamics one can investigate using similar tools. For example, one can consider SQCD with gauge groups $SO(N)$ or $USp(2N)$, or modifications of $SU(N)$ SQCD by the addition of matter in representations other than the (anti)fundamental, e.g. the adjoint, the symmetric or anti-symmetric representations. Several properties that these other theories display are similar to what we have seen here, but some of these theories present also new phenomena that $SU(N)$ SQCD does not enjoy, like generalizations of Seiberg dualities, existence of abelian Coulomb phases, phase transitions between Higgs phase and confinement phase, etc.... We are not going to discuss any of these variants and refer the interested reader to the bibliography at the end of the chapter.

10.6 Exercises

Exercise 10.1. Consider SQCD with one flavor and show that giving a mass to it, upon holomorphic decoupling one ends-up with the pure SYM superpotential, eq. (10.68).

Exercise 10.2. Consider SQCD with $F = N$ with superpotential

$$W = A\left(\det M - B\widetilde{B} - a\Lambda^{2N}\right) + mQ^N\widetilde{Q}_N \tag{10.135}$$

By integrating out the massive flavor, show that one recovers the ADS superpotential for $F = N - 1$ SQCD if and only if $a = 1$.

Exercise 10.3. Check 't Hooft anomaly matching for SQCD with $F = N$ along the baryonic branch, $M = 0, B = -\widetilde{B} = \Lambda^N$.

Exercise 10.4. Check 't Hooft anomaly matching for SQCD with $F = N + 1$ at the origin of the moduli space.

Exercise 10.5. Consider mSQCD for $F = 3(F - N) - \epsilon(F - N)$ with $\epsilon \ll 1$, and find the BZ perturbative fixed point, *i.e.* the values of the dual gauge coupling $g_{\rm m}$ and of the cubic superpotential coupling h, eq. (10.118), for which the corresponding β-functions vanish.

Exercise 10.6. Show that the addition of a mass term in mSQCD corresponds to higgsing in SQCD.

References

[1] M. J. Strassler, *On confinement and duality*, Section 1, ICTP Spring School on Superstrings and Related Matters, ICTP Lect. Notes Ser. 7 (2002) 105.

[2] D. Tong, *Lectures on Gauge Theory*, Sections 2.1 and 2.5, https://www.damtp.cam.ac.uk/user/tong/gaugetheory/gt.pdf

[3] M. E. Peskin, *Duality in supersymmetric Yang-Mills theory*, Sections 2, 3, 4, 7 and 8, TASI 96, arXiv:hep-th/9702094.

[4] N. Dorey, T. J. Hollowood, V. V. Khoze and M. P. Mattis, The calculus of many instantons, Section VII, *Phys. Rept.* **371** (2002) 231 [arXiv:hep-th/0206063].

[5] R. Argurio, G. Ferretti and R. Heise, An introduction to supersymmetric gauge theories and matrix models, Section 3, *Int. J. Mod. Phys. A* **19**, 2015 (2004) [arXiv:hep-th/0311066].

[6] K. A. Intriligator, R. G. Leigh and N. Seiberg, Exact superpotentials in four-dimensions, *Phys. Rev. D* **50** (1994) 1092 [hep-th/9403198].

[7] K. A. Intriligator, Integrating in' and exact superpotentials in 4-d, *Phys. Lett. B* **336** (1994) 409 [hep-th/9407106].
[8] N. Seiberg, Electric — magnetic duality in supersymmetric nonAbelian gauge theories, *Nucl. Phys. B* **435** (1995), 129-146 [arXiv:hep-th/9411149 [hep-th]].
[9] K. A. Intriligator and N. Seiberg, Duality, monopoles, dyons, confinement and oblique confinement in supersymmetric SO(N(c)) gauge theories, *Nucl. Phys. B* **444** (1995), 125–160 [arXiv:hep-th/9503179 [hep-th]].
[10] K. A. Intriligator and P. Pouliot, Exact superpotentials, quantum vacua and duality in supersymmetric SP(N(c)) gauge theories, *Phys. Lett. B* **353** (1995), 471–476 [arXiv:hep-th/9505006 [hep-th]].
[11] K. A. Intriligator and N. Seiberg, Lectures on supersymmetric gauge theories and electric-magnetic duality, *Nucl. Phys. Proc. Suppl.* **45BC** (1996) 1 [arXiv:hep-th/9509066].
[12] J. Terning, *Modern supersymmetry: Dynamics and duality*, Chapters 9, 10 and 11, Oxford University Press (2006).
[13] M. J. Strassler, *The Duality cascade*, Section 1, TASI 2003 [hep-th/0505153].
[14] M. A. Shifman, Nonperturbative dynamics in supersymmetric gauge theories, Sections 1, 2 and 3, *Prog. Part. Nucl. Phys.* **39** (1997) 1 [hep-th/9704114].

Chapter 11

Dynamical Supersymmetry Breaking

Equipped with what we have learned in the previous two chapters, we can now go back to supersymmetry breaking, and finally discuss models where supersymmetry is broken by strong coupling effects, aka *dynamical supersymmetry breaking* (DSB). As already emphasized, besides their intrinsic interest, these models can in principle be used as hidden sectors within gravity or gauge mediation scenarios (or any of their possible variants).

We will first focus on models where supersymmetry is broken dynamically into *stable* vacua, that is at absolute minima of the potential. Later we will discuss dynamical supersymmetry breaking into *metastable* vacua.

The rough general picture for DSB models into stable vacua is as follows (with well motivated exceptions, as we will see):

- The supersymmetric theory at hand is a (asymptotically free) gauge theory. This is because gauge degrees of freedom are the only ones having some chance of generating non-perturbative contributions to the superpotential. As we already discussed, in models of chiral superfields only, the superpotential is tree-level exact.
- The tree level superpotential W_{tree} does not break supersymmetry but lifts all flat directions. Since the superpotential is polynomial in the fields, this typically gives a potential which vanishes at the origin and grows for large field VEVs. Since the superpotential is classically exact in perturbation theory, supersymmetry is preserved at all orders, perturbatively.
- Strong coupling effects generate a non-perturbative superpotential which provides a contribution to the potential which is strong at the origin

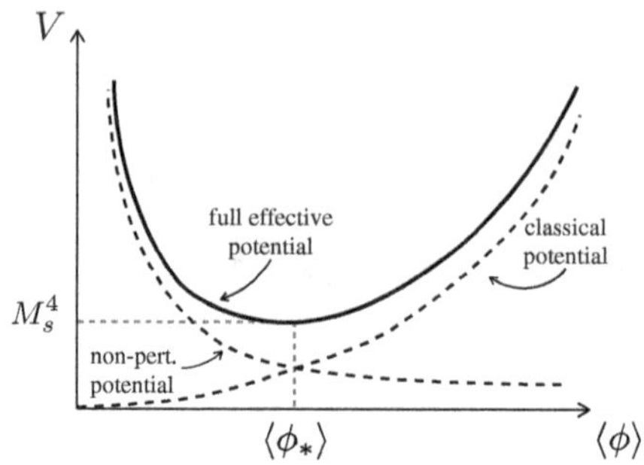

Fig. 11.1 A schematic view of dynamical supersymmetry breaking conspiracy. The classical and non-perturbative contributions to the effective potential sum-up and give a stable supersymmetry-breaking minimum at $\langle\phi\rangle = \langle\phi_*\rangle$.

of field space but decreases for large field VEVs (recall that the large field VEVs region corresponds to the classical region, where quantum corrections are negligible). An instance of such a potential is the effective potential of $F < N$ SQCD.

DSB arises because of the interplay between the classical contribution and the non-perturbatively generated one, as shown in Figure 11.1. Generically, supersymmetry will be broken and the exact potential will display a stable non-supersymmetric minimum at finite distance in field space.

11.1 Calculable and non-calculable models: generalities

Once a stable non-supersymmetric minimum is found, one would like to study quantum fluctuations around it. The spectrum around such vacuum is not supersymmetric and hence quantum corrections are not protected by supersymmetry. Moreover, besides the superpotential, the knowledge of the Kähler potential is also important, if one wants to make any sort of quantitative statement. A knowledge of the Kähler potential is needed to know the exact point in field space where the supersymmetry breaking vacuum sits, the values of the vacuum energy, *i.e.* the supersymmetry breaking scale $M_s \sim (V_{\text{min}})^{1/4}$, and the masses and interactions of light fields. In other words, one should know the structure of the effective Lagrangian. It is in general a difficult task to control the form of the Kähler potential, since K is corrected at all orders in perturbation theory (and non-perturbatively). There is then a problem of *calculability* around a non-perturbatively generated supersymmetry breaking vacuum, in general.

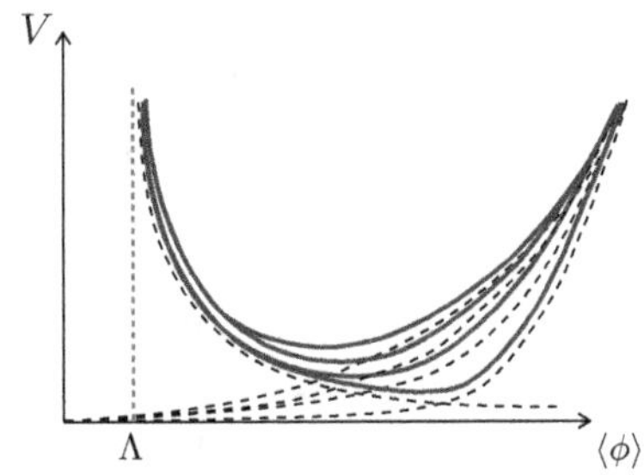

Fig. 11.2 Decreasing the perturbative coupling h the DSB minimum becomes smaller and moves towards large field VEVs. For the red curve the minimum, namely the supersymmetry breaking scale M_s, is smaller than the dynamical scale Λ.

Looking at Figure 11.1, calling h the generic tree level coupling(s), it should be clear that if we decrease h the tree-level potential becomes less and less steep, and the supersymmetry breaking minimum is pushed more and more towards the large VEVs region, where the theory is weakly coupled, see Figure 11.2.

There are three basic reasons why making the tree-level superpotential couplings smaller, calculability is increased.

The smaller h the smaller M_s, too. Eventually, it might become even smaller than Λ (see the red curve in Figure 11.2). This is useful since at energies lower than Λ gauge degrees of freedom can be safely integrated out giving rise to simpler models (of O'R-like type, so to say). Hence, the analysis of the low energy effective theory around supersymmetry breaking vacua might be simpler, the potential having only F-terms contributions

$$V(\phi, \phi^\dagger) = \left[K''(\phi, \phi^\dagger)\right]^{-1} \left|\frac{\partial W}{\partial \phi}\right|^2 . \tag{11.1}$$

Second, as we have discussed at length in the previous chapter, most progresses in understanding supersymmetric theories at the non-perturbative level regard the deep IR, $E < \Lambda$ (structure of vacua, lowest lying state excitations around them, etc...). Hence, having $M_s < \Lambda$ is a welcome feature.

Finally, one more reason why having h small increases calculability has to do with the very possibility of computing the Kähler potential. While the effective superpotential W_{eff} can often be determined exactly, the Kähler potential is in general more difficult to calculate, since it is not protected by holomorphy. Having supersymmetry breaking vacua at large VEVs has the advantage that the theory is more and more classical (*i.e.*, weakly coupled) there. Therefore, one can in principle determine the Kähler potential of the low energy effective fields just by projecting the UV-fields canonical Kähler potential on such operators (which are typically some gauge and flavor invariant combinations of UV fields), getting a correct result up to corrections which, in such semi-classical region, are weak.

According to this general picture, DSB models can be roughly divided into three classes, with increasing level of calculability.

- The worst case scenario is a situation where one cannot get any information on the full potential due to the incapacity of computing both the effective superpotential and the Kähler potential. That supersymmetry is broken can be concluded based on indirect arguments, as those we discussed in Chapter 7 (like R-symmetry and/or global symmetry arguments). In these cases one can reasonably say that supersymmetry is broken and that $M_s \sim \Lambda$, but nothing can be said about the lightest excitations around the supersymmetry breaking vacua nor on the effective Lagrangian describing their dynamics.
- A better situation occurs when one can compute the effective superpotential and explicitly see that the latter generates some non-vanishing F-terms which were vanishing at tree-level. In these cases one can safely say that supersymmetry is broken and possibly tell which are the low energy degrees of freedom around the supersymmetry breaking vacua. Still, the Kähler potential cannot be determined. Hence, one cannot calculate any property of the ground states nor determine the dynamics around them. DSB models belonging to this class are known as *non-calculable models*.
- Finally, there can exist models where the scenario summarized in Figure 11.2 can be fully realized. There exists a region in parameter space where the minimum of the potential is in a weakly coupled region and therefore one can also compute the Kähler potential, there. In these situations one can get also quantitative information about the low energy effective theory, like the precise value of the supersymmetry breaking scale as well as the structure of the light spectrum and of interactions. Possibly, at an arbitrary high level of accuracy, if supersymmetry breaking vacua can be made parametrically far from the origin of field space. Models of this kind are known as *calculable models*.

In what follows, we will present some concrete examples for each of above three classes.

11.2 The one GUT family SU(5) model

Let us consider a supersymmetric gauge theory with gauge group $SU(5)$, a chiral superfield T transforming in the $\underline{10}$ (*i.e.*, the antisymmetric representation), and a chiral superfield $\widetilde{Q}$ transforming in the anti-fundamental representation, $\underline{\overline{5}}$. This theory is UV-free, the one-loop β function coefficient being $b_1 = 13$.

This theory does not have any classical flat direction since it is impossible to construct holomorphic gauge invariant operators out of T and $\widetilde{Q}$. For the same reason a superpotential cannot be added. Therefore, at the classical level there exists one supersymmetric vacuum, sitting at the origin of field space. In this unique vacuum the gauge group is unbroken. Given that the theory is UV-free and so expected to go to strong coupling in the IR and considering the not so large matter content, one can reasonably argue that the theory confines and so that there are no leftover gauge degrees of freedom in the IR.

At the origin the theory is strongly coupled and it is difficult to perform any reliable computation. However, one can use indirect arguments to conclude that non-perturbative corrections break supersymmetry. First, one can easily check that there exist two non-anomalous global symmetries, an abelian $U(1)_Y$ and the $U(1)_R$ R-symmetry. Matter fields have charges $T \simeq (-1, 1)$ and $\widetilde{Q} \simeq (3, -9)$ under these symmetries, as can be easily seen by computing triangle diagrams with one global current and two gauge currents as shown in Figure 11.3, and imposing anomaly cancellations, which read

$$
\begin{aligned}
a &: 3Y(T) + Y(\widetilde{Q}) = 0 \\
b &: 3(R(T) - 1) + (R(\widetilde{Q}) - 1)Y(\overline{U}) = 0,
\end{aligned}
$$

whose solution, up to an overall (inessential) normalization, is as advocated.

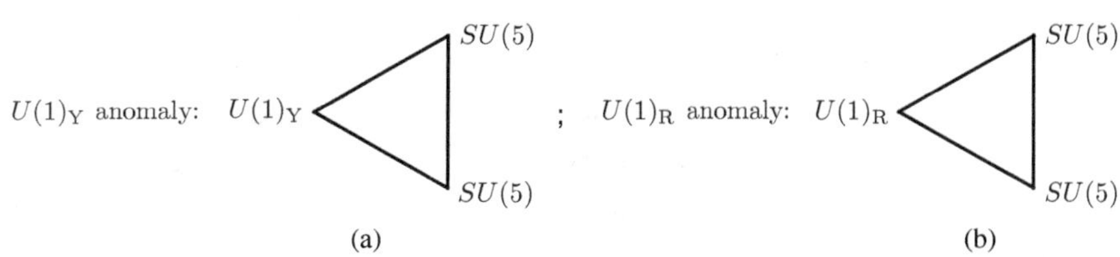

Fig. 11.3 One-loop diagrams contributing to anomalies of the abelian global symmetries of the $SU(5)$ model.

One can now use 't Hooft anomaly matching to argue that the global symmetry group $G_{\rm F} = U(1) \times U(1)_R$ is spontaneously broken. We do not know what the low energy $SU(5)$ invariant degrees of freedom are, but if $G_{\rm F}$ is unbroken, they should reproduce 't Hooft anomalies for $U(1)^3_Y$, $U(1)^2_Y U(1)_R$, etc... of the original theory. One can be as general as possible and allow for a set of putative low energy fields X_i with charges $\simeq (q_i, r_i)$ under $U(1) \times U(1)_R$. One gets four equations for the q_i's and r_i's. Allowing charges not larger than ~50, one needs a least five fields (with rather bizarre charges) to obtain a solution. This sounds quite unnatural (as a comparison: a non-supersymmetric version of this model requires just one massless fermion to match 't Hooft anomalies!). It is therefore quite possible that the system does not admit solutions. So, the global symmetry must be spontaneously broken. But then, since the theory does not have classical flat directions, according to the indirect criteria we have discussed at the beginning of Section 7.6.1, supersymmetry is broken, too.

An independent way to see that supersymmetry most likely is broken is as follows. Add to the $SU(5)$ model one pair of chiral superfields in the 5 and $\overline{5}$ representations, H and $\widetilde{H}$. There are now classical flat directions parametrized by single trace gauge invariant operators written in terms of $T, \widetilde{Q}, H$ and $\widetilde{H}$. These flat directions can be lifted by adding tree-level superpotential terms, including a mass term for $H, \widetilde{H}$, so classically there is a unique vacuum at the origin of field space. A conspiracy between the tree-level superpotential and a dynamically generated one (which one can explicitly show to be generated) along the lines of Figure 11.1 occurs, and supersymmetry is broken. In the limit $m \to \infty$ this theory reduces to the original one, without extra matter. If there are no phase transitions in the limit of large mass, then also the original theory breaks supersymmetry.

The $SU(5)$ model is an instance of the first class of supersymmetry breaking models we discussed in previous section. We do not have direct access to the effective superpotential nor to the Kähler potential, so no quantitative statements can be made. However, symmetry arguments indicate that supersymmetry is most likely broken at the non-perturbative level.

There exist generalizations of this model which break supersymmetry in a similar manner. They are based on a gauge group $SU(N)$, with N odd, $N-4$ chiral superfields transforming in the antifundamental of $SU(N)$, one chiral superfield transforming in the antisymmetric representation of $SU(N)$, and a superpotential which lifts all otherwise present classical flat directions. One can show that at low energy the dynamics of all these models reduces to the one of the $SU(5)$ model described above, and, as the latter, they are therefore expected to break supersymmetry.

11.3 The 3-2 model: supersymmetry breaking by instantons

In what follows, we are going to describe an instance of a calculable model.

Let us consider a supersymmetric theory with gauge group $G = SU(3) \times SU(2)$ and matter as detailed below.

	$SU(3)$	$SU(2)$	$U(1)_Y$	$U(1)_R$
$Q_{\alpha i}$	3	2	1/3	1
$\widetilde{U}^i$	$\bar{3}$	•	−4/3	−8
$\widetilde{D}^i$	$\bar{3}$	•	2/3	4
L_α	•	2	−1	−3

The index i is an index in the (anti)fundamental of $SU(3)$, α in the fundamental of $SU(2)$ and there exist two abelian anomaly-free global symmetries, $U(1)_Y$ and $U(1)_R$. As for the $SU(5)$ model, global symmetry charge assignment is fixed by requiring anomaly cancellation, computing triangle diagrams with one global and two gauge currents insertions as detailed in Figure 11.4.

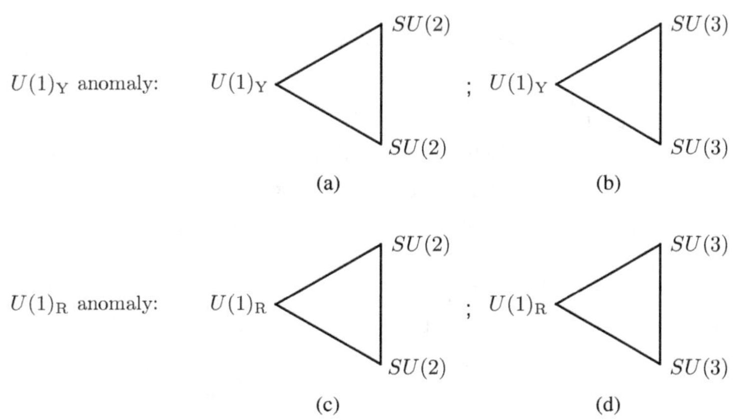

Fig. 11.4 One-loop diagrams contributing to anomalies of the abelian global symmetries, $U(1)_Y$ and $U(1)_R$, of the 3-2 model.

Anomaly-free global symmetries require (letters follow the one-loop diagrams in Figure 11.4)

$$
\begin{aligned}
&a: 3Y(Q)+Y(L)=0\\
&b: 2Y(Q)+Y(\overline{U})+Y(\widetilde{D})=0\\
&c: \frac{1}{2}\left[3(R(Q)-1)+R(L)-1\right]+2=0\\
&d: \frac{1}{2}\left[2(R(Q)-1)+R(\widetilde{U})-1+R(\widetilde{D})-1\right]+3=0.
\end{aligned}
$$

Up to an overall (inessential) normalization, this system of equations admits the solution

$$
\begin{aligned}
&R(Q)=1, \quad R(\widetilde{U})=-8, \quad R(\widetilde{D})=4, \quad R(L)=-3\\
&Y(Q)=1/3, \quad Y(\widetilde{U})=-4/3, \quad Y(\widetilde{D})=2/3, \quad Y(L)=-1,
\end{aligned}
$$

as anticipated. Finally, the theory has a tree-level superpotential

$$
W_{\text{tree}} = h\, Q\widetilde{D}L, \tag{11.2}
$$

which, given the above charge assignment, respects both R and non-R symmetries.

Let us start analyzing this theory at the classical level. The space of D-flat directions has (real) dimension six. This can be seen using the usual parameterization in terms of single trace gauge invariant operators. These read

$$
X_A = Q\widetilde{Q}_A L = Q_{\alpha i}\widetilde{Q}^i_A L_\beta \epsilon^{\alpha\beta}, \; Y = \det\left(Q\widetilde{Q}\right) = \epsilon^{\alpha\beta}\epsilon^{AB}\left(Q_{\alpha i}\widetilde{Q}^i_A\right) Q_{\beta j}\widetilde{Q}^j_B, \tag{11.3}
$$

where $A = 1,2$, $\widetilde{Q}^i_1 \equiv \widetilde{U}^i$, $\widetilde{Q}^i_2 \equiv \widetilde{D}^i$ and $\epsilon^{\alpha\beta}$ is the invariant tensor of $SU(2)$. That these are the correct degrees of freedom to describe the space of D-flat directions can be seen as follows. One can start constructing $SU(3)$ invariants. The only ones are $Q_\alpha\widetilde{Q}_A$ and L_α, which are both $SU(2)$ doublets. Using them to make (single trace) operators which are also $SU(2)$ invariant, one gets the operators X_A and Y.

We should now ask whether the superpotential (11.2) affects this space of supersymmetry preserving vacua, looking for the subspace of D-flat directions where all F-terms also vanish. The F-equation for L_α reads

$$
\frac{\partial W_{\text{tree}}}{\partial L_\alpha} = hQ_\alpha\widetilde{Q}_2 = 0. \tag{11.4}
$$

Contracting with L_α itself this implies that on the moduli space $X_2 = 0$. Similarly, multiplying by $Q_\alpha \widetilde{Q}_1$ so to construct the Y invariant, one can show that also $Y = 0$ on the moduli space. Finally, the F-equation for $\widetilde{Q}_2$ is

$$\frac{\partial W_{\text{tree}}}{\partial \widetilde{Q}_2^i} = hQ_i L = 0. \tag{11.5}$$

Contracting with $\widetilde{Q}_1^i$ one can show that also $X_1 = 0$ on supersymmetric vacua. The conclusion is that because of the presence of the superpotential (11.2) there do not exist classical flat directions but rather one single supersymmetric vacuum at the origin of field space. As for the $SU(5)$ model discussed in the previous section, this implies that if we can prove that some of the global symmetries are spontaneously broken, then we know supersymmetry is broken, too.

Let us start asking whether a dynamical superpotential is generated. In principle, we would expect contributions from $SU(3)$ and/or $SU(2)$ gauge dynamics. Let us choose for now a regime where $\Lambda_3 \gg \Lambda_2$ and $h \ll 1$. In this regime, at scales lower than Λ_3 but bigger than Λ_2, the $SU(2)$ gauge group is weakly coupled while $SU(3)$ confines. Hence, up to subleading corrections, we can consider the $SU(3)$ gauge group as dynamical and the $SU(2)$ gauge group acting as a global symmetry group. Looking at the matter content of the model, we see that from the $SU(3)$ gauge theory view point this is nothing but $F = N - 1$ SQCD, with $N = 3$ and $F = 2$. Hence a non-perturbative superpotential is indeed generated and reads

$$W_{\text{non-pert}} = \frac{\Lambda_3^7}{Y}. \tag{11.6}$$

This is enough to conclude that supersymmetry is dynamically broken! Due to (11.6) the minimum of the potential is certainly at some non-zero VEV for Y. Since $R(Y) = -2$ the R-symmetry is spontaneously broken and since there are no classical flat directions, supersymmetry is broken.

Remarkably, it turns out that in this model we can make more quantitative statements. Summing up the tree-level and non-perturbative superpotential contributions we get for the full effective superpotential

$$W_{\text{eff}} = hX_2 + \frac{\Lambda_3^7}{Y}. \tag{11.7}$$

From the above expression one can easily see that supersymmetry is broken because, in terms of such low energy fields, we have

$$\frac{\partial W_{\text{eff}}}{\partial X_2} = h \neq 0. \tag{11.8}$$

In this derivation we have implicitly assumed that X_1, X_2 and Y are the correct low energy degrees of freedom, and that no other massless fields show up at any point of field space. If this were the case, instead, one could have met singularities in the Kähler metric at such points, invalidating our conclusions. For small enough h, we are safe on this side. The superpotential contribution $W_{\text{non-pert}}$ push the vacuum away from the origin. For $h \ll 1$ the minimum of the potential is certainly in the large $Q, \widetilde{Q}$ region. Since Q and $\widetilde{Q}$ are charged under both gauge groups, in the vacuum the gauge symmetry is completely broken, and (heavy) gauge bosons can be integrated out. This suggests that X_1, X_2 and Y are indeed the correct low energy degrees of freedom and therefore we do not expect singularities in the Kähler potential (which would indicate the presence of extra massless states).

All what we said so far shows that the 3-2 model belongs, at least, to the second class of supersymmetry breaking models we discussed at the beginning of this section, the so-called non-calculable models. But we can do more.

In the regime we chose, $h \ll 1$, $\Lambda_3 \gg \Lambda_2$, the ground states are in a weakly coupled region, and then we are in a situation similar to the red curve of Figure 11.2. Therefore, the Kähler potential can be safely taken to be canonical in terms of UV-fields

$$K = Q^\dagger Q + \widetilde{Q}^\dagger \widetilde{Q} + L^\dagger L. \tag{11.9}$$

We can project this potential onto D-flat directions and get

$$K = 24\, \frac{A + Bx}{x^2}, \tag{11.10}$$

where

$$A = \frac{1}{2}\left(X_1^\dagger X_1 + X_2^\dagger X_2\right), \quad B = \frac{1}{3}\sqrt{Y^\dagger Y}, \quad x = 4\sqrt{B}\cos\left(\frac{1}{3}\arccos\frac{A}{B^{3/2}}\right). \tag{11.11}$$

Plugging these expressions and that for the effective superpotential, eq. (11.7), into eq. (11.1), upon minimization with respect to all scalar fields, one can find the minima, and hence the vacuum energy $E \sim M_s$.

The computation is doable but rather lengthy, so let us first try to get an estimate of the different scales in the problem. The minima will be around a region of field space where the classical and the non-perturbative contributions to the potential are the same order (see Figure 11.1), which is the same as to ask the two contributions to the effective superpotential in eq. (11.7) being roughly comparable. In what follows, we think in terms of fundamental UV fields (an acceptable thing to do, given we are in a weakly coupled region). Calling v the generic VEV of (fundamental) scalar fields at the non-supersymmetric minimum, we get that

$$hv^3 \sim \frac{\Lambda_3^7}{v^4} \quad \text{that is } v \sim \frac{\Lambda_3}{h^{1/7}}. \tag{11.12}$$

This implies that $W \sim \Lambda_3^3 h^{4/7}$ and hence $\partial W/\partial\phi \sim \Lambda_3^2 h^{5/7}$. Therefore, since the potential is proportional to the derivative of the superpotential squared (recall that the Kähler potential is canonical in the UV fields around the supersymmetry breaking vacua), we finally get

$$M_s^4 \sim \Lambda_3^4 h^{10/7} \quad \text{that is } M_s \sim \Lambda_3 h^{5/14}. \tag{11.13}$$

Note that this is a leading order estimate. The Kähler potential receives perturbative and non-perturbative corrections in inverse powers of v. However, in the regime we are considering these are very small, in the sense that v is much larger than any other scale in the theory. Indeed

$$v \sim h^{-1/7}\Lambda_3 \underset{h\ll 1}{\gg} \Lambda_3 \gg \Lambda_2. \tag{11.14}$$

From eq. (11.13) we see that $M_s \ll \Lambda_3$, as well as $M_s \ll \Lambda_2$, if h is small enough. This gives an a posteriori justification of our claim that X_1, X_2 and Y were the correct low energy degrees of freedom. Supersymmetry breaking occurs at an energy scale below the confining scale of both non-abelian gauge groups. Therefore, the low energy effective dynamics is certainly not including light gauge degrees of freedom. The effective Lagrangian should be (and actually is, recall our conclusions below eq. (11.11)) of O'R-like type, with the only complication of a non-canonical Kähler potential. A supersymmetric σ-model, in fact.

As already stressed, our rough estimates do not prevent to compute everything analytically, by means of eq. (11.1). The answer one gets this way is that the minimum of the potential is at $X_1 = 0$, which means that the $U(1)_Y$ symmetry is unbroken (X_2 and Y are uncharged under this symmetry). On the other hand the R-symmetry is broken, since the vacua are at finite value of Y, which is charged under the R-symmetry. This suggests, and confirmed by explicit computations, that the massless spectrum is composed by a goldstino, an R-axion, associated to the breaking of the R-symmetry, and finally a fermionic field with hypercharge $Y = -1$, whose existence can be proved using t'Hooft anomaly matching condition for the unbroken $U(1)_Y$ symmetry. All other fields have masses of order $\sim hv$.

What changes in our analysis if choosing a different regime, namely $\Lambda_2 \gg \Lambda_3$? One can derive an effective superpotential also in this case (which is that of SQCD with $F = N$, now) and show that supersymmetry is still broken (though at a different scale with respect to previous regime). However, generically the theory is strongly coupled and hence the Kähler potential is unknown, regardless how small the superpotential coupling h is. This is because for $F = N$ the effective superpotential is not of runaway type and does not push the vacua towards large field VEVs, where a semi-classical analysis could had been done. More precisely, the quantum constraint sets the VEV of meson and baryon operators to be of order the strong coupling scale Λ_2, regardless the value of the coupling h. So, in this regime the $3-2$ model is non-calculable.

Finally, one can be as general as possible, and consider the two dynamical scales being the same order, leading to a superpotential of the following form

$$W_{\text{eff}} = hX_2 + \frac{\Lambda_3^7}{Y} + A\left(Z - \Lambda_2^4\right) \qquad (11.15)$$

where $Z = \epsilon^{ijk}Q_{i\alpha}Q_{j\beta}\epsilon^{\alpha\beta}Q_{k\gamma}L_\delta\epsilon^{\gamma\delta}$ and A a Lagrange multiplier. The latter is nothing but the gauge invariant expression $\det M - B\widetilde{B}$ for the $SU(2)$ theory, to which we referred to in the previous paragraph and which for $SU(2)$ gauge group has a compact form (mesons and baryons are both quadratic in terms of fundamental fields and can be organized in a single matrix). Classically Z vanishes identically. This shows why in the regime where the $SU(2)$ gauge group is classical, the superpotential reduces to the expression (11.7) we used before. Notice that since Z is classically zero, the $Z^\dagger Z$ term in the Kahler potential is suppressed by some function of Λ_2/v. Restoring canonical normalization for Z kinetic term implies that the mass of Z is enhanced by the inverse of this function. Therefore at low energy, in the regime where the $SU(2)$ group is nearly classical, one can safely integrate Z out and use just X_A and Y as low energy fields, as we did before. Obviously, the analysis in the regime where both $SU(3)$ and $SU(2)$ have a quantum behavior is more complicated but one can again conclude that supersymmetry is broken.

The 3-2 model is the prototype of calculable DSB models, and many interesting generalizations exist, like the so-called $SU(N) \times SU(2)$ and $SU(N) \times SU(N-1)$ models, plus several others. These are discussed in the references at the end of this chapter.

Let us end with an important remark. The 3-2 model is a beautiful instance of a DSB model and provides a natural way to generate a (small) supersymmetry breaking scale dynamically, without the need of having dimension-full parameters put by hand in the theory, as it was the case for the supersymmetry breaking models we discussed in Chapter 7. This holds at *any* point in the parameter space. Calculability, though, does not. As we have seen, the model is fully calculable in the region of the parameter space where the tree-level dimensionless coupling is parametrically small, something not at all generic, from a naturalness point of view.

11.4 The 4-1 model: supersymmetry breaking by gaugino condensation

Let us now consider a model which is similar to the previous one, but differs in that at low energy the theory is not fully higgsed but reduces to pure SYM. In this case supersymmetry breaking will be driven by gaugino condensation, and not by instanton effects as for the 3-2 model.

Let us consider a supersymmetric theory with gauge group $G = SU(4) \times U(1)$ and matter content as detailed below.

	$SU(4)$	$U(1)$	$U(1)_R$
Q_i	4	-3	3
$\widetilde{Q}^i$	$\bar{4}$	-1	-3
A_{ij}	6	2	-2
S	$\bullet$	4	2

The indexes i, j are in the (anti)fundamental of $SU(4)$ and $U(1)_R$ is a non-anomalous R-symmetry. As in previous examples, charges of matter fields under these non-anomalous symmetries can be fixed computing triangle diagrams with one global and two gauge current insertions. The model has also the following tree-level superpotential

$$W_{\text{tree}} = hS\widetilde{Q}Q, \tag{11.16}$$

which has R-charge 2 and sum over gauge indexes is understood.

As usual, let us start analyzing this theory at the classical level. Let first ignore the $U(1)$ dynamics and the superpotential (11.16). The $SU(4)$ D-flat directions can be described by the following gauge invariant operators

$$M = \widetilde{Q}Q, \quad \text{Pf}A = \epsilon^{ijkl}A_{ij}A_{kl}/8, \quad S. \tag{11.17}$$

For later purposes let us notice that M has $U(1)$ charge equal to -4, while PfA and S have both $U(1)$ charge equal to 4.

One can show that along a generic flat direction a $SU(2) \subset SU(4)$ gauge invariance survives under which no matter is charged. At scales below the dynamical scale Λ_2 of the effective $SU(2)$ SYM, the theory confines and glueballs and their superpartners can be integrated out: one is only left with M, PfA and S as low energy degrees of freedom. Gaugino condensation of pure SYM leads to the following non-perturbative generated superpotential

$$W_{\text{non-pert}} = 2\Lambda_2^3 = \frac{2\Lambda_4^5}{\sqrt{M\text{Pf}A}}, \tag{11.18}$$

where the second equality comes from scale-matching.

If we now consider $U(1)$ gauge interactions, we have to project the $SU(4)$ D-flat space onto the subspace which is also $U(1)$ D-flat. The latter is parameterized by two moduli, MPfA and SM, which hence parametrize the classical D-flat directions. Note that the tree-level superpotential (11.16), exactly as in the 3-2 model, lifts them all and leaves only the origin of field space as a supersymmetric vacuum (this can be seen from the F-term equation for S which sets to zero M and hence both MPfA and SM).

Both the $U(1)$ gauge coupling and the superpotential coupling are IR-free, so they would not affect the analysis which led to (11.18). Therefore, we can now consider the full superpotential simply adding up the tree-level and non-perturbative contributions and get

$$W_{\text{eff}} = \frac{2\Lambda_4^5}{\sqrt{M\text{Pf}A}} + hSM. \tag{11.19}$$

This superpotential is essentially the same as that of the 3-2 model, eq. (11.7). Therefore, from this point on the analysis is the same as the one we performed previously. Supersymmetry is broken because of the interplay between the dynamically generated runaway superpotential term (11.18) and the tree-level contribution (11.16). Differently from the 3-2 model, though, the fact that for small enough h, SM and $M\text{Pf}A$ are the correct low energy degrees of freedom, does not follow from complete Higgsing of the gauge group, since on the moduli space there is a surviving $SU(2)$ SYM theory. Still, at energies below Λ_2 the gauge group confines and glueballs and their superpartners can be integrated out. Hence, at low enough energy, the effective theory is indeed given in terms of SM and $M\text{Pf}A$ only.

Similarly to the 3-2 model, one can argue that for small values of the coupling h the model is calculable. This might look strange, given the left-over non-abelian $SU(2)$ gauge dynamics which is strongly coupled. How can that be? One expects non-perturbative strong coupling dynamics associated to $SU(2)$ to give rise to corrections to the Kähler potential in terms of some function of $\sim\Lambda_2/v$, with v the typical scale of a fundamental field VEV. Balancing the two terms in eq. (11.19) and recalling the expression (11.18) one finds

$$v \sim \frac{\Lambda_2}{h^{1/3}} \quad \text{which implies } \frac{\Lambda_2}{v} \ll 1 \quad \text{for } h \ll 1. \tag{11.20}$$

Hence, quantum corrections to the Kähler potential are suppressed in this regime and the model is calculable. Let us stress again how different the dynamics is with respect to the 3-2 model. There, the smallness of h ensures that both gauge groups are fully broken at very high energy, and therefore quantum corrections due to gauge dynamics suppressed. Here, instead, a fully unbroken gauge groups survives at low energy.

The computation of low energy spectrum and interactions goes along similar lines as the 3-2 model, and we do not repeat it here. For instance, also for this model the tree-level superpotential has a R-symmetry which is spontaneously broken in the vacuum. Hence we expect, as for the 3-2 model, a R-axion in the massless spectrum.

Let us summarize, instead, the physical picture one should bare in mind. The theory in the UV is a $SU(4) \times U(1)$ gauge theory. At a scale v this is broken down to $SU(2)$. This left-over non abelian gauge theory confines at a scale $\Lambda_2 \ll v$, below which we have a low energy effective theory with chiral superfields, only. Gaugino condensation gives rise to a superpotential contribution which induces supersymmetry breaking at a scale M_s. Note that in the limit $h \ll 1$ the supersymmetry breaking scale is parametrically smaller than Λ_2 (using the same rationale we used for the 3-2 model, one easily sees that $M_s \sim \Lambda_2 h^{1/6}$ which is well below Λ_2, if h is small). Hence, at the supersymmetry breaking scale all gauge degrees of freedom are heavy and do not contribute to the effective action, which justifies the description in terms of SM and $M\text{Pf}A$, only.

The 4-1 model has several generalizations. The most straightforward ones are theories with gauge group $SU(2l) \times U(1)$ and matter consisting of a chiral superfield transforming in the anti-symmetric representation of $SU(2l)$, $2l-3$ anti-fundamentals $\widetilde{Q}$, one fundamental Q, and $2l-3$ singlets S_i. Supersymmetry breaking is again driven by gaugino condensation of an IR left-over $SU(2)$ gauge group, provided a suitable tree-level superpotential is added which lifts all classical flat directions. The 4-1 model corresponds to $l=2$. More details on these generalizations can be found in the references at the end of the chapter.

11.5 ITIY model: supersymmetry breaking with classical flat directions

Let us now consider an instance of a non-calculable model. While non-calculable, this model is instructive in that supersymmetry is broken even though the theory is non-chiral and admits classical flat directions (the latter get lifted by non-perturbative effects *not* leading to runaway behavior). So, two of the generic properties one expects to be satisfied for a model to break supersymmetry are not met.

Let us consider a gauge theory with group $G = SU(2)$, four fundamental fields Q^i (which correspond to two flavors, since for $SU(2)$ the fundamental and anti-fundamental representations are equivalent) plus six singlets S_{ij} and a superpotential

$$W_{\text{tree}} = hS_{ij}Q^iQ^j \tag{11.21}$$

(notice that the product Q^iQ^j is antisymmetric since what it really means is $Q^i_\alpha Q^j_\beta \epsilon^{\alpha\beta}$ where α, β are $SU(2)$ gauge indexes). This theory admits a $SU(4)$ flavor symmetry group (this enhancement of the global non-R symmetry group from $SU(F)_L \times SU(F)_R \times U(1)_B$ to $SU(2F)$ is always there whenever the gauge group is $SU(2)$). Under this global symmetry the Q^i's transform in the fundamental and the singlets in the anti-symmetric representations, respectively. Hence, the tree-level superpotential (11.21) respects this $SU(4)$ flavor symmetry.

Let us start studying the classical behavior of the theory. The $SU(2)$ D-flat directions can be parameterized by six meson-like operators

$M^{ij} \sim Q^i Q^j$, which transform in the 6 of $SU(4)$ and satisfy the classical constraint of SQCD with $N = F = 2$, that is

$$\text{Pf}M = \epsilon_{ijkl} M^{ij} M^{kl} = 0. \tag{11.22}$$

The indexes i, j should be seen as $SO(4)$ indexes (recall that $SO(4) \simeq SU(2) \times SU(2)$ and notice that for any nonzero value of M the $SU(4)$ global symmetry is broken to $SU(2) \times SU(2)$). The F-flatness condition for S_{ij} sets all Q^i's to zero hence all flat directions are lifted *but* the singlets.

At the quantum level the classical constraint (11.22) is modified and the full effective superpotential reads

$$W_{\text{eff}} = hS_{ij} M^{ij} + A \left(\epsilon_{ijkl} M^{ij} M^{kl} - \Lambda^4 \right). \tag{11.23}$$

The F-equation for S_{ij} still gives $M^{ij} = 0$ but now this is in conflict with the quantum constraint, *i.e.* the F-equation for the Lagrange multiplier A. Therefore, supersymmetry is broken. More precisely, working out the potential from the expression (11.23) one can show that up to symmetry transformations the minimum is at $M_{ij} = \Lambda^2$, $S_{13} = S_{14} = S_{23} = S_{24} = 0$ and $S_{12} = S_{34} \equiv S$. Therefore, there is a pseudoflat direction parametrized by S.

This model is instructive in many respects, which we consider in turn.

Having a flat direction, parametrized by S, one could be worried about where, in field space, the supersymmetry breaking vacua lie, once quantum corrections in the coupling h are taken into account. In principle, there can also be a runaway. A careful analysis, which we refrain to do here, shows that this is not the case: for small enough h and large $h\langle S \rangle$ the Kähler potential for S can be shown to grow logarithmically for large S, hence ensuring that the actual minimum is stabilized at a finite distance in field space.

Notice also that this model is non-chiral. Therefore, one could add a mass term for all fields, lifting all classical flat directions. At low energy one could then integrate all chiral fields out and end-up with pure $SU(2)$ SYM, which does not break supersymmetry (it has two vacua and Witten index equal to 2)! How that can be? The answer comes from a careful analysis of the massless limit.

Let us add a mass perturbation to the superpotential (11.23)

$$W_{\text{eff}} = hS_{ij} M^{ij} + m_{ij} M^{ij} + \frac{1}{2} \widetilde{m}\, \text{Pf}\, S + A \left(\epsilon_{ijkl} M^{ij} M^{kl} - \Lambda^4 \right). \tag{11.24}$$

The F-equations for M^{ij} and S_{ij} set

$$\langle M^{ij}\rangle \sim \epsilon^{ijkl} m_{kl} \left(\frac{\Lambda^4}{\text{Pf}m}\right)^{1/2}$$

$$\langle S_{ij}\rangle \sim \frac{m_{ij}}{\widetilde{m}} \left(\frac{\Lambda^4}{\text{Pf}m}\right)^{1/2},$$

where the square roots get two values, corresponding to the two vacua of pure $SU(2)$ SYM. Take now the limit $\widetilde{m}, m_{ij} \to 0$ with their ratio fixed. This way, $\langle M^{ij}\rangle$ has a finite limit, but $\langle S_{ij}\rangle$ is pushed all the way to infinity. This implies that the supersymmetry preserving vacua are also pushed to infinity and disappear from the spectrum, recovering our previous result.

This is an instance of discontinuous change of the Witten index, which moves from 2 to 0 in the limit of vanishing masses. The reason for that is that the mass terms change the behavior of the Hamiltonian in the large field region. As the limit $\widetilde{m} \to 0$ is taken, the asymptotic behavior of the potential changes since now there are classical flat directions (and the Witten index can and does change, recall our discussion in Section 7.6.2).

The ITIY model (named after Intriligator, Thomas, Izawa and Yanagida) admits many generalizations. An interesting class is based on SQCD with gauge group $USp(2N)$ and $F = N + 1$ flavors. This theory has a $SU(2F) = SU(2N + 2)$ flavor symmetry, and enjoys a quantum deformed moduli space, very much like $SU(N)$ SQCD with $F = N$ flavors. Coupling the quark superfields to a set of gauge singlets transforming in the antisymmetric representation of the flavor symmetry group via a superpotential like (11.21), one can show that supersymmetry is broken in a way identical to that of the original ITIY model (actually, recalling that $SU(2) \simeq USp(2)$, one sees that the ITIY model corresponds to the case $N = 1$ within this class).

11.6 DSB into metastable vacua. A case study: massive SQCD

We now want to consider the possibility that supersymmetry is broken dynamically into *metastable* vacua.

A model of DSB into metastable vacua share some basic properties with ordinary DSB models. The theory should be a gauge theory and should not break supersymmetry at tree level. Only non-perturbative corrections should. The difference is that the non-perturbative dynamics does not lift classical supersymmetric vacua but just ensure that local minima of the potential, whose nature is intrinsically non-perturbative, exist.

On general ground, as we observed towards the end of Chapter 7, due to Witten index arguments, R-symmetry arguments, etc... the landscape of theories admitting metastable supersymmetry breaking vacua is expected to be much larger than those admitting stable ones. This has been known since long, but only more recently it was possible to come-up with the first explicit such construction, when in 2006 Intriligator, Seiberg and Shih (ISS) proved the existence of dynamically generated metastable vacua in the most innocent-looking supersymmetric gauge theory one can imagine: massive SQCD. Note that this is a non-chiral theory, with supersymmetric vacua (a full moduli space, in fact, in the massless limit), non-vanishing Witten index and no R-symmetry (quarks mass terms explicitly break the non-anomalous R-symmetry of massless SQCD). Even more strikingly, the model turns out to be calculable, in the sense that around the metastable vacua one can compute both the superpotential and the Kähler potential, and hence the effective Lagrangian describing the dynamics of light fields.

These results have been extended into several directions, and many interesting applications have been found since then. In what follows, we will review the original model, which represents the core of all these developments.

Summary of basic results

Since the derivation is rather lengthy, let us anticipate the upshot of the analysis we are going to perform. What we will prove is that $SU(N)$ SQCD with massive flavors in the free magnetic phase (that is for $N+1 \leq F \leq \frac{3}{2}N$) admits supersymmetry breaking vacua which, for $m \ll \Lambda$, where m is the scale of quark masses and Λ the dynamical scale of the theory, can be made parametrically long lived. More precisely, the theory admits

- N supersymmetric vacua along the mesonic branch, at

$$\langle M \rangle_{\rm SUSY} = \left(m^{F-N}\Lambda^{3N-F}\right)^{1/N}, \quad \langle B_{i_1 i_2 \dots i_{F-N}} \rangle = 0, \quad \langle \widetilde{B}^{i_1 i_2 \dots i_{F-N}} \rangle = 0. \tag{11.25}$$

- A compact space of metastable supersymmetry breaking vacua along the baryonic branch

$$\langle B_{i_1 i_2 \dots i_{F-N}} \rangle, \langle \widetilde{B}^{i_1 i_2 \dots i_{F-N}} \rangle \neq 0, \quad \langle M \rangle_{\rm meta} = 0, \tag{11.26}$$

with vacuum energy $V_{\rm meta} \sim N|m\Lambda|^2$.

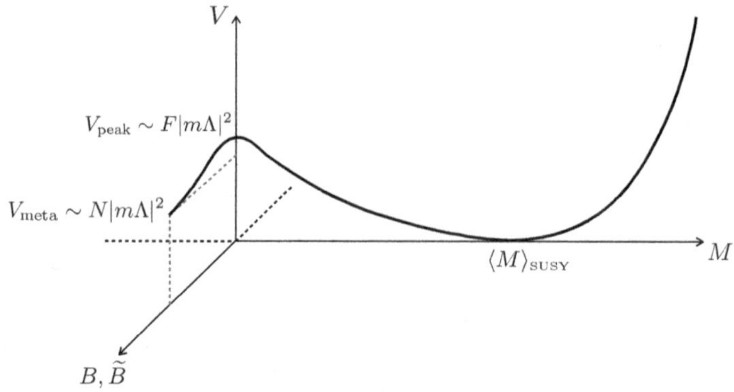

Fig. 11.5 The scalar potential of massive SQCD in the free magnetic phase. On the mesonic branch there are supersymmetric vacua. On the baryonic branch there are supersymmetry breaking vacua, which are parametrically long-lived if $m \ll \Lambda$.

One can also compute the life-time of the metastable vacua and find that

$$\tau \sim e^{S_{\rm B}} \quad \text{where } S_{\rm B} \sim \epsilon^{-4(3N-2F)/N} \quad \text{and} \quad \epsilon = \sqrt{\frac{m}{\Lambda}}, \tag{11.27}$$

with $S_{\rm B}$ the Coleman bounce action. This implies, as anticipated, that for small masses, *i.e.* $\epsilon \ll 1$, the metastable vacua can be made arbitrarily long-lived, and hence potentially viable, phenomenologically. The qualitative of the potential one gets is reported in Figure 11.5.

11.6.1 *Massive SQCD in the free magnetic phase: electric description*

Consider SQCD with gauge group $SU(N)$ in the free magnetic phase, namely for $N+1 \leq F \leq \frac{3}{2}N$. As we have discussed in detail in Chapter 10, this theory has many supersymmetric vacua, actually a full moduli space. Let us add a mass term for all matter fields

$$W = \operatorname{Tr} mQ\widetilde{Q} \equiv \operatorname{Tr} mM. \tag{11.28}$$

where the trace is taken on gauge and flavor indexes. Notice that (11.28) breaks the SQCD R-symmetry explicitly, while the flavor symmetry group is broken to a subgroup H, whose structure depends on the specific form of the mass matrix m (more later).

This theory has two mass scales, the quarks mass, which with a slight abuse of language we refer to as m, which is actually a matrix of masses, and Λ, the dynamical scale of the theory. Let us consider the two obvious possible regimes in turn.

(a) $m > \Lambda$
In this regime one can safely integrate the flavors out before the strong coupling dynamics takes place. The theory at low energy flows to pure SYM with gauge group $SU(N)$ and has N (isolated) supersymmetric vacua. By scale matching, we obtain

$$\Lambda_{\rm L}^{3N} = \det m \, \Lambda^{3N-F} \tag{11.29}$$

which implies

$$W_{\rm eff} = N\Lambda_{\rm L}^3 = N \left(\det m \; \Lambda^{3N-F}\right)^{\frac{1}{N}}, \tag{11.30}$$

an effective superpotential displaying, correctly, the N vacua of pure $SU(N)$ SYM. What's this, really? The mass matrix m and the meson matrix M are Legendre dual variables. The effective superpotential above is nothing but the effective superpotential once the mesons have been integrated out. Hence, using formula (10.79), we get the matrix equation

$$\langle M\rangle_{\rm SUSY} = \left(\det m \; \Lambda^{3N-F}\right)^{\frac{1}{N}} \frac{1}{m} \tag{11.31}$$

which tells where in the moduli space the N supersymmetric vacua sit: they correspond to the N roots of the above equation.

(b) $m < \Lambda$
In this case, which is actually the one we will be interested in, eventually, it is not completely correct to proceed as before since strong coupling dynamics, driven by Λ, enters before being allowed to integrate the massive quarks out. The more correct thing to do, in this case, is to notice that m and M are Legendre dual variables, and integrate M in starting from eq. (11.30). In practice, one should take the determinant of eq. (11.31), solve for $\det m$ and follow the procedure outlined in Section 10.4.3, getting finally

$$W_{\rm eff} = (N-F)\left(\frac{\Lambda^{3N-F}}{\det M}\right)^{\frac{1}{N-F}} + {\rm Tr}\, mM. \tag{11.32}$$

Then we can find eq. (11.31) simply solving the F-equations for M. Notice, however, that strictly speaking $\det M = 0$ for $F \geq N+1$, so one has to go a bit off-shell, so to say, in performing the computation.

The final result, eq. (11.31), is of course a perfectly meaningful on-shell result.

The upshot is that, no matter the value of m/Λ, there exist N supersymmetric vacua on the mesonic branch. That baryon VEVs are vanishing can be easily argued as follows. In the same way as m and M can be thought as Legendre dual variables, one can think of b and $\tilde{b}$ as sources for the baryons B and $\widetilde{B}$, respectively, deforming the theory by $\Delta W = b\,B + \tilde{b}\,\widetilde{B}$ (flavor indexes are suppressed, for the ease of notation). At low energy the theory reduces to pure SYM with gauge group $SU(N)$ and effective superpotential (11.30). Using now eq. (10.79) applied to the Legendre dual variables b and B (respectively $\tilde{b}$ and $\widetilde{B}$) one concludes that $\langle B \rangle = \langle \widetilde{B} \rangle = 0$. Hence the supersymmetric vacua (11.31) have indeed zero baryon number.

In general, m is a matrix transforming under the anti-fundamental of $SU(F)_L$ and the fundamental of $SU(F)_R$. This matrix can always be diagonalized via a bi-unitary transformation and, from here on, we choose for simplicity all entries to be equal, $m_i = m$. The superpotential term hence reads

$$W = m\mathrm{Tr}M, \tag{11.33}$$

where now m is just a number. With this choice, the $SU(F)_L \times SU(F)_R$ flavor symmetry group is broken to $SU(F)_D$, while eq. (11.31) simplifies to

$$\langle M \rangle_{\mathrm{SUSY}} = \left(m^F\,\Lambda^{3N-F}\right)^{\frac{1}{N}}\frac{1}{m} = \left(m^{F-N}\Lambda^{3N-F}\right)^{\frac{1}{N}} = \epsilon^{2\frac{F-N}{N}}\Lambda^2, \tag{11.34}$$

where $\epsilon \equiv \sqrt{m/\Lambda}$.

11.6.2 *Massive SQCD in the free magnetic phase: magnetic description*

So far, we have derived the first part of ISS statement, the easy one. We have obtained, via holomorphic decoupling, the N supersymmetric vacua of massive SQCD, and found they lie along the mesonic branch. In order to find something more interesting, we have to turn to the Seiberg dual description of the theory, *i.e.* mSQCD.

Since we are in the magnetic-free phase we choose, in what follows, $\Lambda_{\mathrm{el}} = \Lambda_{\mathrm{m}} = \mu \equiv \Lambda$ (see the discussion at the end of Chapter 10). The magnetic dual superpotential, including the mass deformation (11.33), reads

$$W = h\mathrm{Tr}\, q\Phi\tilde{q} - m\Lambda\, h\mathrm{Tr}\,\Phi, \tag{11.35}$$

where

$$\Phi = \frac{1}{h\Lambda}M. \tag{11.36}$$

Let us start by recovering, using magnetic variables, the N supersymmetric vacua we have found before. To this aim, let us suppose we give some non-vanishing VEV to the gauge singlet Φ. This provides a mass to dual quarks, q and $\tilde{q}$, which can then be integrated out. The theory reduces to pure $SU(F-N)$ SYM and the effective superpotential one obtains, upon holomorphic decoupling, reads

$$W_{\text{eff}} = -m\Lambda h\mathrm{Tr}\,\Phi + (F-N)\,\Lambda_L^3. \tag{11.37}$$

By matching at dual quarks mass scale we find the relation

$$\Lambda_L^{3(F-N)} = h^F \det\Phi\, \Lambda^{3(F-N)-F} \quad \text{that is } \Lambda_L^3 = \left(h^F \det\Phi\, \Lambda^{2F-3N}\right)^{\frac{1}{F-N}}. \tag{11.38}$$

We can substitute the above relation into the superpotential (11.37) and get

$$W_{\text{eff}} = -m\Lambda h\mathrm{Tr}\,\Phi + (F-N)\left(h^F \det\Phi\Lambda^{2F-3N}\right)^{\frac{1}{F-N}}. \tag{11.39}$$

The F-equation for Φ gives

$$\begin{aligned}\langle h\Phi\rangle_{\text{SUSY}} &= \sqrt{m\Lambda}\,\epsilon^{-\frac{3N-2F}{N}} \gg \sqrt{m\Lambda}\\ &= \Lambda\,\epsilon^{2\frac{F-N}{N}} \ll \Lambda,\end{aligned} \tag{11.40}$$

where the inequalities hold if ϵ is small. The expression in the first line says that the supersymmetric vacua are at a parametrically large distance from the origin of field space in units of $\sqrt{m\Lambda}$. This means that they are located in a very quantum region from mSQCD point of view, since in this regime of parameters mSQCD is an IR-free theory. On the other hand, the expression in the second line shows that they are still well below mSQCD Landau pole, meaning that the above analysis is a meaningful one to do within mSQCD. Notice that, consistently, using the map (11.36) one can easily see that (11.40) is the same as (11.34).

Even using magnetic variables, one can easily conclude that in the supersymmetric vacua baryon VEVs are vanishing. As already observed, on such vacua the dual quarks are massive and can be integrated out, hence their VEVs vanish. The VEVs of the magnetic baryons can be easily computed from that of dual quarks. Indeed, the magnetic theory is IR free and in the supersymmetric vacua the VEV of the product of qs is the same as the product of the VEVs of each q. Therefore

$$\langle b^{j_1 j_2 \ldots j_N} \rangle = 0 \tag{11.41}$$

and similarly for $\tilde{b}_{j_1 j_2 \ldots j_N}$. From the map (10.121) it then follows that $\langle B_{i_1 i_2 \ldots i_{F-N}} \rangle = \langle \widetilde{B}^{i_1 i_2 \ldots i_{F-N}} \rangle = 0$, as anticipated.

Notice that while the ϵ parameter defined here and in the electric description is one and the same, the $\epsilon \to 0$ limit should be understood differently. In the electric description Λ is a dynamical RG-invariant scale and the limit of small ϵ is obtained sending $m \to 0$ keeping Λ fixed. In the magnetic description, Λ is a cut-off scale, above which the theory is not defined. The limit should now be understood as $\Lambda \to \infty$ keeping $\sqrt{m\Lambda}$, the mass scale entering the superpotential (11.35), fixed (notice that $\epsilon = \sqrt{m/\Lambda} = \sqrt{m\Lambda}/\Lambda$). This seemingly pedantic observation will be relevant later.

Let us now come back to the expression (11.35) and analyze the properties of deformed mSQCD more closely. We will do this in steps. Let us forget, for now, that the magnetic group $SU(F-N)$ is gauged. If gauge degrees of freedom are frozen, the vacua of the theory are obtained solving F-equations only. From eq. (11.35) these read

$$\begin{cases} \overline{F}_{\Phi_i^j} = \tilde{q}_a^i q_j^a - m\Lambda\delta_j^i \\ \overline{F}_{q_i^a} = h\Phi_j^i \tilde{q}_a^j \\ \overline{F}_{\tilde{q}_a^j} = hq_i^a \Phi_j^i \end{cases} \tag{11.42}$$

where a are $SU(F-N)$ indexes. We see that the first set of equations cannot be solved. The rank of $\tilde{q}_a^i q_j^a$ is at most $F-N$ while that of δ_j^i is clearly F. Hence we can set to zero at most $(F-N)$ terms of F_Φ-equations: we are left with $F-(F-N) = N$ non-vanishing F-terms. On the other hand, the F-equations for q and $\tilde{q}$'s are easily satisfied. We conclude that supersymmetry is broken, and is so by a *rank condition*. The potential energy gets contribution from the N F-equations that cannot be set to zero

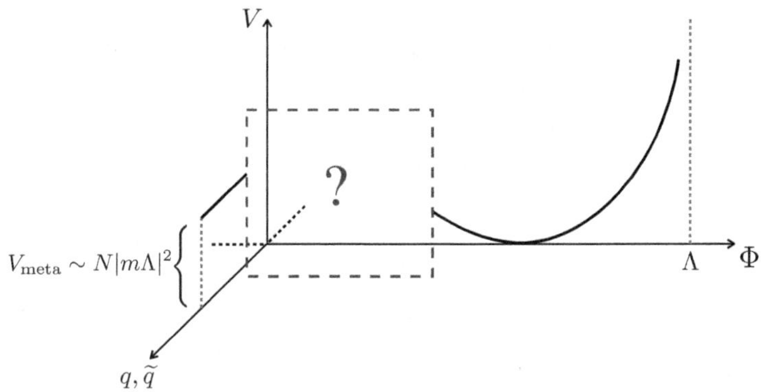

Fig. 11.6 Linearly deformed mSQCD *classical* potential.

and hence reads

$$V_{\rm meta} \sim N|m\Lambda|^2. \tag{11.43}$$

The supersymmetry breaking vacua are at

$$\langle\Phi\rangle = \begin{pmatrix} 0 & 0 \\ 0 & \Phi_0 \end{pmatrix}, \quad \langle q\rangle = \begin{pmatrix} q_0 \\ 0 \end{pmatrix}, \quad \langle\tilde{q}^T\rangle = \begin{pmatrix} \tilde{q}_0 \\ 0 \end{pmatrix} \tag{11.44}$$

where $q_0\tilde{q}_0 = m\Lambda\, \mathbb{1}_{F-N}$, with q_0 and $\tilde{q}_0$ being $F-N\times F-N$ matrices, and Φ_0 an arbitrary $N\times N$ matrix. Therefore, we find a pseudo-moduli space of supersymmetry breaking vacua parameterized by Φ_0, q_0 and $\tilde{q}_0$. If this analysis were correct, the picture we would obtain is what is summarized, schematically, in Figure 11.6.

The question we should now try to answer is to what extent the above results are solid in the full quantum theory. So far, our analysis was classical, both because we have been ignoring $SU(F-N)$ gauge dynamics, and because, even within the ungauged model, we have not taken into account quantum corrections coming from the coupling h. Let us start considering quantum effects due to h. Later, we will consider the role of gauge degrees of freedom and interactions.

Let us first notice that the supersymmetry breaking vacua lie relatively near to the origin, which is the more classical region for mSQCD, which is an IR-free theory. Indeed, as already observed, the scale $\sqrt{m\Lambda}$ is set to be the mass scale entering the mSQCD Lagrangian by the superpotential (11.35), the natural mass unit to measure dimensionfull quantities in the magnetic theory. From eq. (11.43) we see that the energy density of the supersymmetry breaking minima is order one in units of $\sqrt{m\Lambda}$, and so are the values of q_0 and $\tilde{q}_0$ entering eq. (11.44), since $q_0\tilde{q}_0 = m\Lambda\,\mathbb{1}_{F-N}$ (the Φ_0 flat direction does not play any role here since, as we will see momentarily, quantum corrections lift this degeneracy and set $\Phi_0 = 0$). Notice, instead, that from eq. (11.40), first line, it follows that $\langle\Phi\rangle_{\rm SUSY}$ is parametrically large in units of $\sqrt{m\Lambda}$. Since mSQCD is IR-free, the classical region is a neighborhood of the origin in field space so we can safely take the Kähler potential to be canonical in the region where the supersymmetry breaking vacua sit

$$K = \text{Tr}\left(\Phi^\dagger\Phi + q^\dagger q + \tilde{q}^\dagger\tilde{q}\right). \tag{11.45}$$

A second comment regards global symmetries. In the limit where the magnetic gauge group $SU(F-N)$ is taken to be ungauged, mSQCD has a global symmetry group $SU(F-N)\times SU(F)_L\times SU(F)_R\times U(1)_B\times U(1)_{R_0}$. This is broken by the second term in (11.35) to $G = SU(F-N)\times SU(F)_D\times U(1)_B\times U(1)_{R_0}$, where under the non-anomalous R-symmetry $U(1)_{R_0}$ the dual quarks are chargeless and Φ has R-charge $R_0 = 2$, as dictated by the tree-level superpotential (11.35).

On the supersymmetry breaking vacua (11.44) the group G is spontaneously broken. The vacua with maximal unbroken global symmetry sit at (up to unbroken flavor rotations)

$$\Phi_0 = 0, \quad q_0 = \tilde{q}_0 = \sqrt{m\Lambda}\,\mathbf{1}_{F-N}, \tag{11.46}$$

and preserve $H = SU(F-N)_D\times SU(N)\times U(1)_{B'}\times U(1)_{R_0}$ (notice in particular that the R-symmetry is not broken). The preserved baryonic symmetry $U(1)_{B'}$ is the combination of the original baryonic symmetry $U(1)_B$ and of the $U(1)\subset SU(F)_D$ not contained in $SU(F-N)_D\times SU(N)$, under which the lowest part of the dual quark matrices q and $\tilde{q}$ are charged while the upper part is not.

In order to study quantum corrections around the supersymmetry breaking vacua, we can proceed as we did for the O'Raifeartaigh model, and compute the masses of the fluctuations of Φ, q and $\tilde{q}$ as functions of the pseudo-moduli Φ_0, q_0 and $\tilde{q}_0$. It is reasonable to expect that the actual vacuum will sit at a point of maximal symmetry, so as a working hypothesis let us expand around (11.46). We can parametrize the fluctuations as (for ease of comparison we use the same notation of the original ISS paper)

$$\Phi = \begin{pmatrix} \delta Y & \delta Z^\dagger \\ \delta \widetilde{Z} & \delta \hat{\Phi} \end{pmatrix}, \quad q = \begin{pmatrix} \sqrt{m\Lambda} + \frac{1}{\sqrt{2}}(\delta\chi_+ + \delta\chi_-) \\ \frac{1}{\sqrt{2}}(\delta\rho_+ + \delta\rho_-) \end{pmatrix}$$

$$\tilde{q}^T = \begin{pmatrix} \sqrt{m\Lambda} + \frac{1}{\sqrt{2}}(\delta\chi_+ - \delta\chi_-) \\ \frac{1}{\sqrt{2}}(\delta\rho_+ - \delta\rho_-) \end{pmatrix} \tag{11.47}$$

What one finds is that the model looks as N copies of a O'Raifeartaigh-like model. After computing the one-loop effective potential the mass spectrum looks as follows:

- Some fields have (tree-level) mass $\sim |h\sqrt{m\Lambda}|$ from the classical superpotential (11.35).
- Pseudo-moduli are all lifted and get non-tachyonic masses at one-loop $\sim |h^2\sqrt{m\Lambda}|$ from their coupling to massive fields (this gives an posteriori justification for our working hypothesis (11.46)).
- Some fields remain exactly massless. These are the Goldstone bosons associated to the coset G/H, a goldstino, as well as several fermionic partners of Φ_0 pseudo-moduli.

So, after taking into account quantum corrections in the tree-level coupling h we are left with a compact moduli space of stable non-supersymmetric vacua. This moduli space is robust against quantum corrections, because it is protected by symmetries.

What changes of the above analysis if we now switch-on gauge interactions, namely if we let the $SU(F-N)$ group being gauged? Interestingly, not much happens around the supersymmetry breaking vacua (11.46).

First, besides F-equations (11.42) we have now to impose D-equations on the supersymmetry breaking vacua (11.46). These are trivially satisfied, as one can verify plugging the VEVs (11.46) into

$$\sum_A \mathrm{Tr}\left(q^\dagger T_A q - \tilde{q} T_A \tilde{q}^\dagger\right) = 0. \tag{11.48}$$

Hence, the compact space parameterized by (11.46) remains a minimum of the potential (D-terms identically vanish and therefore do not contribute to the vacuum energy).

Second, the $SU(F-N)$ gauge group is completely higgsed in the vacua (11.46), since $\langle q_0 \rangle = \langle \tilde{q}_0 \rangle \neq 0$. Gauge bosons acquire a mass $\sim g\sqrt{m\Lambda}$, eating some of the previously massless Goldstone bosons of the ungauged model. The only change, then, is that the compact moduli space is smaller since global symmetries in the gauged model are less, to start with. In particular we have now that $G = SU(F)_D \times U(1)_B$ and $H = SU(F-N) \times SU(N) \times U(1)_{B'}$. Notice that the R-symmetry of the ungauged model $U(1)_{R_0}$ is now anomalous, while the non-anomalous R-symmetry which mSQCD shares with SQCD is explicitly broken by the mass term in the superpotential (*i.e.* the linear term in Φ, in mSQCD language).

Finally, the gauging does not affect the computation of the one-loop effective potential, either, since the tree level spectrum of massive $SU(F-N)$ fields is supersymmetric and gives no contribution to $\mathrm{Str}\mathcal{M}^2$. This happens because D-terms vanish on the vacua (11.46), and the non-zero expectation values of q and $\tilde{q}$ which provide masses to $SU(F-N)$ gauge fields do not couple directly to any non-vanishing F-term.

So we conclude that, up to a restriction of the compact moduli space, the supersymmetry breaking vacua we found classically in mSQCD survive quantum corrections and are hence supersymmetry breaking vacua of our original theory!

Gauging the $SU(F-N)$ group does have (drastic) consequences on other regions of field space, though. We already know, and we have proved it using both electric and magnetic variables, that massive SQCD has supersymmetric vacua on the mesonic branch. Besides other things, this makes the supersymmetry breaking vacua (11.46) being not absolute minima of the potential. Using magnetic language the effect of gauging is the generation of a non-perturbative superpotential contribution of the form

$$W_{\mathrm{non-pert}} \sim (\det\Phi)^{\frac{1}{F-N}} \sim \Phi^{\frac{F}{F-N}}. \tag{11.49}$$

This contribution is irrelevant near the origin, where supersymmetry breaking vacua sit and the theory is weakly coupled the fields Φ are nearly free, and becomes more and more important the farer we move along the mesonic branch. This operator plays the same role that a mass term for the chiral superfield Φ_2 played in the modified O'Raifeartaigh model discussed in Section 7.3: it brings in supersymmetry preserving vacua. The difference is that everything happens dynamically, here. So we conclude that mSQCD has metastable supersymmetry breaking vacua semiclassically, and non-perturbative restoration of supersymmetry by a a dynamical generated superpotential. On the other hand, in terms of the original SQCD theory, the supersymmetry breaking vacua are far from being classical, since they

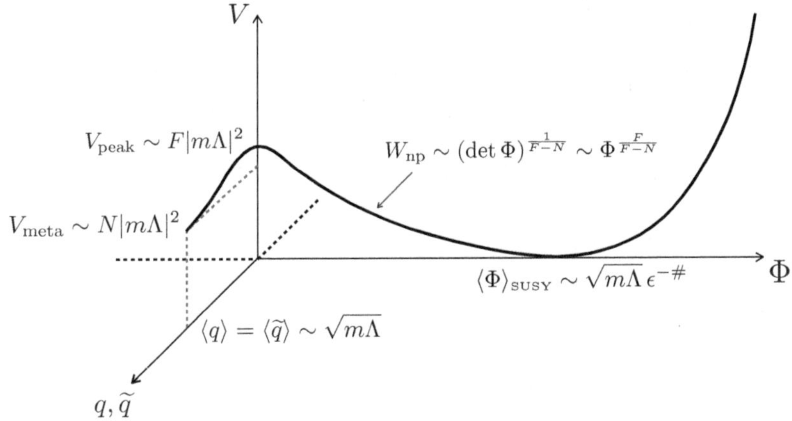

Fig. 11.7 The ISS potential in magnetic dual variables.

sit near the region of field space and this is a highly quantum region, from SQCD view point.

The final picture we obtain confirms what was anticipated in Figure 11.5 or, using magnetic dual language, in Figure 11.7 .

A final comment regards the R-symmetry breaking pattern. Nicely, what we get is consistent with what we learned previously. The SQCD original R-symmetry is explicitly broken by the mass term (11.28). Hence, the theory does not satisfy the necessary condition for supersymmetry breaking, and indeed it has N supersymmetric vacua, and non-vanishing Witten index. On the other hand, the anomalous $U(1)_{R_0}$ R-symmetry is restored, approximately, near the origin. This is more transparent using magnetic variables. The superpotential contribution (11.49) breaks R_0 explicitly, but this operator is irrelevant near the origin and this is why this symmetry arises as an approximate R-symmetry around the vacua (11.46). Therefore, by the (extended version of the) Nelson-Seiberg criterium discussed in Chapter 7.6.1, one would expect metastable vacua to arise there. And this is exactly what happens.

Finally, it is amusing to notice that in the ISS vacuum is plenty of massless fields, so there is no mass gap: a theory with tree-level masses for all matter fields and which enjoys strict confinement... admits vacua without a mass gap!

In all our discussion there is one point that we have overlooked. The magnetic theory has a UV cut-off, Λ. Do our results depend on the physics at scale Λ? Luckily not in the limit we are interested in, namely

$$\epsilon = \sqrt{\frac{m}{\Lambda}} \ll 1 \quad \longrightarrow \quad \begin{cases} \text{SQCD} \quad : m \to 0, \quad \Lambda \ \text{fixed} \\ \text{mSQCD} : \Lambda \to \infty, \quad \sqrt{m\Lambda} \ \text{fixed} \end{cases} \tag{11.50}$$

First, the analysis within the macroscopic theory (*i.e.*, *ungauged* mSQCD) is valid, since this was done at scales of order $\sqrt{m\Lambda} = \epsilon\Lambda$, which are well below the UV cut-off Λ, if ϵ is small. Second, also the supersymmetry preserving vacua can be seen in the magnetic theory: as we have already observed, for small ϵ they are well below the scale Λ

$$\langle \Phi \rangle_{\text{SUSY}} = \Lambda \, \epsilon^{2\frac{F-N}{N}} \ll \Lambda, \tag{11.51}$$

and hence are very weakly affected by any Λ-physics effects. Finally, the one-loop effective potential gives pseudo-moduli mass squares of order $|m\Lambda|$, that is $\sqrt{m\Lambda}\sqrt{\overline{m\Lambda}}$, which is not a holomorphic expression. On the other hand, corrections from Λ-physics are holomorphic in $m\Lambda$ and provide mass contributions of the form

$$\frac{m\Lambda}{\Lambda} \cdot \frac{\overline{m\Lambda}}{\overline{\Lambda}} = |m\Lambda|^2/|\Lambda|^2 = |m\Lambda|\epsilon^2 \ll |m\Lambda|, \tag{11.52}$$

which are then subleading for $\epsilon \ll 1$. A direct way to see this is to note that corrections in Λ would make the Kähler potential (11.45) not being canonical. In particular, to leading order, we would get a contribution as $\delta K = \frac{c}{|\Lambda|^2}(\Phi\Phi^\dagger)^2$, with c a number of order one. This is reminiscent of the Polonyi model with quartic Kähler potential we discussed in Section 7.3. A similar computation as the one there gives a contribution to the pseudo-moduli mass as in eq. (11.52), $\delta m^2 \sim |m\Lambda|^2/|\Lambda|^2$.

The last important check we have to do regards the life-time of the supersymmetry breaking vacua. The life-time can be computed using the Coleman bounce action S_B. Intuitively, the more the two vacua are far in field space in units of the energy difference between them, the more one might expect the life-time to be long. This expectation is confirmed by an explicit computation. It turns out that in the present case we are in a situation in which the so-called thin-wall approximation is valid. In such a situation, up to inessential numerical factors, the bounce action is proportional to the ratio between the fourth power of the distance, in field space, between the supersymmetry breaking and the supersymmetry preserving vacua, and the value of the energy difference between them.

Using previous formulas S_B hence reads

$$S_B \sim \frac{(\Delta\Phi)^4}{V_{\rm meta}} \sim \epsilon^{-4\frac{3N-2F}{N}}, \tag{11.53}$$

which is indeed large for $\epsilon \ll 1$. This ensures that the ISS vacua are parametrically long lived, since $\tau \sim e^{S_B}$. Notice that the largeness of the bounce action is due to different effects depending whether one is working in electric or magnetic variables. From mSQCD view point it is large since $\Delta\Phi$ is parametrically large in units of $\sqrt{m\Lambda}$. From SQCD view point, the bounce action is large because $V_{\rm meta}^{1/4}$ is parametrically small in units of Λ

$$\begin{aligned} &\text{mSQCD} \quad \begin{cases} \Delta\Phi \sim \sqrt{m\Lambda}\epsilon^{-\frac{3N-2F}{N}} \\ V_{\rm meta}^{1/4} \sim \sqrt{m\Lambda} \end{cases} \\ &\text{SQCD} \quad \begin{cases} \Delta\Phi \sim \Lambda\epsilon^{2\frac{F-N}{N}} \equiv \Lambda' \\ V_{\rm meta}^{1/4} \sim \epsilon\Lambda = \left(\Lambda\epsilon^{2\frac{F-N}{N}}\right)\epsilon^{\frac{3N-2F}{NB}} = \Lambda'\epsilon^{\frac{3N-2F}{NB}} \end{cases} \end{aligned} \tag{11.54}$$

where in the electric theory we have defined a rescaled strong coupling scale Λ', to ease the comparison.

11.6.3 *Summary of the physical picture*

Let us summarize the physical picture which emerges from our analysis of massive SQCD in the free magnetic phase.

Since the theory is UV-free, at high energies, larger than the dynamical scale Λ, the theory is weakly coupled, it can be described in terms of electric variables and the gauge coupling $g_{\rm el}$ increases along the flow. The scale Λ is an IR scale for SQCD and a UV cut-off for the IR-free dual magnetic theory. Hence, at scales $E \sim \Lambda$, in order to describe the dynamics of the theory one should better use magnetic language. Below Λ but above $\langle\Phi\rangle$ the theory renormalizes as for an IR-free theory, in the sense that the magnetic gauge coupling $g_{\rm m}$ decreases along the flow. This goes on until the scale $\langle\Phi\rangle$. What happens next depends on the value of such scale. If $\langle\Phi\rangle \neq 0$ at $E \sim \langle\Phi\rangle$ the dual quarks get massive and the theory reduces to pure SYM and leads to N supersymmetric vacua. If instead $\langle\Phi\rangle = 0$ the magnetic theory becomes completely free, gauge degrees of freedom get frozen and one is driven to the supersymmetry breaking vacua at $E \sim \sqrt{m\Lambda}$.

This is what happens for (massive) SQCD in the free magnetic phase, namely in the range $N+1 \leq F \leq \frac{3}{2}N$. A natural question to ask is whether there exist metastable vacua in massive SQCD for different values of F. The short answer is that the existence of such vacua can be rigorously proven only in the free magnetic phase.

For $F \geq 3N$ the dynamics is trivial and there do not exist ISS-like metastable vacua whatsoever.

In the conformal window, $\frac{3}{2}N < F < 3N$, the analysis is not easy since mSQCD is not IR-free. Moreover, one can show that $\langle \Phi \rangle_{\rm SUSY}$ is very near to the origin of field space, hence making metastability difficult to achieve anyway. Finally, the non-perturbative generated superpotential (11.49) is now relevant in the IR, indicating the difficulty in treating separately classical and quantum effects.

For $F < N$ the runaway is too strong and there are simply no tools to say whether local minima develop along the moduli space.

Finally, for $F = N$ the existence of ISS vacua cannot be proven using the magnetic dual theory, which does not exists for $F = N$, but can only be inferred using holomorphic decoupling starting from $F = N + 1$. Even though there are convincing arguments in favor of ISS vacua also in this case, given that the state we are speaking about is non supersymmetric the procedure requires some assumptions which are not fully under control. Hence, the case $F = N$ is not completely understood, in fact. A natural question is therefore whether is it possible to find ISS-like vacua in theories with a quantum deformed moduli space, as $SU(N)$ SQCD with $F = N$ is. The answer is for the affirmative. It has been shown that suitable deformations of the $USp(2N)$ ITIY model we discussed in Section 11.5 allow for dynamically generated metastable vacua, in a theory with a quantum deformed moduli space, as the ITIY model and any of its generalizations actually are. Basically, giving supersymmetric masses to some of the singlets S_{ij} one can show that supersymmetric vacua come in from infinity (because, integrating such massive singlets out, some of the mesonic flat directions are not anymore lifted by the tree-level superpotential). However, dynamically generated non-supersymmetric (local) vacua survive. Remarkably, such supersymmetry breaking vacua can be made parametrically long lived in a region of the parameter space which exactly coincides with the region where Kähler potential corrections are fully under control. So the model is even calculable.

The ISS model admits many generalizations. In particular, at the price of some complications and subtleties which we cannot discuss here, one can generalize the model in order to let the emergent IR R-symmetry to be spontaneously broken in the supersymmetry breaking vacua. This is a feature that the original ISS model does not have, since, as we have seen, quantum corrections stabilize $U(1)_{R_0}$ charged moduli at the origin, $\Phi_0 = 0$. These generalizations are interesting per sé but also from a more phenomenological point if view: if one thinks of the ISS model as a hidden sector in gravity or gauge mediation scenarios, having broken R-symmetry is a necessary condition to let gauginos getting (Majorana) mass.

11.7 Exercises

Exercise 11.1. Consider the $SU(5)$ model and add to it two chiral superfields, H and $\widetilde{H}$, transforming in the 5 and $\bar{5}$ representations of $SU(5)$, respectively. In this extended model there exist flat directions, which can be parametrized by the gauge invariant operators $T\widetilde{Q}\widetilde{H}, TTH, \widetilde{H}H, \widetilde{Q}H$.

- Show that along these flat directions the theory reduces to pure $SU(2)$ SYM (plus 4 singlet chiral superfields) and compute the effective superpotential by scale matching.
- Add now the following tree-level superpotential

$$W_{\text{tree}} = h\,TTH + f\,T\widetilde{Q}\widetilde{H} + m\widetilde{H}H$$

  and show that for $m = 0$ there exists a moduli space of supersymmetric vacua. Show that in the massive case, instead, the F-term equations cannot be solved and hence supersymmetry is dynamically broken.

Exercise 11.2. Consider the 3-2 model discussed in Section 11.3 in the regime $\Lambda_2 \gg \Lambda_3$. Show that also in this regime supersymmetry is broken but, unlike the regime $\Lambda_3 \gg \Lambda_2$, there do not exist a region in the parameter space where the model becomes calculable.

Exercise 11.3. Imposing cancellation of ABJ anomalies, compute the global symmetry charge assignment for the chiral superfields $Q_i, \widetilde{Q}^i, A_{ij}$ and S of the 4-1 model.

References

[1] E. Poppitz and S. P. Trivedi, Dynamical supersymmetry breaking, Section 5, *Ann. Rev. Nucl. Part. Sci.* **48** (1998), 307 [arXiv:hep-th/9803107].

[2] Y. Shadmi, *Supersymmetry breaking*, Section 3, arXiv:hep-th/0601076.

[3] H. Murayama, Studying noncalculable models of dynamical supersymmetry breaking, *Phys. Lett. B* **355** (1995), 187 [arXiv:hep-th/9505082].

[4] Y. Shadmi and Y. Shirman, Dynamical supersymmetry breaking, Sections IV and VI.A, *Rev. Mod. Phys.* **72** (2000), 25 [arXiv:hep-th/9907225].
[5] K. A. Intriligator and S. D. Thomas, Dynamical supersymmetry breaking on quantum moduli spaces, *Nucl. Phys. B* **473** (1996), 121 [arXiv:hep-th/9603158].
[6] K. I. Izawa and T. Yanagida, Dynamical supersymmetry breaking in vector — like gauge theories, *Prog. Theor. Phys.* **95** (1996), 829 [arXiv:hep-th/9602180].
[7] K. A. Intriligator and N. Seiberg, Lectures on Supersymmetry Breaking, Sections 2.7, 2.8, 3.3, 3.7 and 4.3, *Class. Quant. Grav.* **24** (2007), S741 [arXiv:hep-ph/0702069].
[8] K. A. Intriligator, N. Seiberg and D. Shih, Dynamical SUSY breaking in meta-stable vacua, *JHEP* **0604** (2006), 021 [arXiv:hep-th/0602239].
[9] S. R. Coleman, The Fate Of The False Vacuum. 1. Semiclassical Theory, *Phys. Rev. D* **15** (1977), 2929 [Erratum-ibid. D **16** (1977), 1248].
[10] M. J. Duncan and L. G. Jensen, Exact tunneling solutions in scalar field theory, *Phys. Lett.* **B291** (1992) 109.
[11] M. Bertolini, L. Di Pietro and F. Porri, Dynamical completions of generalized O'Raifeartaigh models, *JHEP* **1201** (2012), 158 [arXiv:1111.2307 [hep-th]].

Chapter 12

Supersymmetric Gauge Dynamics: Extended Supersymmetry

In this final chapter we will focus on asymptotically free gauge theories with extended supersymmetry and try to understand their quantum dynamics, in analogy with what we did for theories with $\mathcal{N} = 1$ supersymmetry in Chapter 10.

Asymptotically free gauge theories can enjoy different phases at low energy. In the case of $\mathcal{N} = 1$ supersymmetry, we were able to understand a great deal about the exact vacuum structure and the phases such field theories can enjoy. This can obviously be done also for theories with extended supersymmetry. However, the beautiful thing about theories with extended supersymmetry is that one can also derive the low energy effective action (in $\mathcal{N} = 1$ setups this can be done only in very special circumstances, e.g. for SQCD in the free magnetic phase, but this is not generic at all). The main purpose of this lecture is to show how this is possible.

12.1 Low energy effective actions: classical and quantum

Before entering into any detail, we would like to make some general comments about low energy effective actions in four space-time dimensions, independent from supersymmetry. Suppose to start from some matter coupled, asymptotically free gauge theory. At low energy its dynamics will be described by some (non-renormalizable) effective action whose degrees of freedom will be in general very different from UV ones. What is the structure one would expect for such an action?

Let us assume that in the vacuum we want to expand the theory about, the potential vanishes, $V = 0$. This is not a restriction, since the minimum of the potential can always be chosen to vanish via a constant shift in the

Lagrangian. Moreover, in the context of supersymmetric theories, which is what we are eventually interested in, this is not even a choice but a necessary condition, as far as supersymmetric vacua.

The leading dynamics around these vacua is governed by light fields, eventually only massless ones, as we take the cutoff energy characterizing the low energy effective action to be lower than any scale in the theory. In this limit, the physics is, by definition, scale invariant. However, the nature of the corresponding IR fixed point is not unique. If no massless fields are present (like in the case of strict confinement) there are no propagating degrees of freedom in the limit $E \to 0$, so the IR theory is empty and the IR fixed point a trivial one. The theory is gapped. If massless fields are present, instead, the theory can be in a free or an interacting phase.

A necessary condition for having an interacting conformal field theory at low energy is that massless non-abelian gauge fields are present in the effective action (at least if one assumes that a local Lagrangian description can exist). Indeed, by the Coleman-Gross theorem, in four space-time dimensions any theory of scalars, spinors and abelian gauge fields flows in the IR to a free (or trivial, if everything get a mass) theory. In this sense, the distinction we made between abelian free phase and abelian Coulomb phase for massless and massive SQED in Section 10.2.1 is to some extent semantic. Both theories are actually IR-free, the only difference being that in the latter case the massless spectrum contains only free photons (and gaugini) while in the former also free massless chiral superfields are present. In presence of massless non-abelian gauge fields, instead, one can either end up with, say, confinement, hence a trivial IR fixed point, or an interacting conformal field theory (an example being $\mathcal{N} = 1$ SQCD with $3/2\,N < F < 3N$). There are no general tools to describe strongly coupled, interacting conformal field theories, for which, regardless of supersymmetry, typically one cannot easily derive an effective Lagrangian (recall we are assuming that the UV theory is asymptotically free so we are excluding the case in which there are enough massless charged matter fields to make the β function being IR-free to start with, as e.g. $\mathcal{N} = 1$ SQCD with $F \geq 3N$). Sometime duality can help, like for $\mathcal{N} = 1$ SQCD with $N + 1 < F < 3/2\,N$, whose dynamics can be described by a dual, IR-free, magnetic theory. But this is clearly non generic.

Since our aim is to discuss low energy effective actions and make quantitative statements, in what follows we will focus on effective theories where scalars, spinors and abelian gauge fields enter, only. One may think this is a big restriction we are imposing on the class of theories we want to study. Remarkably, it is not. As we will discuss momentarily, for both $\mathcal{N} = 2$ and $\mathcal{N} = 4$ supersymmetric gauge theories, the moduli space happens to enjoy precisely such an IR-free phase.

A few more comments are in order.

In absence of supersymmetry, one expects the minima of the potential to be isolated, and hence the space of vacua to be a set of isolated points in the space of scalar field VEVs (there might exist a classical pseudo-moduli space which, however, is typically lifted once quantum corrections are taken into account). If this is the case, while scalar fields VEVs do parametrize the space of vacua, no scalar fields can be actually massless. Hence, having truly massless scalar fields in the low energy spectrum, implies the existence of a moduli space of vacua on which the potential vanishes identically, $V = 0$ (this includes also the case of spontaneously broken global symmetries, a scenario which can occur also in non-supersymmetric setups, albeit in this case the moduli space, parametrized by goldstone bosons, is compact). Supersymmetric theories typically admit moduli space of supersymmetric vacua. Hence, in the following, we will assume we are in such a situation, and hence we allow massless scalar fields to be present in the low energy effective action. These scalar fields, or better their VEVs, parametrize the moduli space.

In writing down the most general form of an IR-free effective action, an important simplification occurs. Suppose a charged massless field is present in the theory. Due to one-loop running, the abelian gauge coupling τ under which the massless field is charged vanishes in the far IR, that is $\text{Im}\tau \to \infty$. Hence, the abelian gauge field associated to it is decoupled and does not participate to the low energy effective dynamics. Notice, further, that a charged massless scalar field cannot parametrize the moduli space. Indeed, a non vanishing VEV would Higgs the $U(1)$ and thereby give the field a mass, as the gauge field itself. They would both disappear from the low energy effective action. If, on the contrary, all charged fields are massive, they do not appear in the low energy effective action to start with. Therefore, in the limit $E \to 0$ the low energy effective action just contains massless *neutral* fields and abelian gauge fields (plus fermions).

To sum up, the low energy effective action would be something like

$$\mathcal{L} = g_{ij}(\phi)\partial_\mu\phi^i\partial^\mu\phi^j + \frac{1}{2}\,\text{Im}[\tau_{IJ}(\phi)\mathcal{F}^I_{\mu\nu}\mathcal{F}^{J\,\mu\nu}] + \text{fermions}, \tag{12.1}$$

where i, j run on (neutral and massless) scalar fields, and I, J on (abelian) gauge fields. The complexified gauge coupling matrix τ_{IJ} and field strength are defined, respectively, as

$$\tau_{IJ} = \frac{\theta_{IJ}}{2\pi} + \frac{4\pi i}{g_{IJ}^2}, \quad \mathcal{F}^I_{\mu\nu} = F^I_{\mu\nu} + \frac{i}{2}\,\epsilon_{\mu\nu\rho\sigma}F^{I\rho\sigma}. \tag{12.2}$$

The σ-model metric $g_{ij} = g_{ij}(\phi)$ is the metric on the moduli space $\mathcal{M}$, whose coordinates are the massless scalar fields VEVs. Solving the theory boils down to compute the exact expression of the metric g_{ij}, the gauge coupling matrix τ_{IJ} and the coefficient functions of fermion kinetic terms which also depend on the scalar fields ϕ^i.

So far, we have been rather qualitative. In what follows we will show that for theories with extended supersymmetry one can actually be quantitative and understand a great deal about actions as (12.1) and their quantum dynamics.

12.1.1 $\mathcal{N} = 2$ *effective actions*

Let us first focus on theories with $\mathcal{N} = 2$ supersymmetry and suppose to start from some asymptotically free $\mathcal{N} = 2$ renormalizable action. Such an action is fully specified by the gauge group G and matter content, see Section 6.1.

If supersymmetry is preserved and a moduli space exists, the lightest excitations are massless. Hence, for low enough energy, lower then any scale in the theory, the dynamics on the moduli space is described by an effective action including these massless fields, only. This action should preserve $\mathcal{N} = 2$ supersymmetry. Hence, it should be nothing but a special instance of the $\mathcal{N} = 2$ non-linear σ-model discussed in Section 6.1.1. As such, it would be fully determined by knowing the exact expression of the prepotential $\mathcal{F}(\Phi)$, which gives both the special Kähler metric and the generalized complexified gauge coupling, eqs. (6.9) and (6.10), and by knowing the HyperKähler metric describing the hypermultiplets σ-model.

Scalar fields parametrize a complex manifold which, as discussed in Section 6.1.1, has locally the following product structure

$$\mathcal{M} = \mathcal{M}^V \times \mathcal{M}^H. \tag{12.3}$$

$\mathcal{M}^V$ is a special Kähler manifold, whose coordinates are the massless scalar ϕ^I VEVs belonging to vector multiplets, the subspace of $\mathcal{M}$ where hyper-scalar field VEVs are held fixed. $\mathcal{M}^H$ is a HyperKähler manifold, whose coordinates are the massless scalar H^i VEVs belonging to hypermultiplets, the subspace of $\mathcal{M}$ where vector multiplet scalar field VEVs are held fixed (we refer collectively to H_1^i and H_2^i defined in Section 6.1 as H^i in here).

Let us look at the structure of this (classical by now) moduli space more closely.

The first thing is that, in writing (12.3), we have assumed that the σ-model metric is diagonal, *i.e.* that there are no kinetic terms mixing ϕ^I and H^i. That this is the case comes from a $\mathcal{N}=2$ selection rule. If a cross term where there in the Lagrangian, its supersymmetry variation should be canceled (up to total space-time derivatives) by the supersymmetry variation of some other term. Looking at the supersymmetry variations of vector- and hypermultiplet component fields, one can easily see that such a term does not exist. Hence, metric cross terms are zero.

The subspace $\mathcal{M}^V$ where only the complex scalars ϕ^I get a VEV is called Coulomb branch. This is because the scalars belonging to $\mathcal{N}=2$ vector multiplets transform in the adjoint of the gauge group G and, as such, can at most break G down to $U(1)^n$, where $n = \text{rank}\, G$. More precisely, the scalar potential is proportional to $\text{Tr}\,[\phi, \overline{\phi}]$ and its supersymmetric minima are described by the adjoint scalars VEVs being in the Cartan subalgebra of G

$$\langle \phi \rangle = \sum_{I=1}^{n} a^I \, h_I \quad \text{where } h_I \in CSA \text{ of } G. \tag{12.4}$$

For generic a^I the gauge group is broken as $G \to U(1)^n$. To parametrize the moduli space one should bare in mind that a set of a^I fixes the gauge invariance only up to the action of the Weyl group W_G, the group of residual gauge transformations which, while acting on ϕ, do not take it out from the Cartan subalgebra. So, locally, $\mathcal{M}^V = \mathbb{C}^n/W_G$, meaning that the coordinates of the moduli space should be invariant under Weyl transformations.

Taking, for example, $G = SU(N)$, the Weyl group is S_N, the group of permutations of N elements. A natural set of $U(1)^{N-1} \times S_N$ invariant coordinates on the $(N-1)$-dimensional moduli space $\mathbb{C}^{N-1}/S_N$ can be shown to be

$$u_2 = \sum_{i<j} a^i a^j, \quad u_3 = \sum_{i<j<k} a^i a^j a^k, \ldots, u_N = a^1 a^2 \ldots a^N, \quad i,j,k = 1,\ldots,N \tag{12.5}$$

where, in this case, eq. (12.4) is

$$\langle\phi\rangle = \begin{pmatrix} a^1 & & & \\ & \cdot & & \\ & & \cdot & \\ & & & a^N \end{pmatrix} \quad \text{with } \sum_{i=1}^{N} a^k = 0 \text{ because } \mathrm{Tr}\phi = 0. \qquad (12.6)$$

At low energy, the effective Lagrangian describes $n = N - 1$ massless $\mathcal{N} = 2$ abelian vector superfields V^I. The scalar fields ϕ^I belonging to these massless abelian vector superfields are neutral and the gauge couplings τ_{IJ} are frozen at the value corresponding to the (lightest) massive particles (whose masses are in fact proportional to a^I). Hence the theory is in an IR-free abelian Coulomb phase. The (qualitative) behavior of gauge coupling evolution is shown in Figure 12.1, where we refer to a^I collectively as a.

The subspace $\mathcal{M}^H$, where only the scalars H^i get a VEV is called Higgs branch. This is because, for generic gauge group representations (hyper)scalar field VEVs break the gauge group completely, now. So, on the Higgs branch, one does not expect propagating massless gauge degrees of freedom. Here again, the scalars parametrizing the moduli space $\mathcal{M}^H$ are not only massless but also neutral (if this were not the case, they would acquire a mass by Higgs mechanism and should be integrated out for low enough energy, disappearing from the effective action).

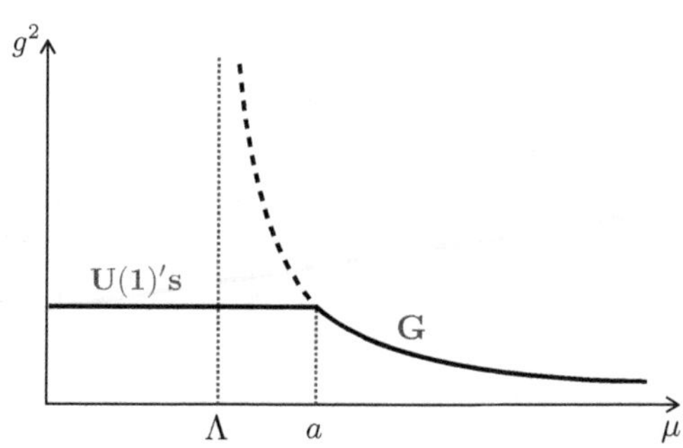

Fig. 12.1 Schematic behavior of gauge coupling running of $\mathcal{N} = 2$ SYM with gauge group $SU(N)$, higgsed to $N-1$ $U(1)$'s at scale $a > \Lambda$ on the Coulomb branch. The gauge coupling stops running at a scale $\mu \simeq a$. The low energy effective theory is described by abelian vector superfields, containing photons and neutral fields (scalars and fermions) and enjoys an IR-free Coulomb phase.

Branches where both ϕ^I and H^i have non-vanishing VEVs are called mixed branches.

So, all in all, we have to deal with a set of massless, neutral scalar fields and $n = \text{rank}\, G$ abelian gauge fields (plus fermionic superpartners). So, we are exactly in a situation as the one advocated in the previous general discussion, see eq. (12.1).

As we have already seen discussing $\mathcal{N} = 1$ theories, the moduli space needs not to be smooth. There can exist singularities where submanifolds of different dimensions meet. For example, classically, at the origin of field space, where the Coulomb and the Higgs branch meet, the theory is fully un-higgsed and the metric of the moduli space is expected to be singular: extra massless degrees of freedom appear and they should be included in the low energy effective description. Figure 12.2 provides a qualitative picture of the $\mathcal{N} = 2$ classical moduli space.

All what we said above is the classical part of the story. How do quantum corrections change it? Answering this question will be the basic goal of this chapter. However, already at this stage a few important facts can be anticipated.

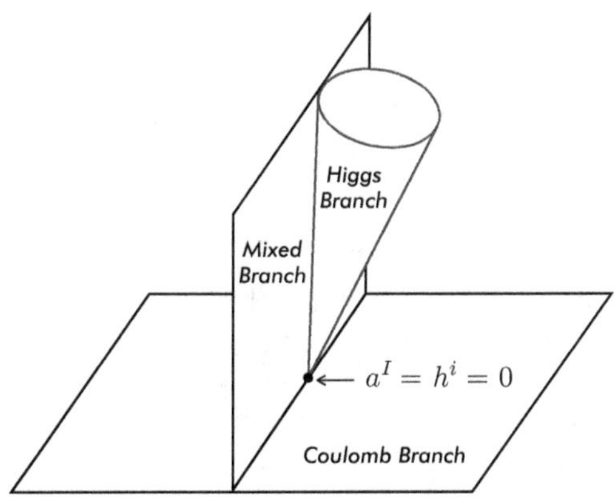

Fig. 12.2 $\mathcal{N} = 2$ classical moduli space. The mixed branch intersects the Higgs branch on a Higgs submanifold and the Coulomb branch along a Coulomb submanifold. The more singular point is the origin where the maximal number of degrees of freedom become massless.

First, the selection rule dictating a direct product for the moduli space $\mathcal{M}$, eq. (12.3), holds also at the quantum level, since it originates from the very structure of the supersymmetry algebra.

Second, $\mathcal{N} = 2$ supersymmetry implies that the special Kähler metric on $\mathcal{M}^V$ and the (imaginary part of the) generalized complexified gauge coupling, τ_{IJ}, are related, see Section 6.1.1. The former is a function of the scalar fields ϕ^I only, and so is the gauge coupling matrix τ_{IJ}. The latter undergoes renormalization, at one loop and non-perturbatively in $\mathcal{N} = 2$, its quantum corrected expression being some (unknown for the time being) function of the strong coupling scale Λ. Since Λ appears in τ_{IJ}, it appears in the Lagrangian in the same way as a VEV of a scalar belonging to a vector multiplet (one can think of Λ as a spurion). Since the metric on $\mathcal{M}^H$ does not depend on vector multiplet scalars, it does not depend on Λ, either. One then concludes that the metric on $\mathcal{M}^H$ is classically exact! The upshot is that the metric on the Coulomb branch can receive quantum corrections, while that on the Higgs branch is classically exact. Therefore, the exact low energy effective action will be described by a quantum corrected Coulomb branch and a classically exact Higgs branch. Solving the quantum theory boils down to determine the geometry on the Coulomb branch. Hence, in what follows, we will mostly focus on Coulomb branches.

One more property which makes $\mathcal{N} = 2$ special is that, unlike for $\mathcal{N} = 1$, in $\mathcal{N} = 2$ theories a moduli space always survives at quantum level. In other words, the classical moduli space can be modified, but never completely lifted (as instead happens in, e.g. SQCD with $F < N$). As for $\mathcal{M}^H$, this is obvious. The HyperKähler manifold is classically exact so, if it exists classically, it persists quantum mechanically. A way to see that this holds also on the Coulomb branch is as follows. For large field VEVs, $a^I \gg \Lambda$, we can use classical intuition where, by ordinary Higgs mechanism, the gauge theory is higgsed to $U(1)^n$ at weak coupling, see eq. (12.4). The corresponding n flat directions can be lifted at the quantum level, if given a mass. However, this cannot occur since in such semi-classical region this can happen only by higgsing, and abelian vector multiplets are neutral and so are the scalar fields ϕ^I, which cannot then Higgs the theory further. Therefore, we conclude that at large fields VEVs the moduli space persists even at the quantum level. But then, by analytic continuation, a moduli space persists also in the strongly coupled region, where $a^I \sim \Lambda$ (complex manifolds can become singular only on complex submanifolds, whose dimension is then at least 2 real dimensions smaller, so there is no obstructions against analytic continuation into a region of strong coupling where classical intuition would fail).

The classical moduli space has singularities of enhanced gauge symmetry and one could wonder if such singularities survive at the quantum level. One of the basic results we will show in the following is that the quantum moduli space does admit singularities where massive particles become massless, but none of them are gauge fields. So, there are no points of enhanced gauge symmetry, and the theory is always in an abelian Coulomb phase. What can exist, instead, are other type of singularities, known as Argyres-Douglas points, where mutually non-local particles, as monopoles and dyons, become *simultaneously* massless. At these singularities the low energy effective dynamics is described by an interacting conformal field theory (which, however, does not admit a Lagrangian description). We will have more to say about this later.

To sum up, apart from special points/curves where a Lagrangian description is not available, the structure of the (bosonic) $\mathcal{N}=2$ low energy effective action is

$$\mathcal{L} = K_I^J(\phi,\bar{\phi})\partial_\mu \overline{\phi}^I\,\partial^\mu \phi_J + \frac{1}{2}\,\mathrm{Im}[\tau_{IJ}(\phi)\mathcal{F}^I_{\mu\nu}\mathcal{F}^{J\mu\nu}] + K_i^j(H,\overline{H})\,\partial_\mu H^i\,\partial^\mu \overline{H}_j, \tag{12.7}$$

where K_J^J is the (special Kähler) metric on the Coulomb branch and K_j^i the (HyperKähler) metric on the Higgs branch. The complexified gauge coupling is related to the prepotential as

$$\tau_{IJ} = \frac{\partial^2 \mathcal{F}}{\partial\phi^I\partial\phi^J}. \tag{12.8}$$

We have slightly changed normalizations with respect to previous chapters. With present normalizations, eq. (6.12), which relates the special Kähler metric to the complexified gauge coupling matrix, reads $K_I^J = \mathrm{Im}\,\tau_{IJ}$. Defining

$$\phi_{DJ} \equiv \frac{\partial \mathcal{F}}{\partial \phi^J}, \tag{12.9}$$

we can re-write (12.7) as

$$\begin{aligned}\mathcal{L} &= \mathrm{Im}\left[\tau_{IJ}(\phi)\left(\partial_\mu\overline{\phi}^I\,\partial^\mu\phi^J + \frac{1}{2}\mathcal{F}^I_{\mu\nu}\mathcal{F}^{J\mu\nu}\right)\right] + K_i^j(H,\overline{H})\,\partial_\mu H^i\,\partial^\mu\overline{H}_j\\ &= \mathrm{Im}\left(\partial_\mu\overline{\phi}^I\,\partial^\mu\phi_{DI}\right) + \frac{1}{2}\,\mathrm{Im}[\tau_{IJ}(\phi)\mathcal{F}^I_{\mu\nu}\mathcal{F}^{J\mu\nu}] + K_i^j(H,\overline{H})\,\partial_\mu H^i\,\partial^\mu\overline{H}_j.\end{aligned} \tag{12.10}$$

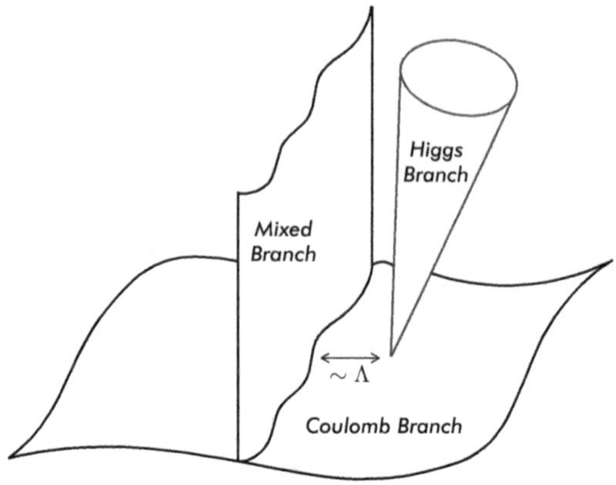

Fig. 12.3 $\mathcal{N} = 2$ quantum corrected moduli space. The (local) product structure remains the same as in the classical limit and the Higgs branch is also unmodified. The same happens for the Higgs directions of the mixed branch, even though they may be deformed in the Coulomb directions. The Coulomb branch is modified at the quantum level, instead (but never completely lifted!). Generically, this may excise (part or all) classical singular submanifolds.

Solving the theory boils down to determine the quantum exact expression of the prepotential $\mathcal{F}$ and, via eqs. (12.8), the Lagrangian (12.10).

A cartoon of the quantum corrected moduli space is depicted in Figure 12.3.

12.1.2 $\mathcal{N} = 4$ *effective actions*

Let us now consider $\mathcal{N} = 4$ supersymmetry. The story here is much simpler. First, there exist only one class of scalar fields, all transforming in the adjoint representation of the gauge group G. So, at a generic point of the moduli space, the low energy dynamics is that of a free $U(1)^n$ $\mathcal{N} = 4$ theory, with $n = \text{rank}\, G$, and the moduli space $\mathcal{M}$ is parametrized by $6n$ neutral real scalars. Hence, at a generic point on the moduli space, we are in an IR-free abelian Coulomb phase. Moreover, as we already discussed in Section 6.2, the gauge coupling does not run, neither perturbatively nor non-perturbatively, and then $\mathcal{M} = \mathbb{R}^{6n}$, meaning that the moduli space is classically exact. This, however, is not boring at all. As we will discuss later, $\mathcal{N} = 4$ non-renormalization theorems, which are the strongest possible, let one get very interesting exact results.

12.2 Monopoles, dyons and electro-magnetic duality

Before proceeding, we need to recall a few properties that ordinary gauge theories may enjoy and discuss how these are realized in supersymmetric contexts.

Let us start from a $U(1)$ gauge theory without matter, namely electromagnetism. Maxwell equations in the vacuum, which in differential form notation can be written as

$$d * F = 0, \quad dF = 0 \tag{12.11}$$

are invariant under the transformation $F \to *F$, $*F \to -F$, which corresponds to the exchange of electric and magnetic fields. This transformation is called S-duality transformation. In presence of electric sources Maxwell equations can still be invariant under S duality if one postulates the existence of magnetic sources and the associated current j_m, with the following action of S

$$F \to *F, \quad *F \to -F \quad \text{and} \quad j_e \to j_m, \quad j_m \to -j_e. \tag{12.12}$$

where now Maxwell equations read

$$d * F = j_e, \quad dF = -j_m. \tag{12.13}$$

The exchange of electric and magnetic currents implies, in particular, that under a S duality transformation electric and magnetic charges are also exchanged.

One crucial consequence of the presence of magnetic monopoles is that the electric charge is quantized. More precisely, as shown by Dirac, it turns out that a theory with both electric and magnetic charges, q and p respectively, can be consistently quantized only if the following condition holds

$$q\,p = 2\pi n \quad \text{with } n \in \mathbb{Z}. \tag{12.14}$$

This is the renown Dirac quantization condition, which implies that any electric charge is an integer multiple of an elementary charge $g \equiv (2\pi/p)\, n_0$, for some integer number n_0. Another important consequence of eq. (12.14) is that regimes where the electric charge is small correspond to regimes where the magnetic charge is large and viceversa. Therefore, S duality is a strong-weak coupling duality.

Maxwell equations are not affected if adding to the action a θ-term

$$\frac{\theta\, g^2}{32\pi^2}\, F_{\mu\nu}\, \widetilde{F}^{\mu\nu}. \tag{12.15}$$

However, in presence of magnetic monopoles, a θ-term does have an interesting physical effect. As shown by Witten, in this case the magnetic charge of a particle contributes to its electric charge, too. Specifically, a particle with magnetic charge $p = \frac{4\pi}{g}$ and $U(1)$ electric charge $n_e g$ has the following physical charges

$$p = \frac{4\pi}{g}, \quad q = n_e g - \frac{\theta g^2}{8\pi^2} p = n_e g - \frac{\theta g}{2\pi}. \tag{12.16}$$

In other words, if a θ-term is present a magnetic monopole always carries an electric charge (even if $n_e = 0$) and such electric charge is not a multiple of some basic unit. This is known as *Witten effect.* In the following, with some abuse of language, we will refer to the $U(1)$ charge g as the electric charge.

Dirac quantization condition is generalized in presence of dyons, which are states carrying both electric and magnetic charges, as

$$q_1\, p_2 - q_2\, p_1 = 2\pi n. \tag{12.17}$$

This is known as Dirac-Schwinger-Zwanziger quantization condition. An aspect regarding eq. (12.17) and that will play a relevant rôle later is that only if the right hand side vanishes the corresponding states are local with respect to each other. So, for instance, two electrically charged states are local with respect to each other while an electrically charged state and a magnetic monopole (or a dyon) are not. As such, they cannot be described within one and the same Lagrangian. Therefore, an effective low energy theory where such mutually non-local objects are both present, is believed not to admit a Lagrangian description.

Let us emphasize that the duality transformation (12.12) is not a symmetry of the theory, since it acts on the couplings. Rather, it maps a description of the theory to another description of the same theory. There exists another transformation, known as T-duality transformation, which does not act on the electro-magnetic field but shifts the θ angle by 2π and, as such, is a symmetry of the theory. These two transformations, S and T, generate a full group, $SL(2,\mathbb{Z}) \simeq Sp(2,\mathbb{Z})$, the duality group of electro-magnetism.

As $SL(2,\mathbb{Z})$ 2×2 matrices, S and T are

$$S=\begin{pmatrix}0 & -1\\ 1 & 0\end{pmatrix}, \quad T=\begin{pmatrix}1 & 1\\ 0 & 1\end{pmatrix} \tag{12.18}$$

The way S and T act on $\tau=\theta/2\pi+4\pi i/g^2$, $\tau\to -1/\tau$ and $\tau\to\tau+1$, respectively, shows that the group they generate acts on the complexified gauge coupling τ as a fractional linear transformation

$$\tau\to\frac{a\tau+b}{c\tau+d} \quad \text{where} \begin{pmatrix}a & b\\ c & d\end{pmatrix}\in Sp(2,\mathbb{Z}). \tag{12.19}$$

12.2.1 *The Georgi-Glashow model and its (supersymmetric) generalizations*

In Maxwell theory, magnetic monopoles (or dyons) are introduced by hand as extra degrees of freedom, they are point-like and carry infinite energy. However, monopole (and dyon)-like sources may arise as solitons, *i.e.* localized, finite energy, non-singular solutions of the equations of motion, in the context of spontaneously broken gauge theories. The first and most famous example is the Georgi-Glashow (GG) model, a $SU(2)$ gauge theory coupled to a scalar ϕ transforming in the adjoint of $SU(2)$ and quartic potential

$$V=\frac{\lambda}{4}\left(\operatorname{Tr}\phi^2-a^2\right)^2, \tag{12.20}$$

with a some real number. This theory undergoes Higgs mechanism which breaks $SU(2)\to U(1)$, and admits soliton solutions carrying monopole and/or dyonic charges under the *low energy effective* $U(1)$. More specifically, there exist a magnetically charged soliton, the 't Hooft-Polyakov soliton, with charges $(n_m,n_e)=\pm(1,0)$, and a dyon, found by Julia and Zee, with charges $(n_m,n_e)=\pm(1,-1)$, where n_m and n_e are the units of magnetic charge $p=\frac{4\pi}{g}$ and electric charge g, respectively. The reason for the 4π in place of the 2π for the magnetic charge p, as compared with eq. (12.14) with $n=1$, is just because in this theory we could add fields in the fundamental representation of $SU(2)$, which would carry electric charge $\pm g/2$ and, in terms of such minimal charge, one would get the usual Dirac quantization condition. In other words, in the GG model the elementary electric charge $g=(2\pi/p)n_0$ is realized with $n_0=2$. The existence of soliton solutions with both plus and minus signs, ensure that the low-energy effective $U(1)$ gauge theory is non-anomalous.

For generic values of the parameters (charge g, scalar field VEV a and quartic coupling λ) these solutions are not known analytically. However, there exists a limit in which the equations of motion can be solved exactly. This is the so-called BPS limit (after Bogomolny, Prasad and Sommerfield), which corresponds to take $\lambda \to 0$ with g and a fixed while retaining the boundary conditions on the Higgs field, that should tend towards a at spatial infinity. In this limit, the minimal energy configurations (the so-called BPS states) satisfy the following relation

$$M = \sqrt{2}\,|\frac{a}{g}(q + ip)| \tag{12.21}$$

where M is the mass of the soliton and $q = n_e g$ and $p = n_m \frac{4\pi}{g}$. It is worth noticing that in the BPS limit all particles in the spectrum, including fundamental degrees of freedom (gauge bosons and Higgs field), satisfy the mass formula (12.21) and so belong to the BPS spectrum.

In presence of a θ-term, the analysis that lead to eq. (12.21) can be repeated almost unchanged, the BPS mass formula becoming now

$$M = \sqrt{2}|a\,(n_e + \tau\, n_m)| \quad \text{where } \tau = \frac{\theta}{2\pi} + \frac{4\pi i}{g^2}. \tag{12.22}$$

Due to Witten effect, acting with the transformation $\theta \to \theta + 2\pi n$ on the monopole and dyon solutions, one can get a full tower of solutions with charges $\pm(1, -n)$ and $\pm(1, -n-1)$, $n \in \mathbb{Z}$. Since this transformation is a symmetry of the theory, these solutions are all physically equivalent. Note that the T-duality transformation $\theta \to \theta + 2\pi$ which acts on the charge vector (n_m, n_e) as

$$T : (n_m, n_e) \to (n_m, -n_m + n_e), \tag{12.23}$$

acts on the complexified gauge coupling as $\tau \to \tau + 1$. Plugging this into the BPS mass formula (12.22), we see that the BPS mass formula is left invariant by the action of T. This is consistent with the fact that since masses are physical observables, they should be insensitive to symmetry transformations.

Looking at eq. (12.12), we see that a S-duality transformation, which sends $\tau \to -1/\tau$, should instead act on a charge vector as

$$S : (n_m, n_e) \to (-n_e, n_m). \tag{12.24}$$

If we demand eq. (12.22) to be invariant under S duality, this should then also be accompanied by the shift $a \to a\tau$.

More generally, a matrix $A \in Sp(2,\mathbb{Z})$ transforming $(a\,\tau, a)^T$ as $A \cdot (a\,\tau, a)^T$, should correspond to a change of the vector of electric and magnetic charges as $(n_m, n_e) \cdot A^{-1}$. We will re-derive this important result later.

All above analysis is (semi) classical. In particular, the derivation of the BPS bound and the construction of the monopole and dyon solutions. One might wonder to what extent this still holds at the full quantum level. This is something difficult to check in the GG model since an analytical handling of the quantum/strong coupling regime is not possible in such a non-supersymmetric setup. But, as usual, supersymmetry helps.

Let us consider $\mathcal{N} = 2$ pure SYM with gauge group $SU(2)$. Since we are going to use slightly difference normalizations with respect to previous lectures, let us write down the on-shell Lagrangian explicitly

$$\mathcal{L} = \frac{1}{g^2}\mathrm{Tr}\left[-\frac{1}{4}F_{\mu\nu}F^{\mu\nu} + \frac{\theta}{32\pi^2}g^2 F_{\mu\nu}\widetilde{F}^{\mu\nu} + \overline{D_\mu\phi}D^\mu\phi - \frac{1}{2}\left[\phi,\overline{\phi}\right]^2 + \right.$$
$$\left. - i\lambda\sigma^\mu D_\mu\overline{\lambda} - i\psi\sigma^\mu D_\mu\overline{\psi} + i\sqrt{2}[\overline{\phi},\psi]\lambda + i\sqrt{2}[\phi,\overline{\lambda}]\overline{\psi}\right], \tag{12.25}$$

where, in present normalizations, $D_\mu = \partial_\mu - iA^a_\mu T_a$. This theory has the same bosonic content of the GG model (including a scalar potential which higgses the theory down to $U(1)$ in the vacuum) and shares with it its basic dynamics. As such, it also admits magnetically charged solitons, as monopoles and dyons.

There are, however, some important differences with respect to the GG model.

First, in the GG model the BPS limit is a rather special limit, since it consists in ignoring the quartic Higgs field potential, while retaining the boundary conditions on the Higgs field at spatial infinity. This is automatically the set-up we have in our $\mathcal{N} = 2$ example, since the potential is proportional to $\left[\phi,\overline{\phi}\right]^2$ and vanishes whenever ϕ is in the Cartan subalgebra, e.g. $\phi = a\,\sigma_3$, for *any* value of a (the gauge invariant combination parametrizing the one-dimensional moduli space can be chosen to be $u = \frac{1}{2}\langle \mathrm{Tr}\phi^2\rangle = a^2$). So here the BPS limit is built in, to some extent.

A more important difference regards the BPS mass formula (12.22). This formula is reminiscent of the bound that massive states in the $\mathcal{N} = 2$ spectrum should satisfy and which 1/2 supersymmetry preserving states (short representations) saturate, $M \geq |Z|$. This suggests that, in presence of charged solitons in the spectrum, the $\mathcal{N} = 2$ central charge may be related to their electric and magnetic charges. Witten and Olive showed that this is indeed the case. In present normalizations the $\mathcal{N}=2$ algebra and the corresponding bound read

$$\{Q^1_\alpha, Q^2_\beta\} = 2\sqrt{2}\epsilon_{\alpha\beta} Z, \quad M \geq \sqrt{2}\,|Z|. \tag{12.26}$$

Starting from the Lagrangian of pure $\mathcal{N}=2$ SYM one can compute the corresponding supercurrents $S^1_{\alpha\mu}$ and $S^2_{\alpha\mu}$ by Nöether method. Recalling that $Q^I_\alpha = \int d^3x\, S^I_{\alpha\,0}$ one finds (after dropping contributions which trivially vanish at spatial infinity) that

$$\begin{aligned}\{Q^1_\alpha, Q^2_\beta\} &= \frac{2\sqrt{2}}{g^2}\epsilon_{\alpha\beta}\int d^3x\,\partial_i\left[\left(F^{a0i} - i\widetilde{F}^{a0i}\right)\overline{\phi}_a\right] = 2\sqrt{2}\epsilon_{\alpha\beta} Z \\ \{\overline{Q}^1_{\dot\alpha}, \overline{Q}^2_{\dot\beta}\} &= \frac{2\sqrt{2}}{g^2}\epsilon_{\dot\alpha\dot\beta}\int d^3x\,\partial^i\left[\left(-F_{a0i} + i\widetilde{F}_{a0i}\right)\phi^a\right] = 2\sqrt{2}\epsilon_{\dot\alpha\dot\beta} Z^*,\end{aligned} \tag{12.27}$$

where in last equalities we have used eq. (12.26). The electric and magnetic charges of the $U(1)$ low energy effective theory in the GG model (and similarly herein) are

$$q = -\frac{1}{ag}\int d^3x\,\partial_i\left(F^{0i}_a\phi^a\right) = g n_e \ , \ p = -\frac{1}{ag}\int d^3x\,\partial_i\left(\widetilde{F}^{a0i}\overline{\phi}_a\right) = \frac{4\pi}{g} n_m. \tag{12.28}$$

So we finally find (after taking into account the effect of a non-trivial θ-term) that

$$\mathrm{Re}\, Z = a n_e, \quad \mathrm{Im}\, Z = a\tau n_m \tag{12.29}$$

and hence eq. (12.22). This shows that in presence of monopoles (and dyons) the supersymmetry algebra must be modified with the addition of a non-trivial central charge and the latter measures their electric and magnetic charges (as in the original GG model in the BPS limit, also in this supersymmetric version fundamental degrees of freedom satisfy the BPS mass formula). From a geometric viewpoint this should not come as a surprise. Supersymmetry charges are space integrals. In calculating their anticommutators one has to deal with surface terms, which one usually neglects. However, as shown in eqs. (12.28), in presence of electric and magnetic charges these surface terms are non-zero and give rise, consistently, to a non-vanishing Z.

Note that unlike the GG model here the relation between masses and charges of BPS states, eq. (12.22), does not come from a (semi) classical analysis but it is dictated by the supersymmetry algebra. Hence, it cannot be spoiled quantum mechanically and should remain valid even when perturbative and non-perturbative corrections are taken into account. BPS states are in short representations and quantum corrections cannot generate the extra degrees of freedom needed to convert a short multiplet in a long one. So, BPS saturated states remain so also at the quantum level in a supersymmetric theory. They are protected operators.

That eq. (12.22) persists quantum mechanically does not mean that the quantities therein do not undergo renormalization. For one thing, in $\mathcal{N} = 2$ we know that the gauge coupling τ runs (at one loop and non-perturbatively). Therefore, upon taking into account renormalization effects, while by its very definition the central charge Z is still a linear combination of conserved (electric and magnetic) abelian charges, the coefficients multiplying n_m and n_e will be replaced by some (holomorphic) functions, which we dub a and a_D, of the strong coupling scale Λ and field VEVs

$$Z = a\, n_e + a_D\, n_m. \tag{12.30}$$

In the classical limit $a^2 = \frac{1}{2}\langle \mathrm{Tr}\phi^2\rangle$ and $a_D = a\tau$. But one expects this not to be true at the full quantum level. In particular, of a_D in terms of a could be different. Seiberg and Witten proposed the following exact relation between a and a_D

$$a_D \equiv \frac{\partial \mathcal{F}}{\partial a} \quad \text{that is } \tau = \frac{da_D}{da}, \tag{12.31}$$

with $\mathcal{F}$ the prepotential. We will provide evidence for the proposal (12.31) later. Here, just notice that this way, in the semi-classical limit, where $\mathcal{F}_{cl} = \frac{1}{2}\tau a^2$, eq. (12.30) correctly reduces to (12.29). But, unlike (12.29), it is by construction renormalization group invariant.

One of our main goals in the following will be to check the proposal (12.31) and compute the exact expression of a and a_D in terms of the scalar field VEVs and Λ. Given this information, the masses of all BPS states (fundamental fields as well as magnetic monopoles and dyons) will be known exactly in terms of the moduli parameters. More importantly, finding the exact expressions of a and a_D amounts to find the exact expression for τ and hence, by (12.10), the full effective action!

This discussion can be repeated for $\mathcal{N} = 4$ SYM, which is also expected to admit charged solitons in its spectrum. There, however, the relation $a_D = \tau a$ is not renormalized, since in this case τ is classically exact, as so is the moduli space. This has important consequences which we will come back to, when discussing the quantum properties of $\mathcal{N}=4$ SYM.

So far, we have been considering pure $\mathcal{N}=2$ $SU(2)$ SYM. One may want to add matter fields, *i.e.* hypermultiplets. This amounts to add to the Lagrangian the superpotential term

$$\sum_{i=1}^{F}\left(\sqrt{2}H_1^i\Phi H_2^i + m_i H_1^i H_2^i\right) + h.c.. \tag{12.32}$$

For equal masses, the theory has a $U(F)$ global symmetry, which is broken to $U(1)^F$ for generic values of m_i. One can repeat previous computations, calculate the contribution of H_1 and H_2 to the supercurrent and, in turn, to the central charge mass formula (12.30). The end result is

$$Z = a\, n_e + a_D\, n_m + \sum_{i=1}^{F}\frac{1}{\sqrt{2}}m_i S_i, \tag{12.33}$$

where S_i are global conserved $U(1)$ charges under which H_1^i and H_2^i have charges $+1$ and -1, respectively.

There exist generalizations of this story. The GG model can be generalized to a gauge theory with gauge group G spontaneously broken to a (not necessarily abelian) subgroup H by some Higgs-like field transforming in the adjoint representation of G. Thanks to the topological nature of soliton solutions, it turns out that an analysis on their existence can be carried out in the context of homotopy theory. In particular, inequivalent solutions are classified by the homotopy group $\Pi_2(G/H)$. This is isomorphic to $\Pi_1(H)_G$, the subgroup of closed paths in $\Pi_1(H)$ which can be contracted to a point when H is embedded in G. If G is simply connected, $\Pi_1(H)_G$ is isomorphic to $\Pi_1(H)$ and non-trivial soliton solutions are hence classified by $\Pi_1(H)$. For example, in the original GG model we have $\Pi_2(G/H) = \Pi_2(SU(2)/U(1)) = \Pi_1(U(1)) = \mathbb{Z}$, and one family of magnetic monopoles with integer charge is indeed present. The same happens in GUT theories. Taking, e.g. $G_{GUT} = SU(5)$ one has $\Pi_2(G/H) = \Pi_2(SU(5)/SU(3)\times SU(2)\times U(1)) = \Pi_1(SU(3)\times SU(2)\times U(1)) = \Pi_1(U(1)) = \mathbb{Z}$, so again magnetic monopoles are expected to exist. This is not the case in the Standard Model, where the gauge group is not simple, $G = SU(2)\times U(1)_Y$ and, more importantly, $\Pi_1(H)_G = 0$. This is because the generator of the unbroken electromagnetic $U(1)$ gets contributions both from the generator of $U(1)_Y$ and from the Cartan of $SU(2)$. This implies that any closed path in $U(1)$ may be deformed to lie completely in $U(1)_Y$, which, unless this path is trivial, cannot be deformed to a point in G. This means that $\Pi_1(H)_G = 0$ and hence magnetic monopoles do not exist. Interestingly (recall the discussion around eq. (12.14)), that the electric charge happens to be quantized can be seen as an evidence in favor of the existence of monopoles and, in turn, of GUT theories.

Exactly as for the original GG model, this more general story finds a natural embedding in supersymmetric contexts. One such situation is nothing but the low energy effective theories describing the $\mathcal{N} = 2$ and $\mathcal{N} = 4$ Coulomb branches we are actually interested in. There, we have a gauge theory with gauge group G broken to its Cartan subalgebra $H = U(1)^n$, where $n = \operatorname{rank} G$. From previous general analysis, it follows that magnetically charged solitons are present in the spectrum. Most of what we said about the supersymmetric generalization of the GG model holds unchanged. In particular, the IR-free effective theory is form-invariant under electro-magnetic duality transformations which are the natural generalization to $n > 1$ of eq. (12.19), and act on the couplings as

$$\tau_{IJ} \to \left(A_I^L \tau_{LM} + B_{IM}\right)\left(C^{JN}\tau_{NM} + D_M^J\right)^{-1} \tag{12.34}$$

where now $M \equiv \left(\begin{smallmatrix} A & B \\ C & D \end{smallmatrix}\right) \in Sp(2n, \mathbb{Z})$. The vector of electric and magnetic charges is now a $2n$-component row vector (n_m^I, n_{eJ}). The corresponding BPS mass formula which generalizes (12.30) is

$$Z = a^I\, n_{eI} + a_{D,I}\, n_m^I = a \cdot n_e + a_D \cdot n_m, \tag{12.35}$$

where, in the second step, matrix multiplication is understood and (12.31) is now $a_{DI} \equiv \partial\mathcal{F}/\partial a^I$. Finally, the addition of (massive) flavors changes the central charge formula in a way similar to eq. (12.33).

Let us conclude this section with a comment which will be relevant later. As already noticed, electro-magnetic duality transformations are not symmetries of the theory. They just express the equivalence of abelian theories coupled to massive sources under a $Sp(2n, \mathbb{Z})$ redefinitions of electric and magnetic charges. It is a redundancy of the effective Lagrangian description. The point, though, is that when a moduli space is present (which is the case we will actually be concerned with), this redundancy can capture important features of the theory. That this is the case, can be seen as follows.

Suppose there is a moduli space of vacua and that the effective dynamics on this moduli space is described by n abelian gauge fields and a bunch of massless, neutral scalars, collectively dubbed ϕ, which parametrize $\mathcal{M}$. Upon traversing a closed loop in $\mathcal{M}$ the physics must be the same at the beginning and at the end of the loop. However, the Lagrangian does not need to: it is enough for it to be invariant modulo an electro-magnetic duality transformation. Geometrically, this corresponds to say that the matrix of couplings τ_{IJ} is a section of a $Sp(2n,\mathbb{Z})$ bundle. In matrix notation, this means that upon making a circle in the ϕ moduli space, the matrix τ should transform as

$$\tau(e^{2\pi i}\phi) = (A \cdot \tau(\phi) + B)\ (C \cdot \tau(\phi) + D)^{-1} \,. \tag{12.36}$$

The element $M = \left(\begin{smallmatrix} A & B \\ C & D \end{smallmatrix}\right) \in Sp(2n,\mathbb{Z})$ is called *monodromy* around the loop. If the closed loop does not encircle any singularity, the monodromy is the identity element of $Sp(2n,\mathbb{Z})$. If it does, the monodromy is instead a non-trivial element of $Sp(2n,\mathbb{Z})$.

As we discussed at length, singularities on the moduli space are associated to massive particles becoming massless, there. The monodromy matrix tells about the nature of such particle. Since masses are physical observables, the BPS mass formula (12.35) should be invariant under monodromies. Hence, as already emphasized for the case $n = 1$, a matrix $A \in Sp(2n,\mathbb{Z})$ transforming $(a_D, a)^T$ as $A\cdot(a_D, a)^T$, should correspond to a change of the vector of electric and magnetic charges as $(n_m, n_e)\cdot A^{-1}$. This means that, in general, the action of the monodromy changes the quantum numbers of charged states.

Now, the state of vanishing mass at a given singularity on the moduli space should be invariant under the action of the monodromy associated to the singularity itself (it is the properties of such massless state which determine the monodromy matrix). That is to say, it should be a left eigenvector of the matrix A with unit eigenvalue. Let us specify, for simplicity, to the case $G = SU(2)$ for which the duality group is $Sp(2,\mathbb{Z})$. It is not difficult to see that the $Sp(2,\mathbb{Z})$ matrix for which the vector (n_m, n_e) is an eigenvector with unit eigenvalue is

$$A(n_m, n_e) = \begin{pmatrix} 1 + 2n_m n_e & 2n_e^2 \\ -2n_m^2 & 1 - 2n_m n_e \end{pmatrix}. \tag{12.37}$$

This means that at a singularity with monodromy matrix of the form above, a state with charges $\pm(n_m, n_e)$ becomes massless.

In fact, any state with charges $l(n_m, n_e)$ with $l \in \mathbb{Z}$ is left invariant by the action of (12.37). However, stable dyons require ln_m and ln_e to be relatively prime, which is the case only for $l = \pm 1$. One way to see it is the following. Suppose to start from a BPS saturated state with charges (N_m, N_e) and mass $M = \sqrt{2}|Z|$. Such state can decay into states whose sum of masses should be less or equal M. For each of these states we have $Z_i = a \cdot n_e^i + a_D \cdot n_m^i$ and $M_i \geq \sqrt{2}|Z_i|$. Since charge conservation implies that $Z = \sum Z_i$, it follows from triangle inequality that $|Z| \leq \sum_i |Z_i|$ which in turn implies that

$$M \leq \sum_i M_i. \tag{12.38}$$

In order for the decay to occur the above bound should be saturated, which implies $|Z| = \sum_i |Z_i|$ (so also the states with charges (n_m^i, n_e^i) should be BPS). This can happen if and only if the vectors (N_m, N_e) and (n_m^i, n_e^i) are proportional, that is if N_m and N_e are not relatively prime, $(N_m, N_e) = l(n_m, n_e)$. If they are, instead, the decay cannot occur.

12.3 Seiberg-Witten theory

Let us now come back to our original problem. We would like to look at asymptotically free $\mathcal{N} = 2$ gauge theories and try to see what can we say about their low energy effective dynamics. As anticipated, we will focus on the Coulomb branch, which is the only part of the moduli space which can be modified at the quantum level. What this boils down to is determine the exact expression of the prepotential $\mathcal{F}$, more specifically of the generalized complexified gauge coupling matrix, whose imaginary part is the metric on the Coulomb branch.

Our starting point is some UV-free $\mathcal{N} = 2$ matter-coupled Lagrangian. This means that if, say, the gauge group is $SU(N)$ and matter multiplets transform in the (anti)fundamental representation of $SU(N)$, we must require that $F < 2N$, since the one-loop coefficient (which captures the full perturbative expression for the β function) is proportional to $2N - F$, in this case.

One thing which will play an important role later is the R-symmetry breaking pattern. Let us first focus on pure SYM. Besides a compact component, $SU(2)_R$, under which all bosons in the $\mathcal{N} = 2$ vector multiplet are singlets and the two gaugini transform as a doublet, there is also a $U(1)_R$ symmetry, under which (both) gaugini have $R(\lambda, \psi) = 1$. This symmetry is anomalous and, following the same discussion we had for $\mathcal{N} = 1$ SYM, one can see that it gets broken as

$$U(1)_R \longrightarrow \mathbb{Z}_{4\,T(Adj)}, \tag{12.39}$$

since the anomaly coefficient is now $A = 2\,T(Adj)$.

The fact that $R(\psi) = 1$ implies that the adjoint scalars ϕ have $R(\phi) = 2$, meaning that on the Coulomb branch the residual symmetry gets further broken. For example, we will see later that for $G = SU(2)$ Coulomb branch vacua preserve a $\mathbb{Z}_4$ subgroup of the full $\mathbb{Z}_8$ and, therefore, each point on the Coulomb branch is paired with its mirror under the residual $\mathbb{Z}_2$, which acts non-trivially on $\mathcal{M}$. For $G = SU(3)$ a $\mathbb{Z}_2$ subgroup survives, only, while for higher ranks the $U(1)_R$ is fully broken.

Interestingly, unlike $\mathcal{N} = 1$, the addition of matter does not restore an anomaly-free $U(1)_R$ symmetry, in general. Indeed, given that $R(\phi) = 2$, from the cubic superpotential term $\sim H_1 \Phi H_2$ one sees that the hyperscalars are neutral. Hence, their fermionic partners ψ_1 and ψ_2 have $R(\psi_1, \psi_2) = -1$. So, if the hypermultiplets transform in the representation $\underline{r}$ of the gauge group G, the $U(1)_R$ is broken at the quantum level as

$$U(1)_R \longrightarrow \mathbb{Z}_{4\,T(Adj)-4F\,T(r)}, \tag{12.40}$$

if adding F hypermultiplets. There can be specific situations where the representation under which the hypermultiplets transform and their number may actually restore the full $U(1)_R$ symmetry. For example, taking $G = SU(N)$ and F hypers in the fundamental representation, one gets $U(1)_R \to \mathbb{Z}_{4N-2F}$. Interestingly, for $F = 2N$ the full $U(1)_R$ R-symmetry is restored, in agreement with the vanishing of the β function and the supposedly conserved R-charge in superconformal field theories.

Let us start considering pure SYM and take, for definiteness, the gauge group to be $G = SU(N)$. Following our general discussion in Section 12.1.1 the low energy Coulomb branch effective (bosonic) Lagrangian looks like

$$\mathcal{L} = \mathrm{Im}(\partial_\mu \overline{\phi}^I\, \partial^\mu \phi_{DI}) + \frac{1}{2}\, \mathrm{Im}[\tau_{IJ}(\phi) \mathcal{F}^I_{\mu\nu} \mathcal{F}^{J\mu\nu}] \tag{12.41}$$

where $I = 1, 2, \ldots, N-1$. Solving the theory amounts to find the exact expression for the prepotential $\mathcal{F}$ (or, which is the same, for the effective abelian gauge coupling matrix τ_{IJ}) as a function of the strong coupling scale Λ and of scalar field VEVs. From non-renormalization theorems we know that the complexified gauge coupling matrix τ_{IJ} gets one-loop and non-perturbative corrections, only, and reads (we refer collectively to a as the common VEV of all scalar fields ϕ^I)

$$\tau_{IJ}(a, \Lambda) = \frac{2N}{2\pi i} C_{IJ} \log \frac{\Lambda}{a} + \sum_{n=1}^{\infty} d_{IJ,n} \left(\frac{\Lambda}{a}\right)^{2Nn}, \tag{12.42}$$

where $2N$ is the one-loop β-function coefficient, C_{IJ} is some constant matrix that can be computed in perturbation theory, $d_{IJ,n}$'s weight n-instanton corrections and we used eq. (9.44). Since the model is Higgsed at a scale a, which can be taken arbitrarily large, these instanton effects can be made arbitrarily small and are calculable. So, in principle, one could compute τ_{IJ}, and hence solve the low energy effective theory exactly, by evaluating all instanton contributions. In practice, this is hard. Seiberg and Witten came-up with a more physical approach to determine τ_{IJ}, which is the one we will follow. We start analyzing the simplest case, $\mathcal{N} = 2$ SYM with gauge group $SU(2)$.

12.3.1 $\mathcal{N} = 2$ $SU(2)$ *pure SYM*

$\mathcal{N} = 2$ SYM with $G = SU(2)$ admits a one-dimensional moduli space. At a generic point of the moduli space the gauge group is broken to $U(1)$. The gauge invariant coordinate on the (classical) moduli space can be chosen to be

$$u = \frac{1}{2}\langle \operatorname{tr} \phi^2 \rangle = a^2, \tag{12.43}$$

where $\langle \phi \rangle = a\sigma_3$ is the (adjoint) scalar field VEV. The u here corresponds to what we called u_2 in eq. (12.5).

The above formula is valid classically. Quantum corrections may change the relation between u and a. In what follows, we will keep on calling u the coordinate on the quantum moduli space but the above equation will be modified as

$$u = \frac{1}{2}\langle \operatorname{tr} \phi^2 \rangle = a^2 + \text{quantum corrections}. \tag{12.44}$$

While classically $a = \sqrt{u}$, quantum mechanically one could expect a more general relation, $a = a(u)$, which only in the classical limit reduces to $a = \sqrt{u}$.

The abelian low energy effective Lagrangian is

$$\mathcal{L} = \operatorname{Im}\left[\tau(\phi)\left(\frac{1}{2}\mathcal{F}_{\mu\nu}\mathcal{F}^{\mu\nu} + \partial_\mu\overline{\phi}\,\partial^\mu\phi\right)\right] = \operatorname{Im}\left[\tau(\phi)\frac{1}{2}\mathcal{F}_{\mu\nu}\mathcal{F}^{\mu\nu} + \partial_\mu\overline{\phi}\,\partial^\mu\phi_D\right], \tag{12.45}$$

and it is univocally determined knowing the exact expression of τ, which is fixed once determining the exact expression for the prepotential $\mathcal{F}$, since $\tau = \partial^2\mathcal{F}/\partial\phi\partial\phi$. Our goal, in the following, will then be to find the exact expression for $\mathcal{F}$ and hence of the analogous of (12.42) which in this case becomes

$$\tau(a,\Lambda) = \frac{1}{2\pi i}\log\left(\frac{\Lambda^4}{a^4}\right) + \sum_{n=1}^{\infty} d_n \left(\frac{\Lambda^4}{a^4}\right)^n . \tag{12.46}$$

The first thing one should readily notice is that $\mathcal{F}$ cannot be a holomorphic function of a all along $\mathcal{M}$; it should be multivalued. Indeed, if this were not the case, $\mathrm{Im}\,\tau(a) = \mathrm{Im}\,\partial^2\mathcal{F}(a)/\partial a\partial a$ (and similarly $\mathrm{Re}\,\tau(a)$) would be a harmonic function. As such, it could not be positive definite everywhere (unless it were a constant, which cannot be since the gauge coupling runs). Hence, there would necessary be regions in the moduli space where $\mathrm{Im}\,\tau(a)$ would be negative, making the effective gauge coupling squared g^2 being negative, too. This would correspond to the propagation of negative norm states, that cannot be. We need $\mathrm{Im}\,\tau > 0$. The way out is to allow for different local descriptions which requires $\mathcal{F}(a)$ to be defined only locally, say in a neighborhood of the classical region $u \to \infty$. In regions where $\tau(a)$ approaches zero, we need a different (but equivalent) description of the theory. Note that this corresponds to regions where the gauge coupling is very large, eventually infinite, so strongly coupled regimes. Geometrically, the moduli space should admit singularities and in the vicinity of such singularities we expect a different coordinate patch with respect to a (in other words, a is not a "good coordinate" on the whole moduli space $\mathcal{M}$).

In order to understand how different local descriptions can emerge, we have to understand how electric-magnetic duality is realized in the low energy effective theory. The action (12.45) can be re-written as

$$\mathcal{L} = \frac{1}{2}\,\mathrm{Im}\left[\tau(\phi)\mathcal{F}_{\mu\nu}\mathcal{F}^{\mu\nu}\right] + \frac{1}{2}\,\partial_\mu \begin{pmatrix}\phi_D\\ \phi\end{pmatrix}^\dagger J\,\partial^\mu \begin{pmatrix}\phi_D\\ \phi\end{pmatrix} \quad \text{where } J = \begin{pmatrix}0 & i\\ -i & 0\end{pmatrix}, \tag{12.47}$$

where $\phi_D \equiv \partial\mathcal{F}/\partial\phi$. The scalar kinetic term is invariant under $Sp(2,\mathbb{R})$ transformations acting on ϕ_D and ϕ as

$$\begin{pmatrix}\phi_D\\ \phi\end{pmatrix} \to M \begin{pmatrix}\phi_D\\ \phi\end{pmatrix} \quad \text{where } M^\dagger J M = J. \tag{12.48}$$

This is the continuum version of the duality group of electro-magnetism previously defined and it is generated by

$$S = \begin{pmatrix}0 & -1\\ 1 & 0\end{pmatrix}, \quad T_b = \begin{pmatrix}1 & b\\ 0 & 1\end{pmatrix} \quad \text{where } b \in \mathbb{R}. \tag{12.49}$$

To see how the duality group acts on the Maxwell kinetic term we should first write the Lagrangian introducing a Lagrange multiplier field $A_{D\mu}$, so to have both the equation of motion and the Bianchi identity emerging as field equations. This is needed, so to put $F_{\mu\nu}$ and $\widetilde{F}_{\mu\nu}$ on equal footing (the Bianchi identity $\partial_\nu \widetilde{F}^{\mu\nu} = 0$ has been solved for in terms of the gauge potential, hence obscuring the duality). Recalling that $\mathcal{F}_{\mu\nu} = F_{\mu\nu} + i\widetilde{F}_{\mu\nu}$, we can write the Maxwell term action as

$$\mathcal{S} = \int \mathrm{Im}\left[\frac{1}{2}\tau(\phi)\left(F_{\mu\nu} + i\widetilde{F}_{\mu\nu}\right)\left(F^{\mu\nu} + i\widetilde{F}^{\mu\nu}\right)\right] + \int A_{D\mu}\partial_\nu \widetilde{F}^{\mu\nu}. \tag{12.50}$$

where $A_{D\mu}$ should be treated as an independent field, with field strength $F_{D\mu\nu} = \partial_\mu A_{D\nu} - \partial_\nu A_{D\mu}$.

In doing the path integral one should now integrate over $F_{\mu\nu}$ and $A_{D\mu}$ independently. Path integrating over A_D we get the equation $\partial_\nu \widetilde{F}^{\mu\nu} = 0$, namely $dF = 0$, which implies $F = dA$ and we are left with the original path integral. But one can path integrate over F, first. Integrating by parts, the second term in eq. (12.50) becomes $2F_{D\mu\nu}\widetilde{F}^{\mu\nu}$ and completing the square one gets

$$\mathcal{S} = \int \mathrm{Im}\left[\frac{1}{2}\tau(\phi)\left\{\left(F_{\mu\nu} + i\widetilde{F}_{\mu\nu}\right) + \frac{1}{\tau(\phi)}\left(F_{D\mu\nu} + i\widetilde{F}_{D\mu\nu}\right)\right\}^2 + \right.$$
$$\left. - \frac{1}{\tau(\phi)}\frac{1}{2}\left(F_{D\mu\nu} + i\widetilde{F}_{D\mu\nu}\right)^2\right], \tag{12.51}$$

where we used the identity $\left(F_{\mu\nu} + i\widetilde{F}_{\mu\nu}\right)\left(F_D^{\mu\nu} + i\widetilde{F}_D^{\mu\nu}\right) = 2i\, F_{D\mu\nu}\widetilde{F}^{\mu\nu}$, which holds because $F_{\mu\nu}F_D^{\mu\nu} = \widetilde{F}_{\mu\nu}\widetilde{F}_D^{\mu\nu}$. Performing the Gaussian path integral over F one gets back the original action (up to an overall normalization) but now in terms of the dual gauge field $A_{D\mu}$ as

$$\mathcal{S} = \int \mathrm{Im}\left[-\frac{1}{\tau(\phi)}\frac{1}{2}\left(F_{D\mu\nu} + i\widetilde{F}_{D\mu\nu}\right)\left(F_D^{\mu\nu} + i\widetilde{F}_D^{\mu\nu}\right)\right]. \tag{12.52}$$

So we see that, as expected, the effect of a S-duality transformation which transforms the gauge coupling as $\tau \to \tau_D = -1/\tau$, is to replace a gauge field, to which electric sources couple locally, by a dual gauge field, to which magnetic sources couple locally. The other generator of $Sp(2,\mathbb{R})$, T_b, does not act on the gauge field but on the coupling, only, shifting the θ angle. In order for it not to change the physics, one should take $b \in \mathbb{Z}$, hence obtaining the actual electro-magnetic duality group, which is $Sp(2,\mathbb{Z})$. We will henceforth call T the generator T_1, to be consistent with conventions in Section 12.2.

In our previous discussion, we argued that whenever $\mathrm{Im}\,\tau(a)$ approaches 0, a different description of the (same) physics should hold. The above discussion suggests what that can be: an S-dual description in terms of a magnetic dual gauge field $A_{D\mu}$, with $\tau \to \tau_D = -1/\tau$ and ϕ and ϕ_D exchanged, see eq. (12.48).

Note, in passing, that the way the duality group acts on (ϕ_D, ϕ), and in turn on (a_D, a), eq. (12.48), provides further evidence for the BPS mass formula (12.30) and the proposed relation (12.31). To see this, let us couple the low energy effective theory to a charged hypermultiplet with charge n_e. Its coupling to the adjoint chiral superfield Φ is fixed by $\mathcal{N} = 2$ supersymmetry to be

$$\sqrt{2}\, n_e H_1 \Phi H_2. \tag{12.53}$$

On the moduli space this induces a mass for the (BPS!) hypermultiplet whose corresponding central charge would then be $Z = n_e a$. By a S-duality transformation, which acts on the adjoint scalar as eq. (12.48), it is clear that for a magnetic monopole with magnetic charge n_m we would have $Z = n_m a_D$ (with $a_D = \partial\mathcal{F}/\partial a$) and, for dyons, the more general formula (12.30).

Singularities and monodromies

We now have all ingredients to understand the singularity structure of the moduli space and the physical meaning of such singularities.

Let us start looking at the (semi)classical region, namely $u \to \infty$. There one can safely use the classical relation $u = a^2$ and the one-loop expression for the prepotential

$$\mathcal{F}_{\text{one-loop}} = \frac{i}{2\pi} a^2 \log \frac{a^2}{\Lambda^2}. \tag{12.54}$$

From this expression we can compute a_D which is

$$a_D = \frac{\partial \mathcal{F}}{\partial a} = \frac{i}{\pi} a \left(\log \frac{a^2}{\Lambda^2} + 1 \right). \tag{12.55}$$

Let us take a counterclockwise contour in the u plane, say $u \to e^{2\pi i} u$, with very large $|u|$. Since in such semiclassical region $u = a^2$ we see that a transforms as $a \to -a$. For a_D, instead, using (12.55), we get

$$a_D \to \frac{i}{\pi}(-a) \left(\log \frac{e^{2\pi i} a^2}{\Lambda^2} + 1 \right) = -a_D + 2a. \tag{12.56}$$

So, there is a non-trivial monodromy, which acts on the vector (a_D, a) as

$$\begin{pmatrix} a_D \\ a \end{pmatrix} \to M_\infty \begin{pmatrix} a_D \\ a \end{pmatrix} \quad \text{where } M_\infty = \begin{pmatrix} -1 & 2 \\ 0 & -1 \end{pmatrix} \tag{12.57}$$

Note that, consistently with previous general discussion, $M_\infty \in Sp(2,\mathbb{Z})$. More specifically, $M_\infty = -T^{-2}$, with T the generator previously defined.

The log term in a_D and the non-trivial monodromy show that a and a_D are multivalued functions: there is a branch cut extending from infinity, due to the log term in the one-loop running. Given the singularity at $u = \infty$, there must be singularities also somewhere else on the u plane, with their associated monodromies M_i.

Since, as shown in Figure 12.4, a contour circling around infinity can be deformed (it is topologically equivalent) to a contour circling around all other singularities, say we have k of them, the following consistency relation should hold, in general

$$M_\infty = M_1\, M_2\, \dots\, M_k. \tag{12.58}$$

Now, how many singularities, besides that at $u = \infty$, do we have on the u plane? The R-symmetry breaking pattern helps, here. As already discussed, the $U(1)$ R-symmetry of the original theory is anomalous and broken to $\mathbb{Z}_8$ at the quantum level. Since ϕ has R-charge 2, on the moduli space, parametrized by $u \sim \langle \mathrm{Tr}\,\phi^2\rangle$, this is further broken to $\mathbb{Z}_4$. The residual $\mathbb{Z}_2$ symmetry acting on the moduli space changes u as $u \to -u$. Therefore, singularities should come in pairs on the moduli space, but at the fixed points of the $\mathbb{Z}_2$ action $u = \infty, 0$. We conclude that if we had one only more singularity beside the one at infinity, this should be at $u = 0$.

But this cannot be. If there were just one singularity at $u = 0$, because of (12.58) we would have $M_0 = M_\infty$. But then, since a^2 is left invariant by M_∞, $u = a^2$ would be a good global coordinate on the full moduli space, not just in the classical region. Then, $\mathcal{F}(a)$ would be a holomorphic function of a and so $\mathrm{Im}\,\tau(a)$ a harmonic function. But then, the latter could not be positive definite (unless it was a constant, which we know it is not).

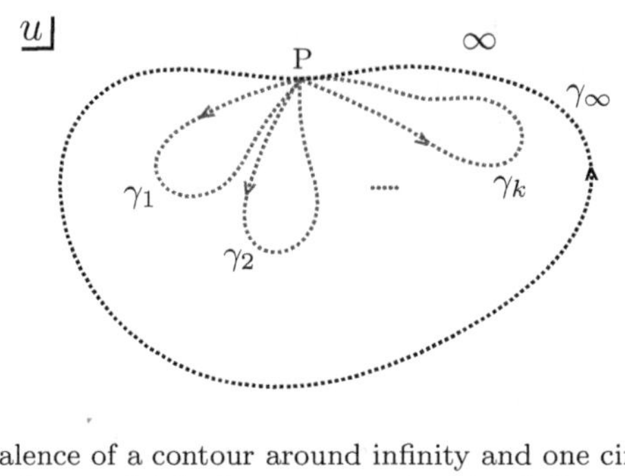

Fig. 12.4 The equivalence of a contour around infinity and one circling all singularities on the u-plane. P is a base point for loops γ_∞ and γ_i, $i = 1, 2, \dots, k$.

So, we conclude that there must be at least two singularities besides that at infinity, located at, say $u = \pm u_0$. If this is this case, $u = 0$, which is a singular point on the classical moduli space, will not be a singular point anymore in the quantum theory. In order to have a singularity at $u = 0$ one should have at least three singularities on $\mathcal{M}$. As we will see, this cannot be either. Having two (and only two) singularities on $\mathcal{M}$, located at $u = \pm u_0$, happens to be the only consistent possibility.

A natural question to ask is what the nature of the particles becoming massless at $u = \pm u_0$ is. Interestingly, unlike to what happens classically, singularities in the quantum moduli space are not associated to enhanced gauge symmetry, namely to extra massless gauge bosons. This can be understood as follows. If an interacting non-abelian Coulomb phase were there in the IR, a conserved R-symmetry should be present (the superconformal R-symmetry). As we have already discussed, singularities on $\mathcal{M}$ occur at $u \neq 0$. Hence, if conformal invariance should be preserved, then the dimension of u (more precisely, the dimension of the operator U of which u is the VEV) at the singularity should be zero. In a SCFT the dimension of an operator is proportional to its R-charge, which for the operator u is $R(u) = 4$ since $R(\phi) = 2$. Therefore, at the singularity the operator u would have non-vanishing scaling dimension and a VEV would break conformal invariance. This suggests that a non-abelian Coulomb phase cannot emerge in the IR. The extra massless degrees of freedom cannot be gauge bosons.

Notice how different the situation is at a arbitrary (smooth) point of the moduli space. At a smooth point of the moduli space the low energy effective theory is an abelian gauge theory. This is also a SCFT, although a non-interacting one. The scalar field u has still R-charge 4 and has a non-vanishing scaling dimension, but is part of an abelian vector multiplet and as such is uncharged under the $U(1)$ gauge group. Therefore, one can shift its VEV to 0 at no cost, without breaking the gauge symmetry (this cannot be done if the gauge symmetry were non-abelian, since then u would have been charged). Therefore, the previous argument does not apply now, since in the new coordinate system $\langle u \rangle = 0$ and this does not break conformal invariance of the (abelian) SCFT.

Coming back to the singularities at $u = \pm u_0$, the analysis of the previous section suggests what the other possibility could be. The only other states in the spectrum (at least that we know from the semi-classical analysis) are monopoles and dyons. For example, magnetic monopoles are very heavy at weak coupling, because of the BPS mass formula (12.30) and tend to become light at strong coupling. So it might very well be that these are the states becoming massless at the strong coupling singularities. Note that if this is the case, by the reasoning in the previous paragraph we conclude that they cannot sit in vector multiplets (which would include spin 1 particles). So they should correspond to hypermultiplets. Indeed, in the $\mathcal{N}=2$ version of the GG model, this was explicitly shown to be the case!

As a corollary, one would expect that the singularity at $a=0$ of the classical moduli space, where extra massless gauge bosons did become massless, should disappear at quantum level. From the exact expression we will eventually get for $a=a(u)$, we will see that this is indeed the case: the point $a=0$ (which would correspond to $\langle\phi\rangle=0$) does not belong to the moduli space, at quantum level (similarly to what happens for $\mathcal{N}=1$ $SU(N)$ SQCD with $F=N$).

Let us then focus on the strong coupling singularities at $u=\pm u_0$. Note that u_0 should be proportional to Λ^2, since in the classical limit, $\Lambda\to 0$, one should recover the (only one) singularity at $u=0$. Hence, from now on, without loss of generality, we will take $u_0=\Lambda^2$. The structure of the moduli space, with punctures and associated monodromies, is depicted in Figure 12.5.

To find the structure of the monodromy matrices M_{Λ^2} and $M_{-\Lambda^2}$ notice that they should have a form like (12.37) in terms of the (integer) electric and magnetic charges (n_m,n_e), (n'_m,n'_e) of the corresponding massless states. Imposing the consistency relation $M_\infty = M_{\Lambda^2}M_{-\Lambda^2}$ and using (12.57) one finds that the unique solution (modulo physically equivalent solutions, remember the comment after eq. (12.22)) is

$$(n_m,n_e)=\pm(1,0),\quad (n'_m,n'_e)=\pm(1,-1) \tag{12.59}$$

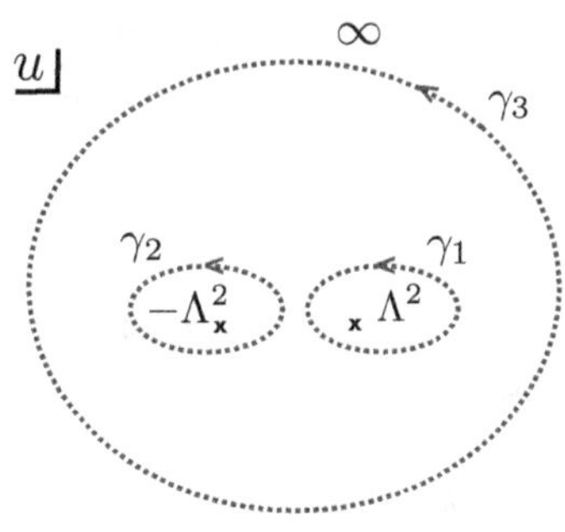

Fig. 12.5 The u plane with the three singularities at $\infty,\Lambda^2,-\Lambda^2$. The monodromies associated to the three cycles γ_i must satisfy the consistency relation $M_\infty = M_{\Lambda^2}M_{-\Lambda^2}$.

corresponding to monodromy matrices

$$M_{\Lambda^2} = \begin{pmatrix} 1 & 0 \\ -2 & 1 \end{pmatrix}, \quad M_{-\Lambda^2} = \begin{pmatrix} -1 & 2 \\ -2 & 3 \end{pmatrix}. \tag{12.60}$$

So, we finally see what the nature of the singularities is: at $u = \Lambda^2$ a monopole with charge $\pm(1,0)$ becomes massless and at $u = -\Lambda^2$ a dyon with charge $\pm(1,-1)$ does. Note, in passing, that $M_{\Lambda^2} = ST^2S^{-1}$ and $M_{-\Lambda^2} = TST^2S^{-1}T^{-1} = TM_{\Lambda^2}T^{-1}$, which nicely agrees with the fact that the residual $\mathbb{Z}_2$ symmetry connecting $u = \Lambda^2$ and $u = -\Lambda^2$ shift the θ angle by 2π.

One might wonder if there can be more than two singularities on $\mathcal{M}$. For this to be the case, one should be able to solve an equation like (12.58) with $k > 2$, with M_∞ given by (12.57) and the M_k's having a structure as (12.37) with integers (n_m^i, n_e^i). While a general proof is not available, one can explicitly show, for not too large values of k, that there are no solutions for $k > 2$. Further evidence suggesting that $k = 2$ is the correct answer will be provided shortly.

Seiberg-Witten curve

Given the knowledge of the singularity structure of the moduli space and its monodromies, we want now to construct holomorphic functions $a = a(u)$ and $a_D = a_D(u)$ satisfying the monodromies (12.57) and (12.60), and from them obtain the exact expression for the complexified gauge coupling $\tau(u)$. A holomorphic function is univocally determined by its singularities. Therefore, if we are able to find a function with the correct monodromies around $u = \infty, \Lambda^2, -\Lambda^2$, we can be sure we get the correct answer. In principle, this can be done, following the so-called differential equation approach, but we will follow a different, more geometric pattern. This was the approach originally pursued by Seiberg and Witten and also the one which makes it easier and more natural to understand generalizations to richer theories (*i.e.* theories with more general gauge groups and including charged matter).

The crucial observation comes from the property that we have learned $\tau = \tau(u)$ should have: a complex quantity with positive definite imaginary part and on which the group $Sp(2,\mathbb{Z})$ acts as a fractional linear transformation, eq. (12.19). Such quantities are fundamental in the theory of Riemann surfaces, where they describe their moduli, the positivity condition ensuring regularity of the surface. In the case at hand, the relevant Riemann surface is just a torus, or equivalently, using the language of algebraic geometry, an elliptic curve. This curve can be written as a complex surface

$$y^2 = (x - \Lambda^2)(x + \Lambda^2)(x - u), \tag{12.61}$$

where u parametrizes the modulus τ of the torus and x and y are complex coordinates. Varying u we vary τ and hence eq. (12.61) describes a family of tori. If we associate to any point on the Coulomb branch parametrized by the complex quantity u a holomorphic varying torus, its modulus will have the same properties we expect for the complexified gauge coupling τ: a holomorphic section of a $Sp(2,\mathbb{Z})$ bundle with positive imaginary part, $\operatorname{Im}\tau > 0$.

A way to understand that (12.61) describes a torus is as follows. From eq. (12.61) we see that y is the square root of a polynomial in x so we can look at the x-plane consisting into two sheets with branch points at Λ^2, $-\Lambda^2$, u and ∞, gluing along the branch cuts (the two sheets corresponding to the $\pm y$ branches). One can take one branch cut between $-\Lambda^2$ and Λ^2 and the second one between u and ∞. The two sheets can be thought as spheres and the branch cuts as tubes connecting them. Topologically, this is a torus, see Figure 12.6.

On a torus the are two independent, non-trivial homology one-cycles, the A and the B cycles, which we can take as in the figure. Degenerate tori (that is tori where some cycles shrink to zero size) occur when any two zero's of eq. (12.61) coincide. In other words, when one of the branch cuts disappears. In particular, for $u = \Lambda^2$ the B cycle shrinks to zero size, for $u = \infty$ the A cycle shrinks to zero size and for $u = -\Lambda^2$ a linear combination of the two, $A + B$, does.

The basis of one-cycles is not unique, but defined up to $Sp(2,\mathbb{Z})$ transformations which act as

$$\begin{pmatrix} B \\ A \end{pmatrix} \to M \begin{pmatrix} B \\ A \end{pmatrix} \quad \text{where } M \in Sp(2,\mathbb{Z}). \tag{12.62}$$

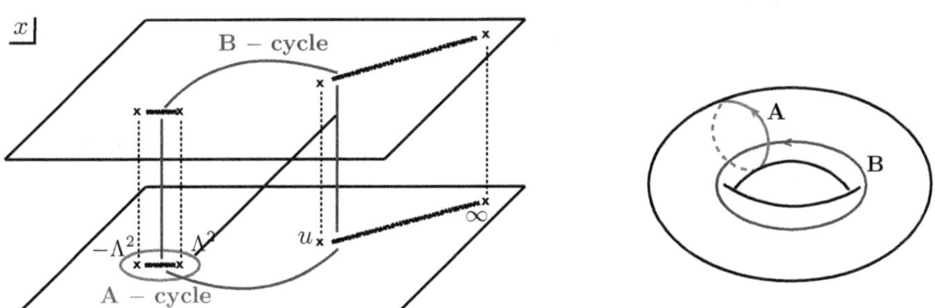

Fig. 12.6 On the left the elliptic curve in the (two sheeted) x plane. A and B are the two one-cycles of the torus. On the right the corresponding torus.

The modulus of the torus $\tau(u)$ corresponds to the ratio of the periods ω and ω_D

$$\tau(u) = \frac{\omega_D}{\omega}, \tag{12.63}$$

the integrals over the A and B cycles of the unique holomorphic (closed) one-form on the torus, $\Omega = dx/y$

$$\omega = \oint_A \frac{dx}{y}, \quad \omega_D = \oint_B \frac{dx}{y} \quad \text{with} \; \frac{dx}{y} = \frac{dx}{\sqrt{(x-\Lambda^2)(x+\Lambda^2)(x-u)}}. \tag{12.64}$$

Note that the periods ω and ω_D inherit from the A and B cycles the transformation properties under $Sp(2,\mathbb{Z})$ and so also the same monodromies at the three singular points $u = \infty, \pm\Lambda^2$, where two branch points collide.

The identification between the $SU(2)$ gauge theory and the above family of tori parametrized by u holds via identifying the modulus of the torus with the complexified gauge coupling, and the periods with the u derivative of a and a_D, that is

$$\tau = \frac{\omega_D}{\omega} \equiv \tau(a) = \frac{\partial a_D}{\partial a} = \frac{\partial a_D/\partial u}{\partial a/\partial u}, \tag{12.65}$$

with the identification

$$\frac{\partial a}{\partial u} = \omega = \oint_A \frac{dx}{y}, \quad \frac{\partial a_D}{\partial u} = \omega_D = \oint_B \frac{dx}{y}, \tag{12.66}$$

up to an overall normalization that we will fix momentarily. Note, in passing, that since u is globally defined, a and a_D have the same monodromies of the periods ω and ω_D, in perfect agreement with the equivalence between transformations (12.62) and (12.48).

Integrating in u on both sides one obtains

$$a = \oint_A d\lambda, \quad a_D = \oint_B d\lambda, \tag{12.67}$$

where the one-form differential $d\lambda$ (also known as Seiberg-Witten differential) can be easily computed to be

$$\frac{\partial d\lambda}{\partial u} = \frac{dx}{y} = \frac{dx}{\sqrt{(x^2-\Lambda^4)(x-u)}} \quad \longrightarrow \quad d\lambda = \frac{(x-u)dx}{y}, \tag{12.68}$$

up to exact forms. Using the above definition of A and B cycles, and deforming them so to lie entirely along the cuts between $-\Lambda^2$ and Λ^2 and between Λ^2 and u, respectively, one can express the integrals (12.67) as

$$a(u) = \frac{\sqrt{2}}{\pi} \int_{-\Lambda^2}^{\Lambda^2} dx \frac{\sqrt{x-u}}{\sqrt{x^2 - \Lambda^4}} \tag{12.69}$$

$$a_D(u) = \frac{\sqrt{2}}{\pi} \int_{\Lambda^2}^{u} dx \frac{\sqrt{x-u}}{\sqrt{x^2 - \Lambda^4}}, \tag{12.70}$$

where the overall normalization has been fixed by requiring that for $u \to \infty$ one recovers the (semi)classical result (12.55).

Using the identity

$$F(\alpha, \beta, \gamma; z) = \frac{\Gamma(\gamma)}{\Gamma(\beta)\Gamma(\gamma - \beta)} \int_0^1 dx\, x^{\beta-1}(1-x)^{\gamma-\beta-1}(1-zx)^{-\alpha}, \tag{12.71}$$

one can finally recast (12.69) and (12.70) in terms of hypergeometric functions

$$a(u) = \sqrt{2}(\Lambda^2 + u)^{1/2}\, F\left(-\frac{1}{2}, \frac{1}{2}, 1; \frac{2}{1 + u/\Lambda^2}\right) \tag{12.72}$$

$$a_D(u) = i\, \frac{\Lambda - u/\Lambda}{2}\, F\left(\frac{1}{2}, \frac{1}{2}, 2; \frac{1 - u/\Lambda^2}{2}\right). \tag{12.73}$$

One can invert (12.72) to obtain $u(a)$ and insert the result into (12.73) to obtain $a_D(a)$. Integrating with respect to a yields $\mathcal{F}(a)$. Equivalently, deriving with respect to a yields $\tau(a)$ and, hence, the exact expression of the low energy effective action (12.45)!

Let us emphasize again that the expression one gets for $\mathcal{F}(a)$ is not globally defined on the moduli space, and different analytic continuations should be used in different patches. For example, near $u = \Lambda^2$, better to use S-dual coordinates, where the role of what is electric and what is magnetic is inverted. This is represented in Figure 12.7.

As a check that the result we got describes the coupling τ entering the effective Lagrangian (12.45), one can expand (12.73) and (12.72) around $u = 0, \Lambda^2$ and $-\Lambda^2$ and show agreement with the expected (singular) behavior for a_D and a, including the monodromies (12.57) and (12.60).

- For $u \to \infty$ we have $a \sim \sqrt{u}$ and $a_D \sim \frac{i}{\pi}\sqrt{u} \log \frac{u}{\Lambda^2} \sim \frac{i}{\pi} a \log \frac{a}{\Lambda}$. This reproduces the (semi)classical result (12.55) and so also the correct monodromy (12.57). Note that in this case there is no choice of (n_m, n_e) giving a vanishing mass (for M_∞ it does not exist a left eigenvector with unit eigenvalue), in agreement with the fact that at $u \to \infty$ there are no extra massless particles in the spectrum.

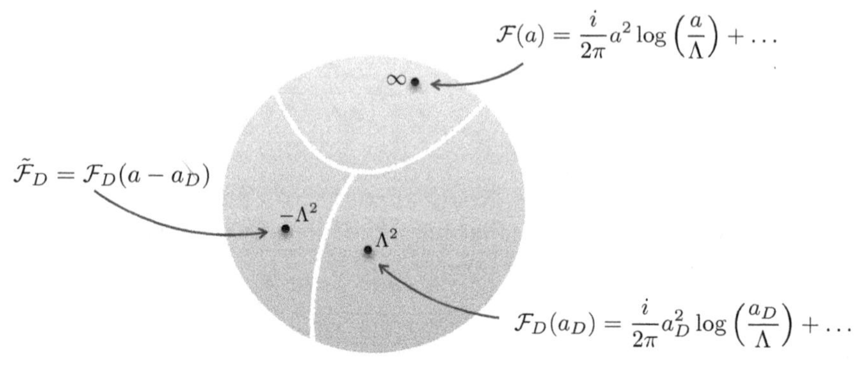

Fig. 12.7 The quantum moduli space $\mathcal{M}_q$ of pure SYM with $G = SU(2)$ represented as a sphere, obtained by adding the point at infinity to the complex u plane. The space is covered by three distinct regions where a local, weakly coupled Lagrangian can be written using appropriate coordinates, *i.e.* the appropriate duality frame. No local Lagrangian exists which would be globally defined on $\mathcal{M}_q$. What are classical and strongly coupled regions is not an invariant concept, since it depends on the coordinate frame.

- For $u \to \Lambda^2$ we have $a \sim \frac{i}{\pi} a_D \log \frac{a_D}{\Lambda}$ and $a_D \sim (u - \Lambda^2)$. So we see that a is singular at $u \sim \Lambda^2$ while a_D vanishes. This is the correct behavior for a magnetic monopole with charge n_m becoming massless at $u = \Lambda^2$, in agreement with what we previously found. Again, the monodromy M_{Λ^2} in eq. (12.60) is correctly reproduced.
- For $u \to -\Lambda^2$ we have $a - a_D \sim (u + \Lambda^2)$ and $a \sim \frac{i}{\pi}(a_D - a) \log \frac{a_D - a}{\Lambda}$. This shows that at $u = -\Lambda^2$ we have a singularity where $a = a_D$, which gives a massless dyon with opposite electric and magnetic charges, $n_e = -n_m$, again in agreement with previous results, including the monodromy $M_{-\Lambda^2}$.

There are other non-trivial checks one can make. For example, one can expand $\tau(u)$, eq. (12.65), in (inverse) powers of u, at large u, and compare with (12.46), using $u = a^2$. This gives perfect agreement with the instanton coefficients d_1 and d_2, which have been independently calculated.

As anticipated, an inspection of the exact solution (12.72) shows that for no values of u the scalar field VEV a becomes 0. So, the point $a = 0$ is not part of the quantum exact moduli space, as anticipated. This is consistent with the claim that nowhere on the moduli space extra massless gauge bosons arise.

To sum-up, at a generic point of the moduli space the effective theory is a $\mathcal{N} = 2$ abelian free theory. At two special points, $u = \Lambda^2, -\Lambda^2$ the effective theory is $\mathcal{N} = 2$ SQED with one massless flavor. Just, to have a local description one should use a dual frame, since monopoles couple locally to an effective Lagrangian written in terms of the dual gauge field $A_{D\mu}$ and dyons with charge $(1, -1)$ couple locally to an effective Lagrangian written in terms of the gauge field $A_{D\mu} - A_\mu$. In this sense the word "monopole" or "dyon" is just conventional and adapted to the large u region, where the theory is semiclassical and local in terms of the gauge field A_μ, in which monopoles and dyons appear as non-perturbative states (for instance, in the $A_{D\mu}$ frame it is an electron which looks as a non-perturbative state).

12.3.2 *Intermezzo: confinement by monopole condensation*

Before discussing generalizations of this model, there is one (very nice) consistency check one can do.

Let us start from $\mathcal{N} = 2$ $SU(2)$ SYM and add a mass m to the chiral superfield Φ belonging to the $\mathcal{N} = 2$ vector multiplet, that is $W = m\,\mathrm{Tr}\Phi^2$. This breaks explicitly $\mathcal{N} = 2$ to $\mathcal{N} = 1$. For $m \gg \Lambda$ we can use the UV Lagrangian, integrate Φ out and end-up with pure $\mathcal{N} = 1$ $SU(2)$ SYM at low energy, which admits two isolated supersymmetric vacua with charge confinement and mass gap. As we discussed in Chapter 10, by supersymmetry this same scenario should hold even if $m \ll \Lambda$. In this regime, the low-energy $\mathcal{N} = 2$ effective description we discussed before should be approximately valid and we should use it, adding to it the small mass perturbation. But how the moduli space can be lifted giving back just two isolate (gapped) vacua? How can the otherwise massless photon get a mass since, following our discussion in Section 12.1, we know that there are no light charged fields?

The addition of a mass term with $m \ll \Lambda$ makes the effective theory becoming an $\mathcal{N} = 1$ abelian gauge theory with a massive (neutral) chiral multiplet Φ. Let us dub $U = \mathrm{Tr}\Phi^2$ the chiral superfield whose lowest component VEV u parametrizes the (original) $\mathcal{N} = 2$ moduli space. At a generic value of u, there are no massless (or nearly massless) chiral superfields other than U so we easily see that the F-term equation we have to impose on the space of D-flat directions cannot be satisfied since

$$\frac{\partial W}{\partial U} = m \neq 0, \tag{12.74}$$

and the moduli space is lifted. But then, what about the two $\mathcal{N} = 1$ supersymmetric vacua we expect to survive? The results we got in the previous section contain the answer. We have learned that there exists two special points on the complex u-plane where extra massless degrees of freedom arise. One such points is $u = \Lambda^2$ where a massless magnetically charged hypermultiplet is present and should hence be included in the effective theory. Let us describe our theory near $u = \Lambda^2$. There, better to use S-dual variables, for which the superpotential reads

$$W = \sqrt{2}\, H_1 \Phi_D H_2 + m\, U, \tag{12.75}$$

where Φ_D is the S-dual of Φ and U should be thought of as a function of Φ_D, now. The D-term equations from the coupling to the (magnetic dual) $U(1)$ gauge field imply that $|H_1| = |H_2|$ (recall that H_1 and H_2 have conjugate internal quantum numbers and hence are oppositely charged under the $U(1)$ gauge symmetry), while the F-term equations read

$$\sqrt{2} H_1 H_2 + m\, \frac{du}{da_D} = 0, \quad a_D H_1 = a_D H_2 = 0. \tag{12.76}$$

Since $du/da_D \neq 0$ (u is a good global coordinate on u!) we get the following answer

$$\begin{aligned} m = 0 : H_1 &= H_2 = 0, \quad a_D = \text{any} \\ m \neq 0 : H_1 &= H_2 = \left(- \frac{m}{\sqrt{2}} \frac{du}{da_D}\bigg|_{a_D=0} \right)^{1/2}, \quad a_D = 0. \end{aligned} \tag{12.77}$$

For $m = 0$ we recover (tautologically) the $\mathcal{N} = 2$ moduli space. For $m \neq 0$, since H_1 and H_2 are (magnetically) charged, their VEVs break the $U(1)$ gauge symmetry and give a mass to the abelian gauge field (to all the $\mathcal{N} = 1$ gauge multiplet, in fact). So, we end up with a supersymmetric vacuum with a mass gap. This same reasoning holds also at $u = -\Lambda^2$, where the role of the massless magnetic monopole is played by a massless dyon. So, for $m \neq 0$ the $\mathcal{N} = 2$ moduli space is fully lifted but at two points, where there are supersymmetric vacua with mass gap. Exactly what we expect. The exact answer we got for the original $\mathcal{N} = 2$ model is just right to give what should hold for the $\mathcal{N} = 1$ massive theory under study!

This is nice, but, superficially, there is still a point of concern. The two $\mathcal{N} = 1$ supersymmetric vacua are confining ones. Here, instead, the dynamics is more of a Higgs-like mechanism, where a scalar monopole or a dyon field condense. The point is that the Higgs mechanism taking place is not the usual condensation of an electrically charged field, but of a magnetically (or dyonically) charged one.

To understand what that means let us recall some basics of the usual Higgs mechanism, where electrically charged fields condense.

The condensation of the electric charge has the effect that any background electromagnetic field gets screened. This implies that electric sources in the theory are (almost) free, since their electric fields can be absorbed by the vacuum condensate and their interaction energy drops off exponentially. Magnetic charges behave very differently. The magnetic field lines have no condensate where to end on. The result is that magnetic field lines tend to be expelled from the vacuum (this is the well-known Meissner effect taking place in superconductors). The minimum energy configuration is for the magnetic field to be confined to a thin flux tube connecting opposite magnetic charges. Therefore, in the Higgs mechanism, electric charges are screened and magnetic charges are confined (Note: this is strict confinement).

In the model above what condenses (let us focus, momentarily, on the vacuum at $u = \Lambda^2$) is not an electric charged state, but a magnetically charged one. By electro-magnetic duality it follows that here magnetic charges are screened, while electric charges are confined. So, eventually, we do have a confining vacuum (due to a magnetic dual of the Meissner effect)! This is a concrete realization of an old idea, due to 't Hooft and others, that confinement in non-Abelian gauge theories maybe due to monopole condensation. The point $u = -\Lambda^2$ where a dyonic field condense is just related to the latter by a different electro-magnetic duality rotation. There, both electric and magnetic charges are confined but dyonic charges proportional to $(1, -1)$ won't, they will just be screened. This is known as *oblique* confinement, also proposed by 't Hooft long ago.

The result we got is beautiful from several point of views. First, it shows that the presence of magnetically charged solitons becoming massless somewhere on the moduli space is necessary to match $\mathcal{N} = 2$ dynamics with $\mathcal{N} = 1$ via holomorphic decoupling, one of our guiding principles. Second, it gives an a posteriori consistency check about the claim that two and only two singularities should be there on $\mathcal{M}$. Finally, it shows (at least in this softly broken $\mathcal{N} = 2$ model) that confinement is due to monopole condensation, providing a concrete realization of the old idea that this could in fact be the way electric charge gets confined.

12.4 Seiberg-Witten theory: generalizations

Till now we have been focusing on pure $\mathcal{N} = 2$ SYM with gauge group $SU(2)$. The story can be generalized to gauge groups with higher rank and/or coupled to matter fields.

We are not going to discuss these generalizations in detail and refer the interested reader to the references at the end of this lecture. Still, we want to make a few remarks and discuss in some detail one (very instructive) example.

For $G = SU(2)$ and no matter we learned that the moduli space, whose complex dimension is one, is the complex plane u with two singularities at $u = \pm u_0$ (beside the singularity at infinity), exchanged by the residual $\mathbb{Z}_2$ R-symmetry. At these two singularities magnetically charged objects, a monopole and a dyon, respectively, become massless. The metric on the moduli space can be described via an auxiliary elliptic curve, a Riemann surface $\Gamma(u)$ of genus $n = 1$ (a torus), whose modulus τ can be explicitly computed and corresponds (in fact, its imaginary part does) to the metric itself, the only unknown in the low energy effective action.

Gauge groups with higher ranks mean moduli spaces $\mathcal{M}$ with complex dimension n, locally $\mathbb{C}^n/W_G$, where W_G is the Weyl group of the gauge group G, n the gauge group rank and u_I, $I = 1, \ldots, n$, gauge invariant coordinates on $\mathcal{M}$, see e.g. eqs. (12.5). As already discussed, these theories are generalizations of the ($\mathcal{N} = 2$ supersymmetric version of the) GG model. As such, they admit several types of charged soliton-like solutions which, in the BPS limit, satisfy the BPS mass formula (12.35). The low energy effective action is form-invariant under electro-magnetic duality rotations, which are generated by $Sp(2n, \mathbb{Z})$. Again, one finds that on the quantum exact moduli space there are singularities where magnetically charged states become massless. The prepotential is not globally defined and different charts (duality frames) should be used. In order to extract the metric on the moduli space, one can again make use of auxiliary elliptic curves $\Gamma(u_I)$, generalizations of the curve (12.61), which are genus n Riemann surfaces, now, and can be described as a double-sheeted x plane with $n + 1$ branch cuts. The period matrices τ of these genus n two-dimensional surfaces, which are defined in analogy to the modulus of a torus and whose imaginary part is in fact positive, get identified with the gauge coupling matrices τ. Hence, they determine the metric on the moduli space and, in turn, the exact low energy effective action (12.41).

A genus n Riemann surface, Figure 12.8, can be characterized in terms of n pairs of homology cycles A^I and B_I and, correspondingly, period integrals defined as

$$\omega_{IL} = \oint_{A^I} \Omega_L, \quad \omega_D{}^J_L = \oint_{B_J} \Omega_L \quad I, J, L = 1, \ldots, n, \tag{12.78}$$

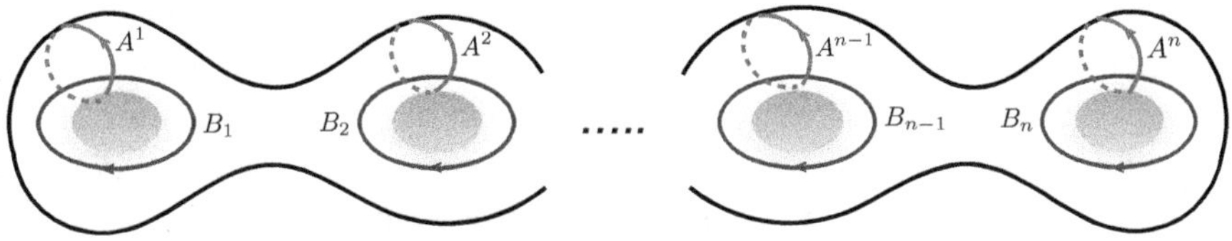

Fig. 12.8 A genus n Reimann surface with the basis of homology cycles A^I, B_J.

where Ω_L are n independent holomorphic one-forms, generalizations of the unique holomorphic differential Ω defined on genus 1 surfaces, eqs. (12.64). The period matrix is given by

$$\tau = \omega_D \cdot \omega^{-1} \tag{12.79}$$

(to be understood as a matrix equation), which ensures that $\mathrm{Im}\,\tau > 0$. The identification goes as before. In particular, the period matrix (12.79) gets identified with the gauge coupling matrix and the periods with the u derivative of a_I and a_D^J, that is

$$\omega_{IL} = \frac{\partial a_I}{\partial u_L} = \oint_{A^I} \frac{\partial d\lambda}{\partial u_L}, \quad \omega_D{}^J_L = \frac{\partial a_D^J}{\partial u_L} = \oint_{B^I} \frac{\partial d\lambda}{\partial u_L} \tag{12.80}$$

and

$$a_I = \oint_{A^I} d\lambda, \quad a_D^J = \oint_{B_J} d\lambda, \tag{12.81}$$

where $d\lambda$ is again the Seiberg-Witten differential.

What about adding matter? The presence of matter fields opens-up the possibility for Higgs branches. However, in what follows we will focus on the Coulomb branch only, since, as already emphasized, that is the only component of the moduli space which gets modified at the quantum level. Note, also, that since we want to keep the theory UV-free, matter is constrained, there cannot be too much. For example, adding matter to $SU(2)$ SYM one can add up to three hypermutiplets in the fundamental representation or, if allowing a vanishing β function, four hypermultiplets in the fundamental representation or one in the adjoint (the latter case corresponds to $\mathcal{N} = 4$ SYM).

While matter fields do not change the dimension of the Coulomb branch (and so the genus of the corresponding Seiberg-Witten curve), they do change its singularity structure. One way to see this is to notice that hypermultiplets enjoy two contributions to their (effective) mass: the bare mass m_i and, whenever the adjoint chiral superfield Φ gets a VEV, the one inherited from the $\mathcal{N} = 2$ supersymmetry preserving cubic coupling, see eq. (12.32). So, one expects new singularities on the moduli space wherever the two mass contributions cancel each other and the (charged) hypermultiplets become effectively massless. Notice, further, that in order to understand the monodromy associated to these singularities, one should use the generalized BPS mass formula

$$Z = a \cdot n_e + a_D \cdot n_m + \sum_{i=1}^{F} \frac{1}{\sqrt{2}} m_i S_i, \tag{12.82}$$

in place of (12.35).

In order to construct the curves $\Gamma(u_I)$ concretely, one should follow the same logical steps we discussed for the $SU(2)$ theory, some of the guiding principles being matching their singularity structure with the appearance of massless particles in the $\mathcal{N} = 2$ theory spectrum, R-symmetry (and when hypermultiplets are present also flavor symmetry) considerations, agreement, by holomorphic decoupling or scale matching, with curves with less flavors or smaller gauge groups. In any event, one ends up with equations like

$$y^2 = f(x, u_I, m_i, \Lambda), \tag{12.83}$$

where u^I, the moduli space coordinates, parametrize the period matrix τ of the curves, eq. (12.79), m_i are hypermultiplet bare masses (where $i = 1, \ldots, F$) and Λ is the strong coupling scale (we are assuming, for simplicity, that the gauge group G is simple).

It should be remarked that different parametrizations can be used to represent the curve associated to the moduli space of a given theory. Some may be more useful than others, depending what one wants to look at. For this reason, in the literature (and in the references at the end of this lecture) different parametrizations can be found. Needless to say, they are all physically equivalent, as they should. We will see one such example shortly.

Admittedly, despite the clear logical steps and guiding principles one can follow, some amount of educated guesswork is usually needed to find the correct curves. That is to say, there is no general recipe for constructing the Seiberg-Witten curve for an arbitrary theory.

Interestingly, a systematic way to construct a large class of Seiberg-Witten curves does exist and relies on M-theory, where a physical meaning can be given to the Riemann surfaces (the curves) themselves.

Four-dimensional $\mathcal{N} = 2$ theories can be engineered from $M5$ branes wrapped on suitably chosen two-dimensional compact surfaces. At low energy, smaller than the typical size of the surface, the theory becomes effectively four-dimensional and preserves $\mathcal{N} = 2$ supersymmetry. Such Riemann surfaces are nothing but the Seiberg-Witten curves! This makes several properties of the low energy effective theory having a geometrical interpretation. This helps and indeed a plethora of interesting results have been obtained following this approach.

As already stressed, there are a number of consistency checks one can make on the curves (12.83). For example, by making one hypermultiplet massive, that is taking its mass m_i large, eventually $m_i \to \infty$, one can integrate the hypermultiplet out and end on a theory with one flavor less, and then show that the limit of the corresponding curve agrees with the curve with one less hypermultiplet. Similarly, by letting one of the vacuum expectation values becoming large, one obtains a limit in which the gauge group is higgsed at high energy, e.g. $SU(N) \to SU(N-1)$. Again, the corresponding limit of the $SU(N)$ curve should agree with the curve for $SU(N-1)$. Etc....

12.4.1 *A case study: $\mathcal{N} = 2$ $SU(2)$ SQCD with one flavor*

To make the above discussion a bit more concrete, we will now discuss in some detail one of the simplest generalizations of the original Seiberg-Witten model, namely $\mathcal{N} = 2$ SQCD with gauge group $SU(2)$ and one massive hypermultiplet in the fundamental representation. The proposed corresponding Seiberg-Witten curve reads

$$y^2 = x^2(x-u) - \Lambda_1^6 + 2m\Lambda_1^3 x, \tag{12.84}$$

where, with obvious notation, Λ_1 is the strong coupling scale.

Let us first check that upon holomorphic decoupling one recovers the curve for pure $SU(2)$ SYM. If we send $m \to \infty$ keeping $m\Lambda_1^3$ fixed, the flavor decouples and, using the scale-matching relation $\Lambda^4 = m\Lambda_1^3$, one ends up with the curve

$$y^2 = x^2(x-u) + 2\Lambda^4 x. \tag{12.85}$$

This is indeed the curve of pure $SU(2)$ SYM, although in a different parametrization (and normalization) with respect to eq. (12.61). Let us see how the curves (12.61) and (12.85) are related.

First, as previously noticed, in a normalization where charged fields transforming in the adjoint representation of the gauge group have integer charges, those of fields transforming in the fundamental are half integers. Hence, in making a comparison between gauge theory with and without matter, it is convenient to first change the normalization we used to treat $SU(2)$ pure SYM and multiply, in eq. (12.30), n_e by 2, so to ensure n_e to still be an integer, and divide a by 2, so to ensure that (12.30) is unchanged. This change of conventions corresponds to the following transformation on the vector (a_D, a)

$$\begin{pmatrix} a_D \\ a \end{pmatrix} \to \begin{pmatrix} 1 & 0 \\ 0 & 1/2 \end{pmatrix} \begin{pmatrix} a_D \\ a \end{pmatrix}, \tag{12.86}$$

which changes monodromy matrices as

$$\begin{pmatrix} a & b \\ c & d \end{pmatrix} \to \begin{pmatrix} 1 & 0 \\ 0 & 1/2 \end{pmatrix} \begin{pmatrix} a & b \\ c & d \end{pmatrix} \begin{pmatrix} 1 & 0 \\ 0 & 2 \end{pmatrix} = \begin{pmatrix} a & 2b \\ c/2 & d \end{pmatrix}. \tag{12.87}$$

With these normalizations, the monodromy matrices (12.57) and (12.60) become

$$M_\infty = \begin{pmatrix} -1 & 4 \\ 0 & -1 \end{pmatrix}, \quad M_{\Lambda^2} = \begin{pmatrix} 1 & 0 \\ -1 & 1 \end{pmatrix}, \quad M_{-\Lambda^2} = \begin{pmatrix} -1 & 4 \\ -1 & 3 \end{pmatrix} \tag{12.88}$$

The corresponding elleptic curve reads

$$y^2 = x^2(x-u) + \frac{1}{4}\Lambda^4\, x, \tag{12.89}$$

which, by a constant rescaling of Λ, coincides with (12.85).

As compared to (12.61), the above expression makes it less transparent the points on the complex plane u where singularities arise, namely where two branch points collide and the curve degenerates. To this aim, regardless the parametrization one is using, it suffices to compute the discriminant of the x-poynomial, $\Delta = \prod_{i<j}(\alpha_i - \alpha_j)^2$ (where α_i are the roots of the polynomial), and find the values of u such that some roots coincide. For a cubic polynomial of the form

$$x^3 + bx^2 + cx + d, \tag{12.90}$$

we have $\Delta = b^2c^2 - 4c^3 - 4b^3d + 18bcd - 27d^2$ which applied to eq. (12.89) gives

$$\Delta = \frac{1}{16}\Lambda^8(u^2 - \Lambda^4), \tag{12.91}$$

which vanishes at $u = \pm\Lambda^2$, in agreement with our previous analysis. More generally, the roots α_i of the x-polynomial (*i.e.* the branch points) are functions of the parameters of the theory which, in the pure SYM $SU(2)$ are just u and Λ, $\alpha_i = \alpha_i(u, \Lambda)$. The explicit form of these functions (and even their number) depends on the parametrization chosen to describe the elliptic curve. For example, in the parametrization (12.61) they are $\alpha_1 = \Lambda^2, \alpha_2 = -\Lambda^2, \alpha_3 = u$, while in the parametrization (12.89) they are $\alpha_1 = 0, \alpha_2 = 1/2(u + \sqrt{u^2 - \Lambda^4}), \alpha_3 = 1/2(u - \sqrt{u^2 - \Lambda^4})$. What does not (and cannot) change, instead, are the number and the locations of the singularities on the u plane (namely the values of u for which two or more roots coincide), nor the nature of the particles becoming massless there, since these are physical information.

Let us now go back to the curve (12.84), and look at the singularity structure of the moduli space of $\mathcal{N} = 2$ SQCD with $F = 1$ more carefully. As compared to pure SYM, we expect now three singularities at finite distance in the u plane. This can be seen as follows. Let us first suppose the mass m to be very large. Then, at roughly the scale $|m|$ the hypermultiplet decouples and, below that scale the theory behaves, effectively, as $SU(2)$ SYM. This suggests that the structure of the moduli space in the region $|u| \ll |m^2|$ should be the same of the pure $SU(2)$ theory, with two singularities at $u \simeq \pm\Lambda^2$. Moreover, following our general discussion, we expect a third singularity where the hypermultiplet becomes effectively massless. For m large, this could only happen in the large VEV region where $u = a^2$. One can then easily see that a balance between the bare mass and the one coming from the cubic term in the superpotential, eq. (12.32), will occur at $u \simeq m^2$. Consistently with holomorphic decoupling, sending $m \to \infty$ this third singularity is pushed all the way to infinity and one recovers, correctly, the pure $SU(2)$ SYM moduli space of Figure 12.5.

Let us now consider the other extreme case, namely $m = 0$. While a mass term for the hypermultiplet completely breaks the R-symmetry, in the massless case there is a preserved $\mathbb{Z}_6$ R-symmetry at the quantum level. The u coordinate has R-charge 4 and one can then easily see that each point on the u plane preserve a $\mathbb{Z}_2$ symmetry. Hence, in the massless case we expect three (strong coupling) singularities, related by the broken $\mathbb{Z}_3$ generators.

These expectations can be checked analytically from the curve (12.84), by computing the discriminant and the three roots expanding the result for large, respectively small m. The discriminant of the x-polynomial (12.84) is

$$\Delta = 4m^2u^2\Lambda_1^6 - 32m^3\Lambda_1^9 - 4u^3\Lambda_1^6 - 27\Lambda_1^{12} + 36um\Lambda_1^9. \tag{12.92}$$

Let us consider first the large mass regime, $m \gg \Lambda_1$, eventually $m \to \infty$ keeping $m\Lambda_1^3 = \Lambda^4$ fixed. One can compute Δ at $u \simeq m^2$ and find $\Delta = 0 + \mathcal{O}[(\Lambda_1/m)^3)]$. Similarly, at $u^2 = \pm 8\Lambda^4$, one finds $\Delta = 0 \pm \mathcal{O}[(\Lambda_1/m)^{3/2})]\Lambda^4$, as expected. In the massless case, $m = 0$, the discriminant reduces to $\Delta = -4u^3\Lambda_1^6 - 27\Lambda_1^{12}$ and one easily sees that the three singularities are instead located at

$$u = \left(-\frac{27}{4}\Lambda_1^6\right)^{1/3}, \tag{12.93}$$

which shows, as anticipated, that they all are in the strong coupling region and get transformed one another by $\mathbb{Z}_3$ rotations. Clearly, one can interpolate between these two extreme cases by continuously increasing (decreasing) m. Figure 12.9 shows how the moduli space changes as we vary the hypermultiplet mass.

Interestingly, the nature of the hypermultiplet becoming massless at the three singularities depends on m. When m is large, the two strong coupling singularities (and the associated monodromies) are basically the same as the pure theory, and the corresponding massless particles are a monopole and a dyon, respectively. At $u \simeq m^2$ the (elementary) hypermultiplet becomes massless, which is an electrically charged object. So, for $m \to \infty$ the charges of the massless particles are $(1,0), (1,-1), (0,1)$ respectively. As we decrease m the three singularities become closer and closer and more and more similar to those of the massless theory, which are related by $\mathbb{Z}_3$ rotations. So there is less and less a clear distinction between hypermultiplets coming from solitons or from elementary objects. In fact, when $m = 0$ the charges of the massless particles can be easily computed to be $(1,0), (1,1), (1,2)$,

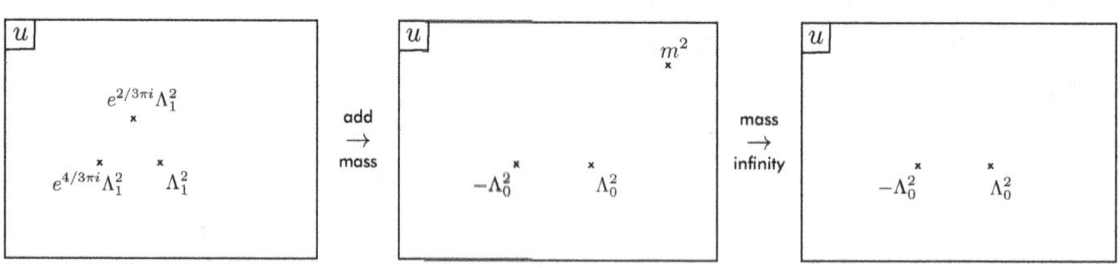

Fig. 12.9 The moduli space of $\mathcal{N} = 2$ SQCD with $F = 1$ as the hypermultiplet mass is varied. In the limit of infinite mass one recovers the pure $SU(2)$ moduli space, rightmost figure.

which is very different from the massless spectrum when m is large. This does not hurt. The mass is a UV parameter and, as we change it, we do change the theory. Hence, that the physics changes should not come as a surprise.

Notice, finally, that at any given singularity, regardless the value of m, there always exists a dual frame where the massless particle looks "electric" and the effective theory is then nothing but $\mathcal{N} = 2$ SQED with one massless flavor.

Argyres-Douglas theories

In the theory we have just discussed there is a choice of UV parameters which makes one of the singular points of the u-plane special. Let us first recap. The roots of the x-polynomial are functions of the Coulomb branch parameter u, the strong coupling scale Λ_1 and the hypermultiplet bare mass m, $\alpha_i = \alpha_i(u, \Lambda_1, m)$. There exists three different ways to collide two branch points, which correspond to the three singularities of the u-plane discussed above, where two roots coincide, meaning that a non-trivial cycle of the torus collapses and one hypermultiplet (no matter its nature) becomes massless. This is what happens for generic values of m. Can something more singular happen?

Let us choose the mass m to be $m = (3/2)\,\omega\Lambda_1$ with $\omega^3 = 1$. With this choice the Seiberg-Witten curve (12.84) becomes (we set for simplicity $\Lambda_1 = 1$ and $\omega = 1$)

$$y^2 = x^2(x - u) + 3x - 1, \tag{12.94}$$

and the discriminant (12.92) reads

$$\Delta = 9u^2 - 108 - 4u^3 + 54u - 27. \tag{12.95}$$

Singularities on the u-plane occur at those values of u for which the discriminant vanishes. In the present case it so happens that two singularities on the u-plane merge. There is one singularity at $u = -15/4$ where two roots of the x-polynomial coincide and another one at $u = 3$ where *all* three roots coincide. Indeed, when $u = -15/4$ we get $y^2 = (x + 2)^2(x - 1/4)$ while for $u = 3$ the Seiberg-Witten curve becomes $y^2 = (x-1)^3$ and all branch points coincide at $x = 1$. This means that at $u = 3$ the torus becomes double-degenerate two (mutually non local!) hypermultiplets become massless. This can happen in three different ways, depending on the root chosen for $\omega^3 = 1$ (and in turn on the mass m). In terms of the identification of the massless states in the large m regime, these three different ways correspond to the electrically charged hypermultiplet, the magnetic monopole and the dyon becoming massless in pairs).

This is something not specific to $SU(2)$ SQCD with one flavor, but it can happen whenever the gauge group rank is larger than one or, as in the present case, matter is added to pure SYM: there exist special points on the moduli space, known as Argyres-Douglas (AD) points, where mutually non-local objects become simultaneously massless. This means that there does not exist a duality frame in which all (light) fields are electric and that the theory cannot be as simple as SQED coupled to massless flavors. What this effective theory can be?

It is believed that at points where mutually non-local objects becomes simultaneously massless the theory enjoys an interacting (as opposed to free) conformal phase. At first sight this might sound surprising. Coleman-Gross theorem states that in four dimensions any theory of scalars, spinors and abelian gauge fields is IR-free. Our low energy effective theory is abelian and, as we already emphasized, there are no points whatsoever on the moduli space where new gauge bosons, besides those associated to $U(1)^n$, where $n = \text{Rank}G$, become massless. So there seems not to be room for an interacting fixed point.

Actually, as we already observed, what is special about these AD points is that cycles having non-vanishing intersections (like the A and B cycles of the two-torus) simultaneously shrink. Physically, this corresponds to, e.g. a dyon and a monopole, or a dyon and an electrically charged object becoming simultaneously massless. This is a situation where the Coleman-Gross theorem cannot be proven, since the theory lacks a Lagrangian description. As we now discuss, there are strong arguments suggesting that at such points the theory enjoys an interacting abelian Coulomb phase.

Let us consider a (not necessary supersymmetric) CFT in four dimensions and focus on operator scaling dimensions and Lorentz spins. The latter can be represented as (j_+, j_-), the two eigenvalues being $SU(2)$ quantum numbers. Unitarity and conformal symmetry provide lower bounds on the scaling dimensions of various operators. For instance, for an operator annihilated by special conformal generators K_μ and for which either j_+ or j_- vanish, and as such dubbed antichiral or chiral primary operator respectively, the following inequality holds

$$\Delta \geq j_+ + j_- + 1. \tag{12.96}$$

For non-chiral primaries, instead, it follows that

$$\Delta \geq j_+ + j_- + 2. \tag{12.97}$$

Equality in the above equations holds for free fields.

Let us consider the field strength operator $F_{\mu\nu}$. This is the sum of two conformal primary operators, schematically $F^{\pm} = F \pm^* F$, whose Lorentz spins are $(1,0)$ and (0,1), respectively. One can show that the states associated to the conserved currents $J^{\pm}_{\mu} = \partial^{\nu} F^{\pm}_{\mu\nu} = {}^*dF^{\pm}$ satisfy the equation $||J^{\pm}_{\mu}\rangle|^2 = 2(\Delta - 2)$, where Δ is the conformal dimension of $F_{\mu\nu}$. So we see that $\Delta = 2$ if and only if the currents are null vectors, $J^{\pm} = 0$, which are nothing but the Bianchi identity and the equation of motion of a free Maxwell theory. Conversely, if F is not free $\Delta > 2$ and *both* J^+ and J^- are different from zero. Since they are descendants of different primary fields, they are linearly independent and, in turn, this implies that both the electric current $J_e = J^+ + J^-$ and the magnetic current $J_m = J^+ - J^-$ are non zero. So we conclude that in a CFT any interacting field strength must couple both to electric and magnetic charged objects. In other words having both elementary monopoles and electric charges allow an abelian gauge theory to have a non-trivial (*i.e.* interacting) fixed point, while QED without elementary monopoles cannot have a non-trivial fixed point (in agreement with Coleman-Gross theorem).

We have seen that in our theory (for a proper choice of the UV mass parameter) at the $u = 3$ singularity mutually non-local objects become massless and so, by the above argument, the IR dynamics is believed to be described by an interacting fixed point, a so-called Argyres-Douglas theory.

In all this discussion supersymmetry did not play any role. Not surprisingly, however, the extra constraints imposed by $\mathcal{N} = 2$ supersymmetry let one get more clues on the property of this interacting fixed point, for example by providing the exact scaling dimension of some CFT operators. As discussed in Section 6.3, a new feature of the superconformal algebra as compared with the conformal algebra is that there is another symmetry generator under which operators transform, the R-symmetry. In the $\mathcal{N} = 2$ superconformal algebra this is $U(2)_R$ and therefore CFT operators are also characterized by the $U(1)_R$ charge R and $SU(2)_R$ "spin" I. From the superconformal algebra, one can show that a chiral primary operator (a state with $j_- = 0$ and annihilated by the supercharge $\overline{Q}^i_{\dot{\alpha}}$) satisfies the relation

$$\Delta = 2I + \frac{1}{2}R \geq 2I + j_+ + 1. \tag{12.98}$$

Let us now consider the $\mathcal{N} = 2$ vector superfield. In $\mathcal{N} = 2$ superfield formalism this is a scalar superfield U satisfying the chiral constraint

$$\overline{D}_{\dot{\alpha}i}U = 0 \quad \text{where } i = 1, 2. \tag{12.99}$$

Using $\mathcal{N} = 2$ unconstrained superfield formalism one can show that in $\mathcal{N} = 1$ language this has the same field content of a chiral superfield Φ and a gaugino superfield W_α. The lowest component of U, that we dub u, has $I = 0$ and Lorentz spin $(0,0)$, so $\Delta(u) = \frac{1}{2}R(u) \geq 1$. This implies that $\Delta(F^+) \geq 2$ and, when the equality is saturated we have that $dF^+ = 0$ (recall previous discussion). When $R(u) = 2$ (and hence $\Delta(u) = 1$) the superfield U satisfies also the equation

$$D^{\alpha(i}D^{j)}_{\alpha}U = 0. \tag{12.100}$$

which is then equivalent to say that the field is free and there is a null state, $dF^+ = 0$. For an interacting vector multiplet eq. (12.100) does not hold, just (12.99) does, there are no null states and electric and magnetic currents cannot vanish.

Let us now consider the Coulomb branch of $\mathcal{N} = 2$ $SU(2)$ one-flavor SQCD. At a generic point we have a free Maxwell theory described by a free $\mathcal{N} = 2$ $U(1)$ vector multiplet U with $\Delta(u) = 1$. What changes in the IR theory if we add a relevant operator in the UV theory, like a mass term for the one flavor?

If we shift the elementary hypermultiplet UV mass m, the prepotential of the effective theory is modified. The leading order operator generated by such shift at a given point on the moduli space can be obtained by expanding the variation of the prepotential in U: the constant term would not contribute, the linear one, $\int d^4\theta U$, using eqs. (12.99) and (12.100) can be shown to be a total space-time derivative so the leading term is proportional to U^2 and provides a shift of the effective coupling τ. The same holds at singular points of the Coulomb branch where one hypermultiplet becomes massless, since eqs.(12.99) and (12.100) still hold and therefore the linear term in U again does not contribute.

At an interacting fixed point, instead, U is chiral but eq. (12.100) is not satisfied anymore. This implies that at such special point, if it exists, the leading effect is the linear one

$$\int d^4\theta\, mU, \tag{12.101}$$

since this is not anymore a total space-time derivative. Therefore we get

$$\Delta(m) + \Delta(U) = 2, \tag{12.102}$$

which shows that m is a source for the CFT operator U.

Let us now go back to the special point $u = 3, m = 3/2$ of our model where the IR dynamics is a SCFT, and let us expand the curve (12.94) around such point in terms of shifted variables $(M, \tilde{u}, \tilde{x})$ defined as

$$m = \frac{3}{2} + M, \quad u = 3 + 2M + \tilde{u}, \quad x = \frac{1}{3}u + \tilde{x}. \tag{12.103}$$

We get

$$y^2 = \tilde{x}^3 - 2(M + \tilde{u})\tilde{x} - (\tilde{u} + \frac{4}{3}M^2) + \cdots, \tag{12.104}$$

where the dots are higher order terms in M and $\tilde{u}$, which are small in a neighborhood of the fixed point. Notice that the possibility of freely shifting the VEV of u to 0, $\langle \tilde{u} \rangle = 0$, follows the same argument we used when discussing smooth point of the moduli space (see the paragraphs after eq. (12.58)), since also in this case u is uncharged under the (abelian) gauge symmetry.

From the above equation we see that in order to see the cubic singularity at $\tilde{u} = M = 0$, meaning that nothing should dominate against $\tilde{x}^3$ in (12.104), we have to assign the following relative scaling to $\tilde{x}, M$ and $\tilde{u}$

$$\Delta(\tilde{x}) : \Delta(M) : \Delta(\tilde{u}) = 1 : 2 : 3. \tag{12.105}$$

So in eq. (12.104) we can drop $\tilde{u}\,\tilde{x}$ and M^2 terms and get the simplified expression

$$y^2 = \tilde{x}^3 - 2M\tilde{x} - \tilde{u} \tag{12.106}$$

at the SCFT point. Remarkably, from the scaling dimensions of the coefficients of the curve (12.106) one can extract the scaling dimensions of CFT operators. Indeed, from (12.106) it follows that

$$[y] = \frac{3}{2}[\tilde{x}], \quad [\tilde{u}] = 3[\tilde{x}], \quad [M] = 2[\tilde{x}]. \tag{12.107}$$

Recalling that $a \sim (\tilde{u}/y)d\tilde{x}$ gives the mass of BPS particle and should then have scaling dimension one we conclude that

$$[\tilde{x}] = \frac{2}{5}, \quad [\tilde{u}] = \frac{6}{5}, \quad [M] = \frac{4}{5}, \quad [y] = \frac{3}{5}. \tag{12.108}$$

From these relations we learn a few interesting facts:

- We get the exact scaling dimension of $\tilde{u}$, and this is larger than 1, showing that we are at an interacting fixed point, as expected. Notice, further, that since in the UV $u = \frac{1}{2}\text{Tr}\phi^2$ has dimension $\Delta = 2$, we see that u acquires an order one anomalous dimension along the RG flow, in agreement with the idea that we are at a strongly coupled fixed point.

- $\Delta(\tilde{u}) + \Delta(M) = 2$, in agreement with the idea that M is the dual coupling to $\tilde{U}$, and $\int d^4\theta M\tilde{U}$ a deformation out the fixed point theory.
- Since $\Delta(\tilde{u}) < 2$, $\int d^4\theta M\tilde{U}$ is a *relevant* deformation.

As already remarked, Argyres-Douglas fixed points exist not only for the theory we have been considering. For instance, as far as $SU(2)$ SQCD, there exist Argyres-Douglas CFTs for any $F \leq 3$. The same happens for the other minimal generalization of the original Seiberg-Witten model, namely pure SYM with gauge group $SU(3)$ (which is actually the first instance where this phenomenon was discovered) and several generalizations thereof.

12.5 $\mathcal{N} = 4$: Montonen-Olive duality

A central role in understanding the low energy effective dynamics of $\mathcal{N} = 1$ and $\mathcal{N} = 2$ supersymmetric gauge theories we discussed so far, has been played by two strong-weak coupling dualities. Both these dualities were loosely referred to as electric-magnetic dualities, but their nature is quite different. For $\mathcal{N} = 1$ theories, this is an IR equivalence between two different theories, with different gauge groups and matter content. For $\mathcal{N} = 2$ it is an equivalence between different IR descriptions of one and the same theory: the duality acts on the effective abelian action describing vacua on the Coulomb branch.

Increasing further the number of supersymmetries one could wonder if yet another type of duality would emerge. The answer is for the affirmative and, as we will see, it turns out that this is an exact duality, *i.e.* a duality acting not just at the level of the IR effective action. In order to understand what this duality is and how it comes about, we should do a step back.

Let us start again from the Georgi-Glashow model we discussed in Section 12.2.1. In the BPS limit, all states satisfy the BPS mass formula (12.22). Combined with Dirac quantization condition, this implies that states carrying magnetic charge are very heavy at weak coupling and states carrying electric charge are heavy at strong coupling (and viceversa). Therefore, one could imagine that at strong coupling the rôles of electric (fundamental) and magnetic (solitonic) sources are interchanged, and that the theory at strong coupling is a theory of light monopoles. This idea was put forward by Montonen and Olive which suggested that the

GG model could have two completely equivalent descriptions, one in terms of electric sources and one in terms of magnetic sources, the two being exchanged under S duality (which indeed interchanges electric and magnetic couplings).

There are two non-trivial evidences in favor of such a duality in the GG model. The first is that the BPS mass formula (12.22) is invariant under S duality, and this is a necessary condition for Montonen-Olive (MO) duality to hold. The second such evidence has to do with monopole interforce. Starting from the 't Hooft-Polyakov monopole solution, one can show that there is no interaction between two monopoles, while there is a non-vanishing interaction between a monopole and an anti-monopole. If duality is correct, since upon S transformations monopoles and massive gauge bosons are exchanged, the same should hold for the W^+ and W^- bosons, in the GG model. This has been shown to be the case. Basically, in the BPS limit the (massless) Higgs field contributes exactly the opposite to the photon in the interactions between W's with equal charge, which therefore do not exert any force between each other, and exactly the same to the photon in the interactions between W's with opposite charge, and the two add in this case.

These are both quite non-trivial properties that the GG model enjoys, but not enough to conclude that MO duality is realized in there. For example, there is no guarantee that the semiclassical BPS mass formula (12.22) holds at quantum level (differently from theories with extended supersymmetry). So the very existence of monopole-like states cannot be proven in the full quantum theory. Moreover, in order for the duality to hold, monopoles and $W^\pm$ bosons should carry the same spin. This cannot be verified in the GG model. If the 't Hooft-Polyakov monopole has any fixed integer spin it must be spherically symmetric classically, as the explicit soliton solution actually is. So this does not tell what the actual spin of the configuration is. Naive semi-classical quantization suggests that monopole are spinless, but not having an understanding of the full quantum version of GG model (exact spectrum and symmetries), one cannot draw any definitive conclusions about this, either. Finally, beside monopoles, also dyons exists in the semi-classical spectrum of the GG model. What is their role in the all picture? Notice, further, that the gauge coupling is a parameter only classically since it runs quantum mechanically so it is not even completely clear what MO duality actually means in the full quantum theory. More generally, the MO duality is non-perturbative in nature and it cannot be verified in a perturbative framework, unless one has some control over the full quantum dynamics of the system (perturbative and non-perturbative), something we can hardly achieve in a non-supersymmetric framework.

What about the supersymmetric version of this story?

The persistence of (12.22) at the quantum level ensured by the $\mathcal{N}=2$ supersymmetry algebra could suggest that exact S duality could be realized in $\mathcal{N}=2$ theories. However, the other necessary condition, namely that monopoles should have the same quantum numbers of massive gauge bosons, does not hold. It has been explicitly shown that magnetically charged states (the 't Hooft-Polyakov monopole and the Julia-Zee dyon) sit in hypermultiplets, which do not accommodate spin one particles. In fact, we have seen that in $\mathcal{N}=2$ theories a quite different duality is realized, which is not an exact duality but rather an electro-magnetic duality which holds at the level of the IR effective theory.

What about $\mathcal{N}=4$ SYM? Differently from $\mathcal{N}=2$, $\mathcal{N}=4$ SYM is believed to realize exact S duality! There are several facts which suggest this to be plausible.

First, as shown in Chapter 3, it follows from the supersymmetry algebra that in $\mathcal{N}=4$ massive representations cannot be anything but BPS multiplets containing spin one particles. This implies that monopoles sit in vector multiplets, as gauge bosons do, differently from $\mathcal{N}=2$ SYM. So they transform under the same Lorentz representations. In $\mathcal{N}=4$ SYM all physical states (massive or massless) sit in BPS saturated vector multiplets, which is the only possible $\mathcal{N}=4$ representation in theories without gravity.

Second, differently from $\mathcal{N}=2$, in $\mathcal{N}=4$ the BPS bound (12.22) does not only hold true at quantum level, but non-renormalization theorems guarantee that the quantities therein are classically exact. Hence, the $U(1)$ couplings entering the effective Lagrangian (12.1) do not renormalize, implying that the matrix τ_{IJ} is proportional to the exactly marginal (and non-abelian!) UV-coupling τ

$$\tau_{IJ} = C_{IJ}\,\tau, \tag{12.109}$$

where C_{IJ} is a constant matrix (which, in a suitably chosen basis for the Cartan generators, can be made proportional to the Cartan matrix of the gauge group). This suggests that, unlike for $\mathcal{N}=2$ theories, the electro-magnetic duality of the effective theory may propagate all the way to the UV and imply a stronger form of duality. More precisely, the $Sp(2n,\mathbb{Z})$ transformations (12.34) contain transformations which would act on the (non-abelian) UV coupling τ, via eq. (12.109), as $\tau \to \tau+1$ and $\tau \to -1/\tau$, and these generate the group $Sp(2,\mathbb{Z}) \simeq SL(2,\mathbb{Z})$ (this is true for gauge groups whose corresponding Lie algebra is simply laced, generalizations to non simply-laced algebras will be discussed later). This implies that theories with UV couplings τ related by $SL(2,\mathbb{Z})$ transformations should be physically equivalent. That is to say, not only theories where electrically charged, respectively magnetically charged states are the fundamental degrees of freedom are expected to be physically equivalent, as originally proposed by Montonen and Olive: all theories whose fundamental degrees of freedom are dyonic states related to purely electric ones by a $SL(2,\mathbb{Z})$ transformation should. Notice how different this duality is with respect to the $\mathcal{N} = 2$ IR dualities discussed previously, which act at the level of free IR effective abelian theories. Here we are claiming that different *interacting* theories, with different UV couplings, are physically equivalent. It is a non-abelian version of electro-magnetic duality of Maxwell theory, where the gluons of the S-dual (or more generally $SL(2,\mathbb{Z})$-dual) theory are seen as non-abelian monopoles (more generally dyons) of the original theory, and viceversa.

Another argument in favor of MO duality in $\mathcal{N} = 4$ SYM comes from the following observation. If a given theory enjoys MO duality, the BPS spectrum of any theory related to the latter by a $SL(2,\mathbb{Z})$ transformation should be the same. Therefore, not only monopoles and dyons should carry the same Lorentz representations as gauge bosons: the whole spectrum of the theory should be duality invariant. In particular, given that massive gauge bosons are BPS states with charges $\pm(0,1)$, there should also be in the theory all BPS states which can be obtained acting on $\pm(0,1)$ with $SL(2,\mathbb{Z}) \subset Sp(2n,\mathbb{Z})$ transformations (more precisely, there should exist one state for all relatively prime choice of electric and magnetic charges, recall the discussion at the end of Section 12.2.1). An equivalent way to make the same statement is that in theories where MO duality is realized, the charge lattice of BPS states should be invariant under $SL(2,\mathbb{Z})$ transformations. Several evidence and consistency checks were given showing this to be the case for $\mathcal{N} = 4$ SYM.

One could wonder what the connection between the (massive) BPS states of the abelian effective theory and the UV (massless) non-abelian $SL(2,\mathbb{Z})$ invariant spectrum could be. The Coulomb branch of $\mathcal{N} = 4$ SYM is classically exact and therefore at its origin the full non-abelian symmetry is recovered (this is the so-called superconformal phase of $\mathcal{N} = 4$ SYM). Let us start from an arbitrary point of the Coulomb branch whose BPS state mass spectrum is given by

$$M = \sqrt{2}|a^I n_{eI} + a_{D,I} n_m^I|, \tag{12.110}$$

where $a_{D,I} = \tau_{IJ} a^J$ and $\tau_{IJ} = C_{IJ}\tau$, with τ the non-abelian gauge coupling (we assume for simplicity the gauge group to be simple). Note that at fixed τ_{IJ}, the smaller a^I the smaller $a_{D,I}$, and the lighter the BPS mass spectrum. Let us move towards the origin of the Coulomb branch, $a^I = 0$. Since $a_{D,I} = \tau_{IJ} a^J$ is an exact relation, also $a_{D,I} = 0$, there. Hence, under the assumption that the process is smooth, at the origin where the full non-abelian gauge symmetry is recovered, all BPS states become massless. These are nothing but the non-abelian fundamental and monopole-like states predicted by MO duality (actually, since the theory at the origin of field space is a conformal field theory, the notion of particles is ill defined so, strictly speaking, one should better talk about operators belonging to a CFT). Note, in passing, that this argument does not only apply to $\mathcal{N} = 4$ SYM but actually to any exactly conformal supersymmetric theory with a Coulomb branch, so also to $\mathcal{N} = 2$ SCFTs (one such example being $\mathcal{N} = 2$ $SU(2)$ SYM coupled to four hypermultiplets transforming in the fundamental representation).

One independent argument in favor of MO duality in $\mathcal{N} = 4$ SYM comes, finally, from string (and M-) theory. There, it exists an intricate set of dualities between different string theories which implies, as a by-product, MO duality of $\mathcal{N} = 4$. The self-consistency of this web of dualities has passed many tests and it is regarded as yet another indication for the duality of $\mathcal{N} = 4$ SYM.

We said that MO duality can be seen as a non-abelian version of the electric-magnetic duality of Maxwell theory. This is roughly correct but a bit oversimplified, since the non-abelian nature of the gauge group G makes the duality map more involved than for Maxwell theory. In particular, a S-duality transformation has in general a non-trivial action also on the gauge group (and even on the algebra, in some cases), and transforms it into the so-called magnetic dual $\widehat{G}$, as we now review.

The very idea of a magnetic gauge group was originally proposed by Goddard, Nuyts and Olive (GNO) when trying to extend to non-abelian monopoles the Dirac quantization condition of electro-magnetism. For Maxwell theory, Dirac quantization condition comes from the requirement that a theory with both electric and magnetic charges can be consistently quantized. Requiring that Dirac quantization condition is satisfied for non-abelian monopoles determines the magnetic group $\widehat{G}$ (the group to which magnetic monopoles couple electrically) starting from a group G. And the two differ, in general. The group $\widehat{G}$ is often called the GNO-dual of G in the physics literature, while in the mathematical literature is called the Langlands-dual of G, because of its role in the Langlands program.

Let us consider a gauge theory with compact and connected Lie group G and algebra $\mathfrak{g}$. The Lie algebra $\hat{\mathfrak{g}}$ of the magnetic gauge group $\widehat{G}$ is specified by its roots $\hat{\alpha}$ being the co-roots of $\mathfrak{g}$, that is

$$\hat{\alpha} = 2\,\frac{\alpha}{\alpha \cdot \alpha}, \tag{12.111}$$

where α are the roots of $\mathfrak{g}$. From these definitions one concludes that simply-laced Lie algebras, *i.e.* $\mathfrak{su}(N), \mathfrak{so}(2N), \mathfrak{e}_6, \mathfrak{e}_7$ and $\mathfrak{e}_8$ are self-dual, $\hat{\mathfrak{g}} \simeq \mathfrak{g}$, as well as $\mathfrak{f}_4$ and $\mathfrak{g}_2$. The Lie algebras $\mathfrak{so}(2N+1)$ and $\mathfrak{sp}(2N)$, instead, are exchanged between each other.

From eq. (12.111) it also follows that if $\mathfrak{g}$ is not simply laced the S-duality transformation is not $\tau \to -1/\tau$, but actually $\tau \to -1/n_{\mathfrak{g}}\tau$, with $n_{\mathfrak{g}} = 2$ for $\mathfrak{f}_4, \mathfrak{so}(2N+1)$ and $\mathfrak{sp}(2N)$, and $n_{\mathfrak{g}} = 3$ for $\mathfrak{g}_2$ and reads

$$\widetilde{S} = \begin{pmatrix} 0 & 1/\sqrt{n_{\mathfrak{g}}} \\ -\sqrt{n_{\mathfrak{g}}} & 0 \end{pmatrix}, \tag{12.112}$$

which we dub $\widetilde{S}$ to distinguish it from S defined in eq. (12.18). Note that $\widetilde{S} \notin SL(2,\mathbb{Z})$. This implies that for non simply-laced Lie algebras the actual duality group is not $SL(2,\mathbb{Z})$ but a certain infinite discrete subgroup of $SL(2,\mathbb{R})$. More precisely, it is an extension of the group $\Gamma_0(n_{\mathfrak{g}})$ by the generator $\widetilde{S}$, where $\Gamma_0(n_{\mathfrak{g}})$ is the subgroup of $SL(2,\mathbb{Z})$ consisting of the matrices whose lower left entry is a multiple of $n_{\mathfrak{g}}$, and it is generated by T and $ST^{n_{\mathfrak{g}}}S$.

This was about the Lie algebra $\hat{\mathfrak{g}}$. To define properly the magnetic group $\widehat{G}$ we first need to recall a few facts about semi-simple Lie algebras and compact groups.

Given an algebra, a group is not univocally determined. Groups sharing the same algebra are locally isomorphic but may differ by their global structure, which is univocally determined by their center. So, we can define a group G_K starting from the unique simply connected Lie group $\widetilde{G}$ an algebra $\mathfrak{g}$ admits (a simply connected group is a group whose first homotopy group is trivial), via the relation

$$G_K = \widetilde{G}/K \quad \text{where } K \subset Z(\widetilde{G}), \tag{12.113}$$

$Z(\widetilde{G})$ being the center of $\widetilde{G}$. Compact groups of semi-simple Lie algebras have abelian finite (hence discrete) centers, and so is $K \subset Z(\widetilde{G})$ in eq. (12.113). Using that K is abelian, from eq. (12.113) it follows that

$$Z(G_K) = Z(\widetilde{G})/K, \tag{12.114}$$

while from K being discrete one can conclude that

$$\Pi_1(G_K) = K. \tag{12.115}$$

The magnetic group $\widehat{G}$ is defined as the group with algebra $\hat{\mathfrak{g}}$ (12.111) and whose center is related to that of the electric group G via the relation

$$Z(\widehat{G}) = \frac{Z(\widetilde{G})}{Z(G)} = \Pi_1(G), \tag{12.116}$$

where $\widetilde{G}$ is the simply connected group with the same algebra of G, and the second equality follows from eqs. (12.114) and (12.115). From (12.116) we see that S duality exchanges the centre and the fundamental group, since $Z(\widehat{G}) = \Pi_1(G)$. This shows that S duality interconnects topological and algebraic properties of the group, in agreement with the exchange of electric and magnetic degrees of freedom (and of the corresponding currents).

The case of $\mathfrak{su}(N)$: $SU(N)$, $SU(N)/\mathbb{Z}_N$ and their siblings

In order to make the above discussion a bit more concrete, in the following we would like to specialize to $\mathcal{N} = 4$ SYM with gauge group $SU(N)$. In this case the Dynkin diagram is simply-laced so $\hat{\mathfrak{g}} = \mathfrak{g}$ and the duality group is $SL(2,\mathbb{Z})$, with generators S and T. The difference is only at the level of groups. Since $SU(N)$ is simply connected and its center is $\mathbb{Z}_N$, it follows from eq. (12.116) that $\widehat{G}$ is in this case $SU(N)/\mathbb{Z}_N$.

In Section 10.1.1 we have seen how Wilson line operators (which can be thought of as worldlines of external electrically charged particles) can be used to probe the phase of a gauge theory. One can also consider external magnetically charged particles, the corresponding line operators being the so-called 't Hooft lines. Dyons, instead, are associated to mixed Wilson-'t Hooft line operators.

A non-abelian gauge theory without matter fields or with matter in the adjoint representation as it is the case for $\mathcal{N} = 4$ SYM, admits Wilson line operators transforming in any representation of the gauge group. These are in one-to-one correspondence with the weight lattice $\Lambda_{\rm w}$ of $\mathfrak{g}$ (modulo the Weyl group W_G, which for $SU(N)$ is S_N, the group of permutations of N elements). As shown by GNO, for non-abelian gauge theories the consistency of Dirac quantization condition restricts magnetic monopole charges, and hence the spectrum of 't Hooft lines, to $\Lambda_{\text{co-root}}(\mathfrak{g})$. For simply-laced algebra, as $\mathfrak{su}(N)$, this is the same as the root lattice, which is a *subgroup* of the weight lattice Λ_{w}. Therefore, 't Hooft lines are less than Wilson lines in $SU(N)$. More precisely, Wilson lines can sit in any representation of $SU(N)$ (they are so-called genuine line operators and act as order parameters, as we have seen in Section 10.1.1), while 't Hooft lines can only sit in the adjoint or tensor products thereof.

What about the GNO-dual, whose gauge group is $SU(N)/\mathbb{Z}_N$? Here the situation is somewhat reversed. Wilson lines are labelled by representations of the group, which for $SU(N)/\mathbb{Z}_N$ are a subset of those of $SU(N)$ in that they correspond to the root lattice $\Lambda_{\text{root}}(\mathfrak{g}) \subset \Lambda_{\text{w}}(\mathfrak{g})$. For example, Wilson lines in the fundamental representation are not allowed. On the contrary, 't Hooft lines are not anymore restricted by Dirac quantization condition, and they are labelled by the weight lattice $\Lambda_{\text{w}}(\mathfrak{g})$. The genuine line operators which can serve as order parameters are 't Hooft lines, now. In particular, we can have 't Hooft lines in the fundamental representation.

The $SL(2,\mathbb{Z})$ generator T shifts the θ angle by 2π. This is a symmetry of the $SU(N)$ gauge theory, but not of the $SU(N)/\mathbb{Z}_N$ theory, whose θ-periodicity is instead $2\pi N$:

$$SU(N) \ : \ \theta \text{ periodicity } [0, 2\pi), \qquad SU(N)/\mathbb{Z}_N \ : \ \theta \text{ periodicity } [0, 2\pi N).$$

Upon shifting $\theta \to \theta + 2\pi$ nothing happens in $SU(N)$, the shift just permutes line operators and maps their spectrum into itself. The $SU(N)/\mathbb{Z}_N$ theory is not invariant under the same shift, the spectrum of line operators changes (and this happens for any shift $\theta \to \theta + 2\pi n$ with $n = 1, \dots, N-1$). So, starting from $G = SU(N)$, acting with S and T, we get a family of theories which can labelled as $(SU(N)/\mathbb{Z}_N)_0, (SU(N)/\mathbb{Z}_N)_1, \dots, (SU(N)/\mathbb{Z}_N)_{N-1}$. All these theories look the same, locally, but differ by the spectrum of allowed line operators. The order parameters are Wilson lines for $SU(N)$, 't Hooft lines for $(SU(N)/\mathbb{Z}_N)_0$ and, as one can easily guess, mixed Wilson-'t Hoof lines for $(SU(N)/\mathbb{Z}_N)_n$. This is summarized in Figure 12.10.

More generally, in $\mathcal{N} = 4$ SYM with gauge group G, a Wilson line operator is labeled by a representation R of G while a 't Hooft line operator is labeled by a representation $\widehat{R}$ of the magnetic dual group $\widehat{G}$ (and a mixed Wilson-'t Hooft line operator by a suitable $R \times \widehat{R}$ representation).

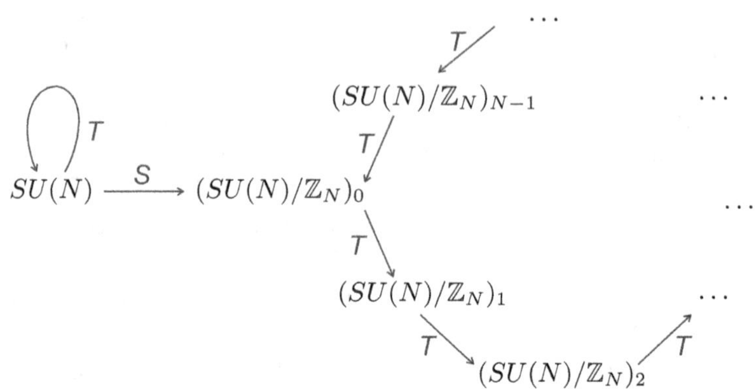

Fig. 12.10 Some of the different gauge theories in the $SL(2,\mathbb{Z})$ duality orbit of $\mathcal{N}=4$ SYM with algebra $\mathfrak{g}=\mathfrak{su}(N)$.

Under the action of S a Wilson line operator in the theory with gauge group G maps to a 't Hooft line operator in the theory with the dual group $\widehat{G}$ and viceversa. Non-trivial evidence for MO duality requires that given a Wilson line operator in the theory with gauge group G, it exists a 't Hooft line operator for the theory with dual gauge group $\widehat{G}$ with precisely the same quantum numbers as the original Wilson line operator. In other words, the charge lattice should go one into the other upon S duality. The duality further predicts that correlation functions of dual operators are also the same. In particular, S duality predicts that the expectation value of a 't Hooft line operator gets mapped to the expectation value of a Wilson line operator in the dual theory and, more generally, to the expectation value of mixed Wilson-'t Hooft line operators in the theory obtained from G acting with the corresponding element of the duality group. Several consistency checks have been done in the literature, showing this to be the case.

In closing, let us remark again that (in flat space) local physics does not depend on global issues. In particular, the Lagrangian depends on the choice of the algebra and not on the gauge group, and so do correlation functions of local operators. Therefore, it does not matter whether the gauge group is $SU(N)$ or $SU(N)/\mathbb{Z}_N$ (in any of its variants). What changes is the spectrum of allowed line operators that one can use to probe the theory. But as far as local physics, the theory is what it is, no matter how one probes it. For example, we have seen in Section 10.1.1 that in YM theory the VEV of the Wilson loop operator is 0 and this is associated to a one-form (electric) symmetry being preserved. Since local dynamics is independent of the global topology of the gauge group, there should not be a difference between, say, $G = SU(N)$ or $G = (SU(N)/\mathbb{Z}_N)_0$. However, for $G = (SU(N)/\mathbb{Z}_N)_0$ the genuine line operator which may serve as order parameter is the 't Hooft line operator. Recalling the discussion in Section 12.3.2, we expect a 't Hooft operator to follow a perimeter law if the theory is in a confining phase and therefore its VEV should be $\neq 0$. This has its counterpart at the level of one-form symmetries: $(SU(N)/\mathbb{Z}_N)_0$ does not have any electric one-form symmetry but a *magnetic* one-form symmetry $\mathbb{Z}_N^{(1)}$, which is spontaneously broken in the confining phase

$$\text{Confinement:} \quad \langle T_f(\gamma)\rangle \neq 0 \quad \text{Broken } \textit{magnetic}\, \mathbb{Z}_N^{(1)} \tag{12.117}$$

where $T_f(\gamma)$ is a 't Hooft operator in the fundamental representation. So, eventually, we see that $SU(N)$ and $(SU(N)/\mathbb{Z}_N)_0$ gauge theories do differ, but do so at the level of one-form symmetries, which have no effect on local dynamics (similar reasonings could be done verbatim for the whole $(SU(N)/\mathbb{Z}_N)_n$ family, with $n = 1, \ldots, N-1$).

What about $\mathcal{N} = 4$ SYM? At the origin of field space the theory is in an interacting (super)conformal phase. The $\mathbb{Z}_N^{(1)}$ one-form is spontaneously broken, no matter the duality frame one describes the theory with: there are no scales in the problem and the VEV of any order parameter, being it a Wilson, 't Hooft or mixed Wilson-'t Hooft operator is constant and different from 0. At a generic point of the moduli space, $\mathcal{N} = 4$ SYM is instead in a Coulomb phase. As discussed in Section 10.1.1, in such a Coulomb phase there is an enhancement of the one-form symmetry as

$$\mathbb{Z}_N^{(1)} \longrightarrow \left[U(1)_e^{(1)} \times U(1)_m^{(1)}\right]^{N-1}, \tag{12.118}$$

$U(1)_e^{(1)}$ and $U(1)_m^{(1)}$ being electric and magnetic one-form symmetries with respect to which Wilson operators and 't Hooft operators are charged, respectively (mixed Wilson-'t Hooft operators will be charged under both). Their VEVs are all $\neq 0$ since at low enough energy the theory is (trivially) conformal. Therefore, the whole continuous one-form symmetry (12.118) is (spontaneously) broken.

The above discussion concerns extended operators and, as we already pointed out, local physics is insensitive to these issues, in flat space. Things become more relevant when putting the theory on compact manifolds, because even local dynamics may be affected. For example, suppose to put the theory on $\mathbb{R}^3 \times S^1$. By wrapping a line operator on S^1 one gets a local operator on $\mathbb{R}^3$. So theories which only differ by the spectrum of line operators in four dimensions, may have different local structure in three dimensions (they could even have a different number of vacua, as for example $\mathcal{N}=1$ SYM with gauge groups $SU(N)$ or $SU(N)/\mathbb{Z}_N$ on $\mathbb{R}^3 \times S^1$).

More details about what we have discussed in this section (including further insights one can achieve rephrasing this story in the language of generalized symmetries) can be found in the references at the end of the chapter.

12.6 Exercises

Exercise 12.1. Consider the $\mathcal{N}=2$ pure SYM Lagrangian (12.25) and compute the supercurrents $S^1_{\alpha\mu}$ and $S^2_{\alpha\mu}$ by Nöether method. Recalling that the supersymmetry charges are related to the supercurrent as $Q^I_\alpha = \int d^3x\, S^I_{\alpha\,0}$ $(I=1,2)$, derive eqs. (12.27).

Exercise 12.2. Consider a rank one $\mathcal{N}=2$ supersymmetric gauge theory and a BPS state with charges (n_m, n_e) on the Coulomb branch. Suppose this state becomes massless at some point u_0 of the moduli space. Derive the expression of the corresponding monodromy matrix, eq. (12.37).

Exercise 12.3. Consider the three following parametrizations of the elliptic curve for $\mathcal{N}=2$ $SU(2)$ SYM

$$y^2 = (x-\Lambda^2)(x+\Lambda^2)(x-u)$$

$$y^2 = x^2(x-u) + \frac{1}{4}\Lambda^4 x$$

$$y^2 = (x^2-u)^2 - \Lambda^4$$

Show that they are physically equivalent, *i.e.* that the number and the locations of the singularities on the u plane as well as the nature of the particles becoming massless there, do not depend on which curve one considers.

References

[1] P. Argyres, *Non-perturbative dynamics of four-dimensional supersymmetric field theories*, https://homepages.uc.edu/~argyrepc/cu661-gr-SUSY/fgilec.pdf

[2] S. Coleman, *The magnetic monopole fifty years later*, Chapters 3 and 4, Springer, 1983.

[3] S. Coleman, *Aspects of Symmetry: Selected Erice Lectures*, Chapter 6, Cambridge University Press, 1985.

[4] E. Witten and D. I. Olive, Supersymmetry Algebras That Include Topological Charges, *Phys. Lett. B* **78** (1978), 97.

[5] N. Seiberg and E. Witten, Electric — magnetic duality, monopole condensation, and confinement in N=2 supersymmetric Yang-Mills theory, *Nucl. Phys. B* **426** (1994) 19 Erratum: [*Nucl. Phys. B* **430** (1994) 485] [hep-th/9407087].
N. Seiberg and E. Witten, Monopoles, duality and chiral symmetry breaking in N=2 supersymmetric QCD, *Nucl. Phys. B* **431** (1994) 484 [hep-th/9408099].

[6] E. D'Hoker and D. H. Phong, *Lectures on supersymmetric Yang-Mills theory and integrable systems*, Sections 2.10, 3 and 4, hep-th/9912271.

[7] M. E. Peskin, *Duality in supersymmetric Yang-Mills theory*, Sections 5 and 6, arXiv:hep-th/9702094.

[8] A. Bilal, *Duality in N=2 SUSY SU(2) Yang-Mills theory: A Pedagogical introduction to the work of Seiberg and Witten*, Sections 3 and 5, hep-th/9601007.

[9] L. Alvarez-Gaume and S. F. Hassan, Introduction to S duality in N=2 supersymmetric gauge theories: A Pedagogical review of the work of Seiberg and Witten, Section 3, *Fortsch. Phys.* **45** (1997) 159 [hep-th/9701069].

[10] P. Di Vecchia, *Duality in N=2, N=4 supersymmetric gauge theories*, Sections 12, 16 and 17, hep-th/9803026.

[11] Y. Tachikawa, N=2 supersymmetric dynamics for pedestrians, *Lect. Notes Phys.* **890** (2014) [arXiv:1312.2684 [hep-th]].

[12] A. Klemm, W. Lerche, S. Yankielowicz and S. Theisen, Simple singularities and N=2 supersymmetric Yang-Mills theory, *Phys. Lett. B* **344** (1995), 169 [arXiv:hep-th/9411048 [hep-th]].

[13] P. C. Argyres and M. R. Douglas, New phenomena in SU(3) supersymmetric gauge theory, *Nucl. Phys. B* **448** (1995), 93 [arXiv:hep-th/9505062 [hep-th]].
P. C. Argyres, M. R. Plesser, N. Seiberg and E. Witten, New N=2 superconformal field theories in four-dimensions, *Nucl. Phys. B* **461** (1996), 71 [arXiv:hep-th/9511154 [hep-th]].

[14] C. Montonen and D. I. Olive, Magnetic Monopoles as Gauge Particles?, *Phys. Lett. B* **72** (1977), 117.

[15] A. Sen, Strong — weak coupling duality in four-dimensional string theory, *Int. J. Mod. Phys. A* **9** (1994), 3707 [arXiv:hep-th/9402002 [hep-th]].
A. Sen, Dyon — monopole bound states, selfdual harmonic forms on the multi — monopole moduli space, and SL(2,Z) invariance in string theory, *Phys. Lett. B* **329** (1994), 217 [arXiv:hep-th/9402032 [hep-th]].

[16] P. Goddard, J. Nuyts and D. I. Olive, Gauge Theories and Magnetic Charge, *Nucl. Phys. B* **125** (1977), 1.

[17] A. Kapustin and E. Witten, Electric-Magnetic Duality And The Geometric Langlands Program, *Commun. Num. Theor. Phys.* **1** (2007), 1 [arXiv: hep-th/0604151 [hep-th]].

A. Kapustin, Wilson-'t Hooft operators in four-dimensional gauge theories and S-duality, *Phys. Rev. D* **74** (2006), 025005 [arXiv:hep-th/0501015 [hep-th]].

[18] P. C. Argyres, A. Kapustin and N. Seiberg, On S-duality for non-simply-laced gauge groups, *JHEP* **06** (2006), 043 [arXiv:hep-th/0603048 [hep-th]].

[19] D. Tong, *Lectures on Gauge Theory*, Section 2.6, https://www.damtp.cam.ac.uk/user/tong/gaugetheory/gt.pdf

[20] O. Aharony, N. Seiberg and Y. Tachikawa, Reading between the lines of four-dimensional gauge theories, Sections 1 and 2, *JHEP* **08** (2013), 115 [arXiv:1305.0318 [hep-th]].

[21] D. Gaiotto, A. Kapustin, N. Seiberg and B. Willett, Generalized Global Symmetries, *JHEP* **02** (2015), 172 [arXiv:1412.5148 [hep-th]].

Index